ELECTRICITY AND ELECTRONICS FUNDAMENTALS 1

Frank D. Petruzella

Technical Director
E. L. Crossley S.S.
Fonthill, Ontario

McGraw-Hill Book Company

New York Atlanta Dallas St. Louis San Francisco
Auckland Bogotá Guatemala Hamburg Lisbon
London Madrid Mexico Milan Montreal
New Delhi Panama Paris San Juan São Paulo
Singapore Sydney Tokyo Toronto

Cartwright.
Booth.
Rumble.
Hahn.
McGuire.
Price.
Mutch.
Jenny.

Booth
Hahn
Rumble
Cartwright

Library of Congress Cataloging-in-Publication Data

Petruzella, Frank D.
Electricity & electronics fundamentals.

Includes index.
1. Electricity. 2. Electronics. I. Crossley, E. L. II. Title. III. Title: Electricity and electronics fundamentals.
TK146.P443 1987 621.3 87-2641
ISBN 0-07-049676-5

Electricity and Electronics Fundamentals, Book 1

1 2 3 4 5 6 7 8 9 0 HALHAL 8 9 4 3 2 1 0 9 8 7

ISBN 0-07-049676-5

Special adaptation of INTRODUCTION TO ELECTRICITY AND ELECTRONICS, Book 1 by Frank D. Petruzella for distribution in the United States only.

Cover Designer: P.L.K. Graphics, Inc.
Cover Photographer: Geoffrey Gove

Acknowledgments

The author would like to thank the following companies and organizations for their assistance in providing photographs and illustrations for this text.

Ontario Hydro
The Cooper Tool Group
Shure Brothers Inc.
Ideal Industries Inc.
Miller Services
Allen-Bradley Co.
Hickok-Electrical Instrument Co.
Centralab, Inc. of North America
Philips Company
Potter and Brumfield
Duracell Inc.
Siemens Electric Limited

CONTENTS 1

PREFACE

ELECTRICITY AND ELECTRONICS FUNDAMENTALS, Book 1 and Book 2, *constitutes a comprehensive introduction to one of the most exciting and challenging fields of modern technology. While we often try to distinguish the two, it is difficult, if not impossible, to say where "electricity" ends and where "electronics" begins. Each embraces content that is unique to one or the other as well as much content that is common to both. With this set of two books, students should gain, through their studies, insight into both fields.*

Electricity and Electronics Fundamentals, Book 1, is designed for students who are new to electricity and electronics. It is devoted to the properties, measurement, and uses of electricity, and it provides both basic theory and applications necessary for further study in the field.

Both career-oriented students and those taking a course out of general interest should find this book readable, clearly laid-out, and informative. Simple two-color illustrations support the text throughout, helping students understand the theory and visualize its practical applications.

Book 1 is written in a logical sequence of units designed for beginners. Each unit is organized in a practical way starting with objectives that specify the goals of the unit, followed by the subject material and suggested practical assignments for learning the important "hands-on" skills. The review section includes a self evaluation test relating very closely to the unit objectives. These questions should help students evaluate their understanding of each unit.

In the interest of focusing on the basics, mathematics is kept to a minimum. Many worked examples of numerical problems are provided. The use of a calculator is recommended for students. Projects and examples are simple and directly related to theory. Projects in most cases can be performed with readily available components, equipment, and instruments.

One of the features of this book is the strong emphasis on teaching the proper use of schematic and wiring diagrams, starting with very basic diagrams for series and parallel circuitry and moving on to more involved circuitry. Students are aided throughout with wiring number sequence charts, which they will learn to formulate themselves as they progress through the series.

It is hoped that students will find this book accessible and interesting.

Frank D. Petruzella

1

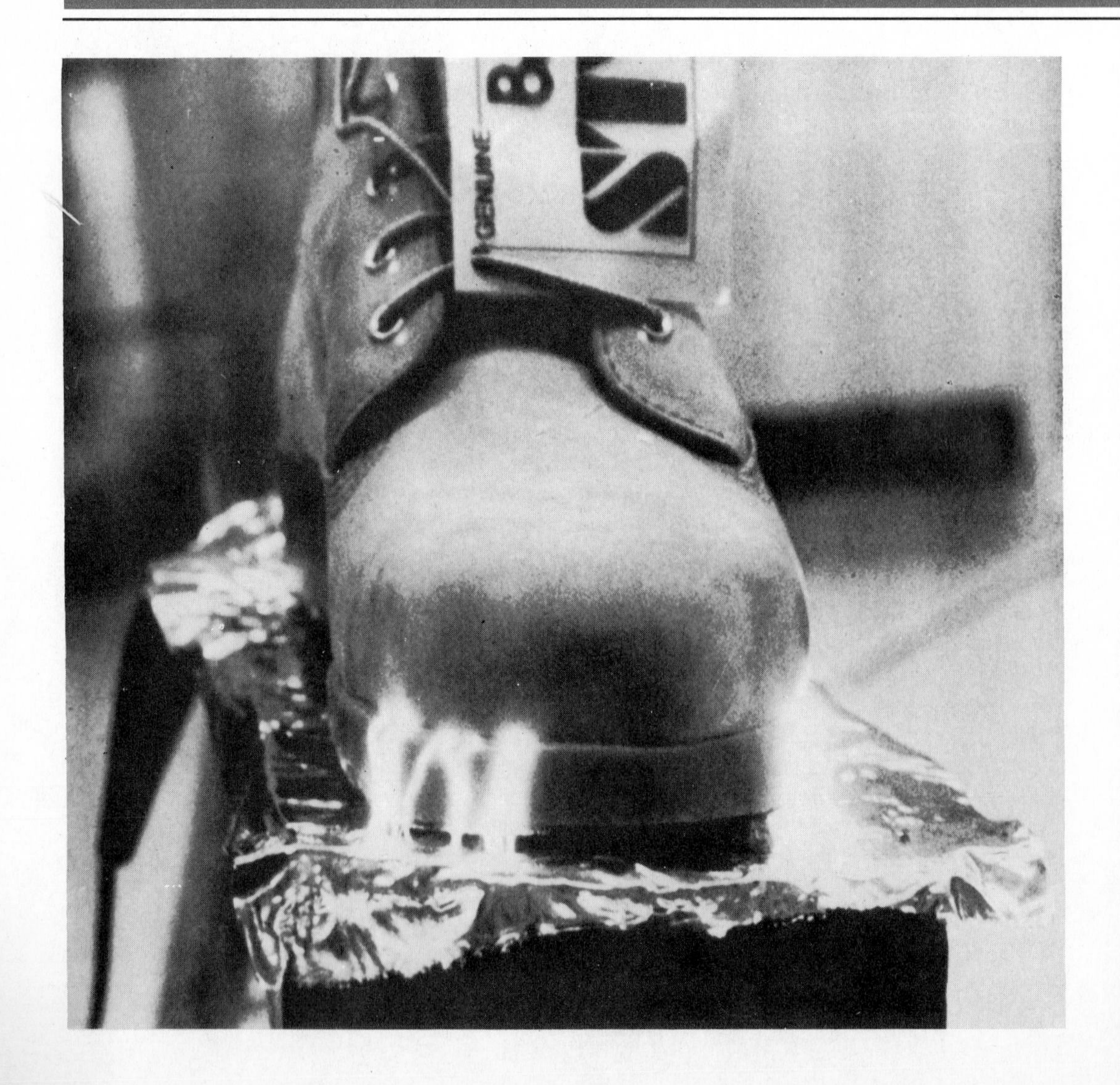

SAFETY

OBJECTIVES

Upon completion of this unit, you will be able to:

- Outline electrical safety rules for the home.
- Outline electrical safety rules for the school shop.
- Describe the electrical factors that determine the severity of an electrical shock.
- State the maximum safe current and voltage values.
- Given the body resistance and applied voltage, calculate the body current flow.
- Define the term "first-aid."
- Outline the first-aid procedure for bleeding, burns and electric shock.
- Outline the mouth-to-mouth method of artificial respiration.
- Outline the procedure to be followed in case of an electrical fire.

1-1 Electrical Hazards in the Home

Electrical equipment used in homes today is safe when properly installed, maintained and used. However, hazards are created when this equipment is improperly used or suitable safety measures are not employed.

Next to motor vehicles, the home accounts for most fatal accidents each year in the United States. The best way to reduce accidents at home is to know the potential hazards and take the necessary precautions to eliminate them. The following list of electrical safety suggestions is designed to increase your awareness of electrical accidents that can occur in the home:

1. Never run extension cords under rugs. They are not designed for this type of rough service nor are they a substitute for permanent wiring.
2. Do not jerk extension cords from electrical outlets.
3. Unused electrical outlets should be covered so that children cannot poke pins, etc. into them.
4. Do not over-fuse a circuit. Never use a penny to make a fuse connection.
5. Always turn off the main electrical switch before replacing a blown fuse.
6. Basement floors around washtubs and machines should be kept dry to help eliminate falls and to reduce the hazards of electric shock (Figure 1-1).
7. Never use water to put out an electrical fire.
8. Replace frayed appliance cords or defective wiring on all appliances as soon as discovered (Figure 1-2).
9. If fuses blow often, have an electrician discover the cause.
10. Do not overload electrical outlets by use of multiple tap-off devices (Figure 1-3).
11. Do not use electric space heaters, radios or appliances in the bathroom, laundry room or near the kitchen sink (Figure 1-4).

FIGURE 1-1 WATCH OUT FOR WET FLOORS AROUND APPLIANCES

FIGURE 1-2 REPAIR FRAYED CORDS

FIGURE 1-3 AVOID USING TAP-OFF DEVICES

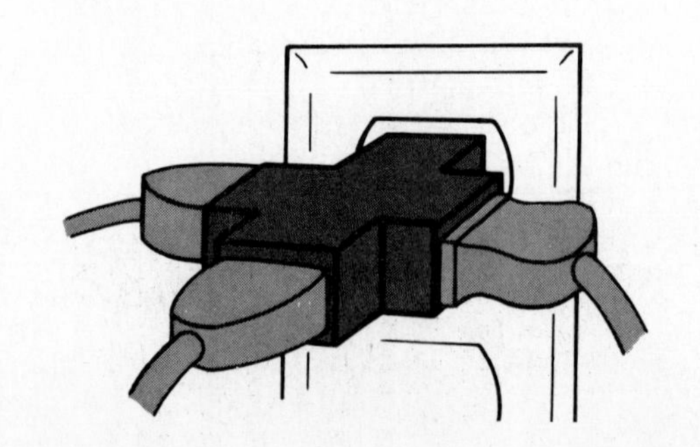

FIGURE 1-4 AVOID THE USE OF SPACE HEATERS IN THE BATHROOM

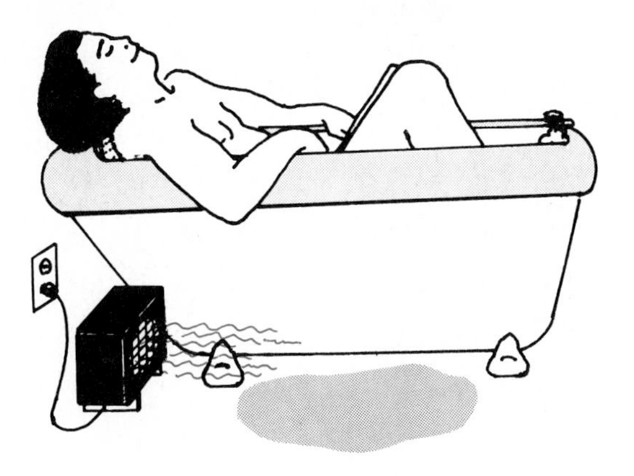

12. Turn off the main electrical switch when checking or replacing switches and outlets (Figure 1-5).
13. Disconnect such appliances as irons and toasters as soon as you have finished using them.

FIGURE 1-5 DON'T GUESS! MAKE SURE THE POWER IS OFF WHEN REPLACING OUTLETS

FIGURE 1-6 AVOID TWO HEAVY LOADS ON THE SAME CIRCUIT

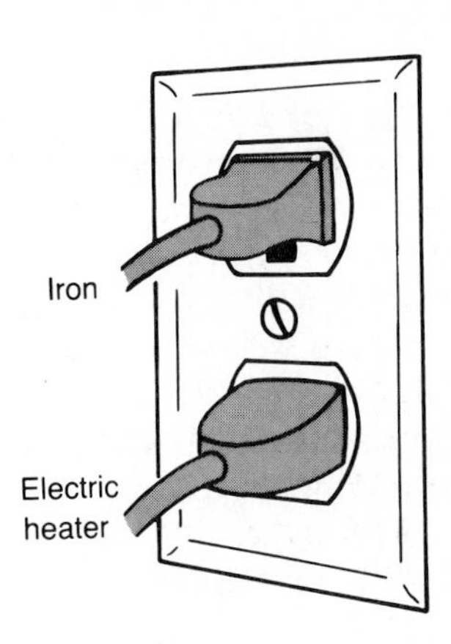

14. Avoid using two or more electric heating appliances on the same circuit (Figure 1-6).

1-2 Safety in the School Shop

Most school shop accidents are caused because safety rules are not observed. The school must ensure safe working conditions for the students. However, you must learn how to protect yourself and others working near you. Use your common sense! The school shop, in particular, is not the place for horseplay or carelessness. You should be aware of the following general school safety rules. Your teacher will point out the specific safety rules that apply to the shop you are working in.

1. Inform your teacher immediately when you become aware of a safety hazard.
2. Always notify your teacher when you are injured in the shop. Have proper first-aid applied.
3. Do not underestimate the potential danger of a 120 V circuit.
4. Work on live circuits ONLY when absolutely necessary and under supervision of the teacher.

5. Stand on dry, non-conductive surfaces when working on live circuits.
6. Never bypass an electrical protective device.
7. If you have a hot soldering iron on your bench, arrange your work so that you never have to reach over it.
8. Keep your work area clean.
9. Always wear safety glasses/goggles when you are operating any kind of power tool or when soldering.
10. Avoid horseplay and practical jokes.
11. Know where the fire extinguisher is and know how to use it.
12. Check all "dead" circuits before you touch them.
13. Never take a shock on purpose.
14. Do not touch two pieces of plugged-in equipment at the same time. An equipment defect could cause a shock.
15. Do not open or close any main switch without permission from the teacher.
16. At all times—if in doubt, ASK your teacher.
17. Make sure all electrical connections are secure before applying a voltage.
18. Always use properly grounded tools. Use only those tools with three pronged plugs or double insulated tools with two pronged plugs.

TABLE 1-1 SKIN CONDITION OR AREA AND ITS RESISTANCE

SKIN CONDITION OR AREA	RESISTANCE VALUE
Dry Skin	100 000 to 600 000 Ω
Wet Skin	1 000 Ω
Internal Body—hand to foot	400 to 600 Ω
Ear to Ear	about 100 Ω

FIGURE 1-7 MEASURING BODY RESISTANCE

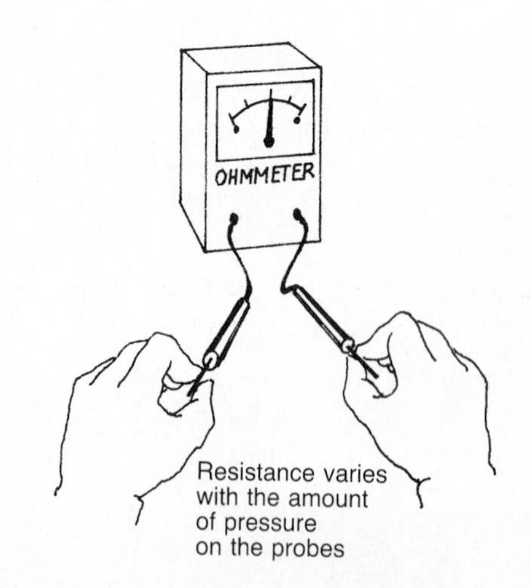

1-3 Effects of Electricity on the Human Body

The three electrical factors involved in an electric shock are: resistance, voltage and current. Electrical resistance (R) is defined as the opposition to the flow of current in a circuit and is measured in ohms (Ω). The lower the body resistance, the greater the potential electrical shock hazard. Body resistance varies with the condition of the skin and the area in contact. Typical body resistance values are listed in Table 1-1. Body resistance can be measured with an instrument called an *ohmmeter* (Figure 1-7).

Voltage (V) is defined as the pressure that causes the flow of electric current in a circuit and is measured in volts. The amount of voltage that is dangerous to life varies with each individual due to variations in body resistance and heart conditions. Generally, any voltage above 30 V is considered dangerous.

Electric current (I) is defined as the rate of flow of electrons in a circuit and is measured in amperes. The amount of current flowing through the body depends on the

FIGURE 1-8 RELATIVE MAGNITUDE AND EFFECT OF ELECTRIC CURRENT

Current in a 100 W lamp can electrocute 20 adults at one time
1 000 mA (1 A)
Current
Heart convulsions usually fatal
50 mA
Painful shock — inability to let go
20 mA
Muscles contract
5 mA
Safe current values
No sensation
1 mA
0

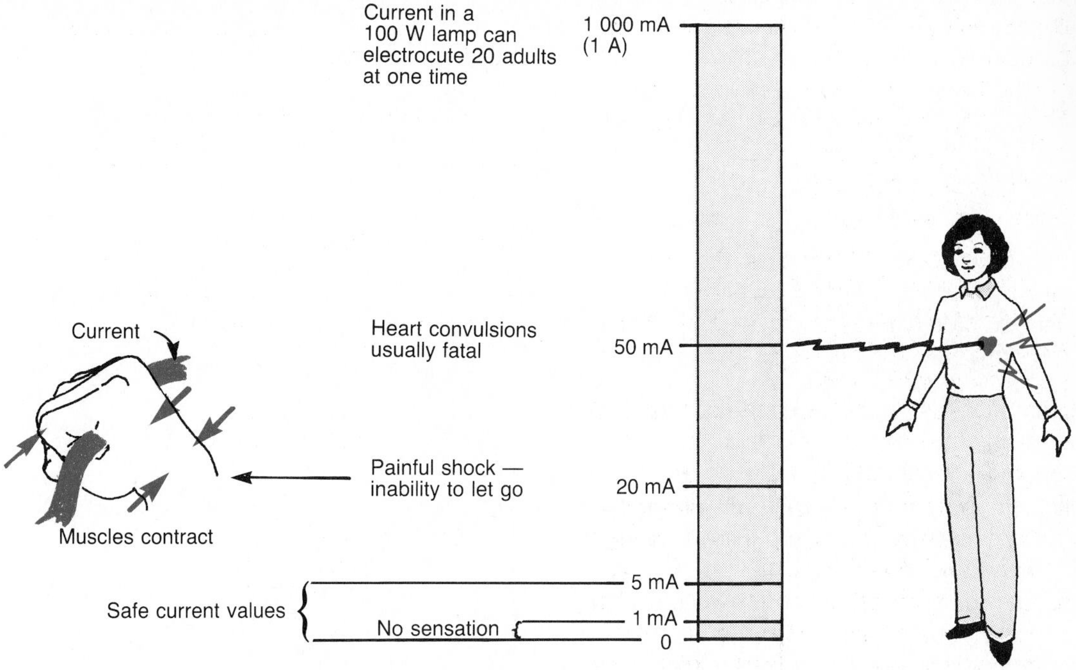

voltage and resistance. Body current can be calculated using the following formula:

$$\text{Current Through Body} = \frac{\text{Voltage Applied to Body}}{\text{Resistance of Body}}$$

$$I\ (\text{amperes}) = \frac{V\ (\text{volts})}{R\ (\text{ohms})}$$

OR

$$I\ (\text{milliamperes}) = \frac{V\ (\text{volts})}{R\ (\text{kilohms})}$$

Generally, any current flow above 0.005 A (Amperes) or 5 mA (milliamperes) is considered dangerous. A flashlight cell can deliver more than enough current to kill a human being. Yet it is safe to handle. This is because the resistance of human skin is high enough to greatly limit the flow of electric current. In lower voltage circuits, resistance restricts current flow to very low values. Therefore, there is little danger of an electric shock.

Higher voltages on the other hand, can force enough current through the skin to produce a shock. The danger of harmful shock increases as the voltage increases. Those who work on very high voltage circuits must use special equipment and procedures for protection.

Figure 1-8 illustrates the relative magnitude and effect of electric current.

1-4 First-Aid

First-aid is the immediate and temporary care given to the victim of an injury or illness. Its purpose is to preserve life, assist recovery and prevent aggravation of the condition. If someone else is hurt, *first* send for help immediately. A few of the basic first-aid procedures are listed on page 6. These are standard practices, but different methods may sometimes apply. An emergency first-aid course, such as that offered by the American Red Cross, is recommended for anyone entering the electrical field.

Bleeding To control bleeding, apply direct pressure on the wound using a clean pad or your hand. Raise the arm, leg, or head above heart level.

Burns Immerse the injured area in cold water or apply cold packs to relieve the pain. *Do not* break blisters. Cover with a clean dressing only.

Electric Shock To treat for electric shock, turn the power off or use a dry board or stick to remove the electrical contact from the victim. *Do not* touch the victim until he or she has been separated from the current. Begin first-aid. Apply artificial respiration if the victim is not breathing. Keep the victim warm; position so that the head is low and turned to one side to encourage flow of blood.

1-5 Artificial Respiration

If breathing stops, you can help the victim by knowing how to apply artificial respiration. The basic mouth-to-mouth method of artificial respiration is as follows (Figure 1-9):

1. Place victim on his or her back immediately! *Turn head, clear throat* area of water, mucus, foreign objects, or food.
2. *Tilt head back* to open air passage.
3. *Lift jaw up* to keep the tongue out of air passage.
4. *Pinch nostrils* closed to prevent air leakage when you blow.
5. *Seal your lips* around victim's mouth.
6. *Blow into the mouth* until you see the chest rise.
7. *Remove your mouth* to allow natural exhalation.
8. *Repeat* 12 to 18 times per minute until natural breathing starts.

1-6 Fire Control

In case of an electrical fire the following procedures should be followed:

1. Trigger the nearest fire alarm to alert all personnel in the workplace as well as the fire department.
2. If possible, disconnect the electrical power source.
3. Use a carbon-dioxide or dry-powder fire extinguisher to put out the fire. Under no circumstances use water, as the stream of water may conduct electricity through your body and give you a severe shock.
4. Ensure that all persons leave the danger area in an orderly fashion.
5. Do not re-enter the premises unless advised to do so.

1-7 Practical Assignments

1. Working in pairs, measure and record (using an ohmmeter) the body resistance between each of the following points for both you and your partner.
   (i) hand to hand
   (ii) hand to foot

2. Using the recorded body resistance values, calculate the amount of voltage required to produce a current flow of 100 mA.

1-8 Self Evaluation Test

1. Describe three safety rules that deal with the use of extension cords.

2. Describe two safety rules that deal with the replacement of fuses.

3. Since water is such a good conductor of electricity we must observe certain safety rules. Describe two of them.

4. List five safety rules that apply directly to your work in the shop.

FIGURE 1-9 ARTIFICIAL RESPIRATION — MOUTH-TO-MOUTH

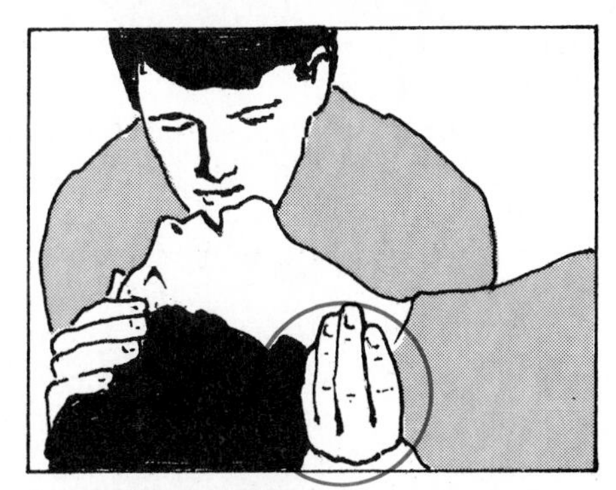

A. Tilt head — Clear throat — Lift jaw

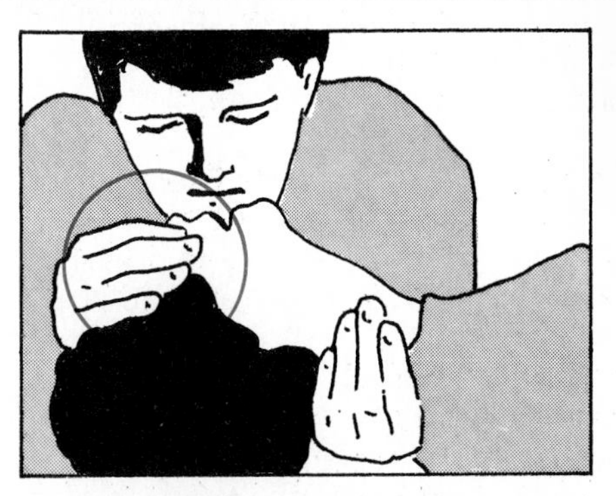

B. Pinch nostrils

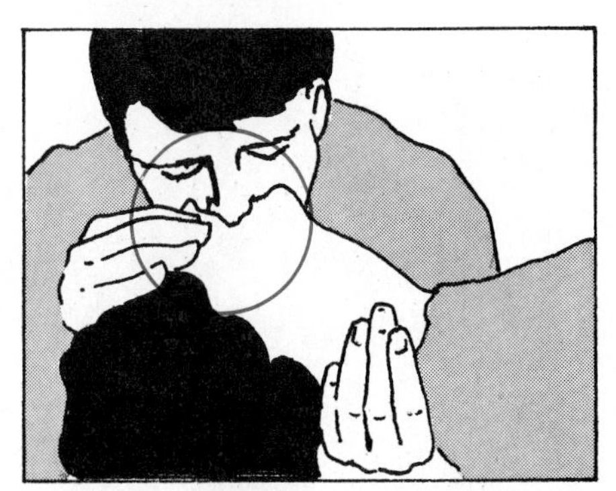

C. Make a tight seal— blow into mouth

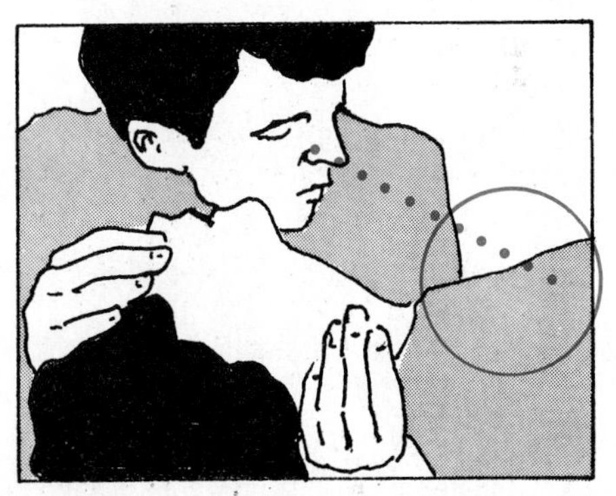

D. Watch to see chest rise and fall — repeat 12 to 18 times per minute

5. State if the severity of an electrical shock increases or decreases with each of the following changes:
   (a) a decrease in voltage
   (b) an increase in current
   (c) an increase in body resistance.

6. (a) What is the maximum safe current value?
   (b) What is the formula for finding the amount of current flow through a body?

7. Voltages are generally considered to be dangerous when they are above what value?

8. Calculate the body current flow of an electrical shock victim that comes in direct contact with a 120 V energy source. Assume a body contact resistance of 1000 Ω.

9. Define the term "first-aid."

10. Outline the basic first-aid procedure for:
    (a) bleeding
    (b) burns.

11. List the important steps to be followed when applying mouth-to-mouth artificial respiration.

12. What important rescue procedure should be followed in case of an electrical accident involving a live electrical circuit?

13. In case of an electrical fire, what type of fire extinguisher should be used?

2

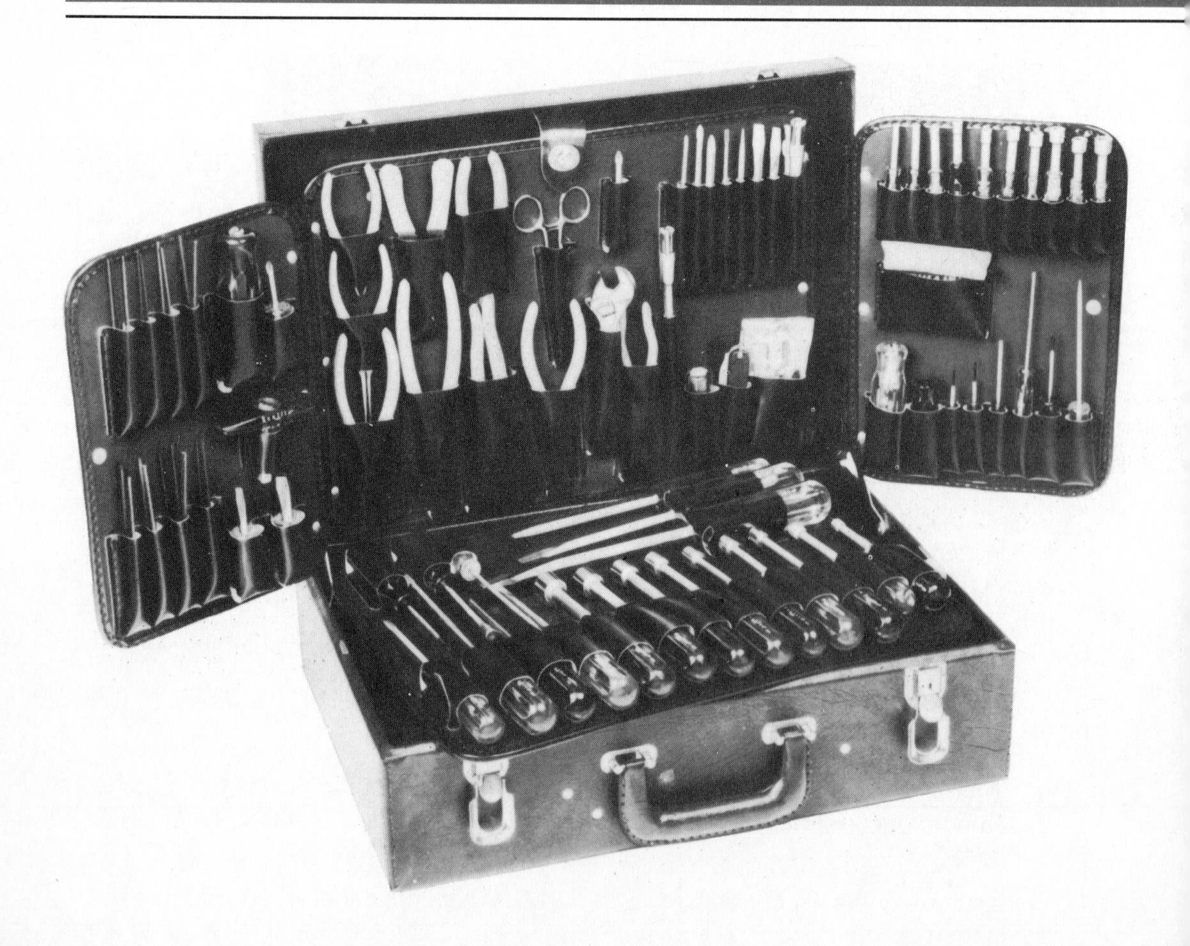

TOOLS OF THE TRADE

OBJECTIVES

Upon completion of this unit you will be able to:

- Identify and state the use for common measuring devices used in the electrical trade.
- Identify and state the use for common tools used in the electrical trade.
- Outline the procedure to be followed for the proper care and use of tools.

FIGURE 2-1 VOLTMETER AND AMMETER

Courtesy Bach-Simpson Limited

2-1 Measuring Devices

There are many kinds of instruments used by the electrician and technician for the measurement of electrical quantities. Some of the more important ones are discussed in more detail in other units of this text. Most electrical measuring devices are fairly expensive and delicate. When using any electrical measuring device you should:

1. handle it with care,
2. be sure it is properly connected to the circuit,
3. be sure not to exceed the voltage or current rating of the device.

For experimental work in the school, separate voltmeters, ammeters and ohmmeters are often used (Figure 2-1). They accurately measure voltage, current and resistance respectively.

On the job, a single *multimeter* is usually used to accurately measure voltage, current, or resistance (Figure 2-2).

The *voltage tester* (Figure 2-3) is often used by the electrician to measure approximate circuit operating voltages. Its rugged construction makes it ideally suited for rough on-the-job handling.

FIGURE 2-2 MULTITESTER

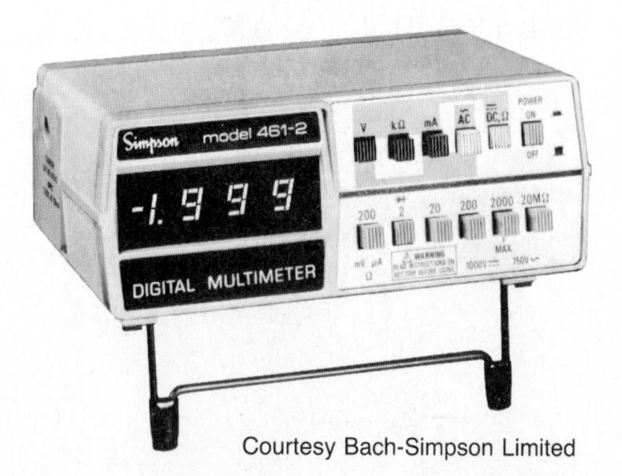

Courtesy Bach-Simpson Limited

The *clip-on ammeter* (Figure 2-4) is often used by the electrician to measure current flow. This device is able to measure current without any direct electrical contact with the circuit.

FIGURE 2-3 VOLTAGE TESTER

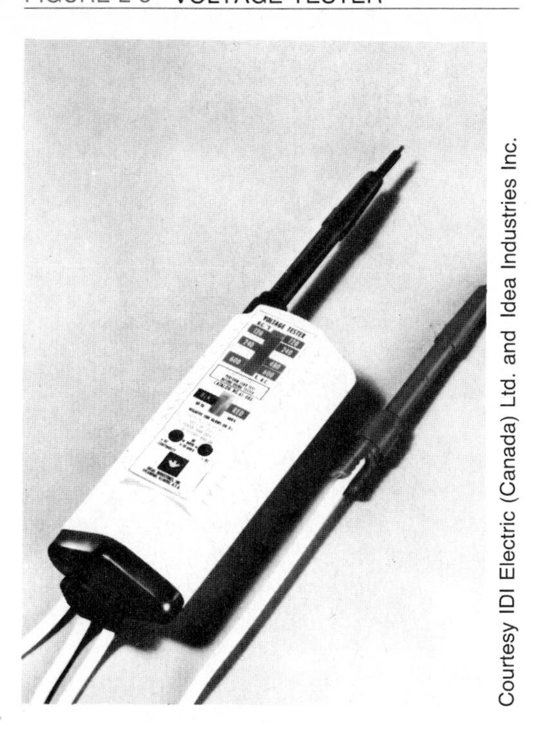

Courtesy IDI Electric (Canada) Ltd. and Idea Industries Inc.

FIGURE 2-4 CLIP-ON AMMETER

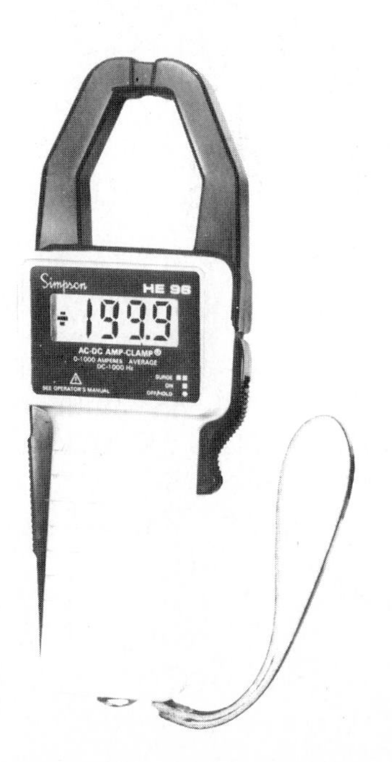

Courtesy Bach-Simpson Limited

FIGURE 2-5 NEON TEST LIGHT

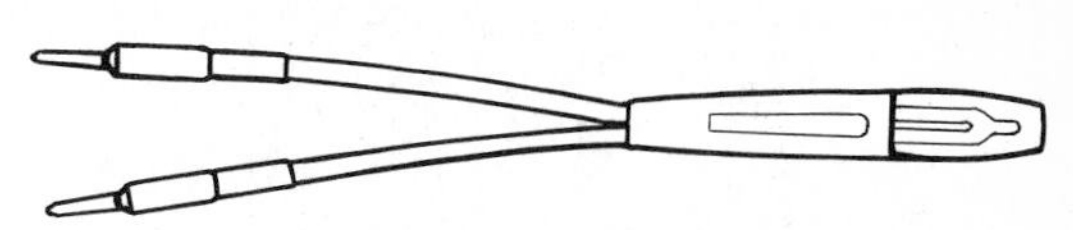

The *neon test light* (Figure 2-5) is an inexpensive device that can be used by the homeowner to indicate the presence of a voltage.

2-2 Hand Tools

The skilled electrician or technician must be familiar with the proper use of the tools of his or her trade. **Remember:** always use the right tool for the job.

Screwdrivers The screwdriver is a tool designed to loosen or tighten screws. Screwdrivers are identified by the shape of their head. A *slot-head* or *standard* screwdriver (Figure 2-6A) is designed for use on screws with slotted heads. This type of screw is often used on the terminals of switches, receptacles and lampholders.

FIGURE 2-6 SCREWDRIVERS

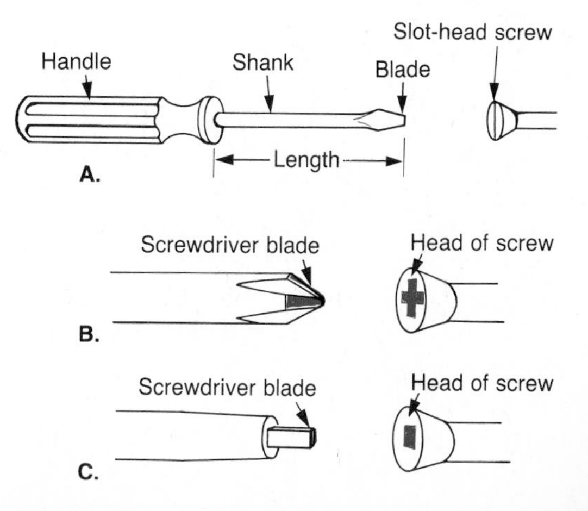

FIGURE 2-7 TYPES OF PLIERS

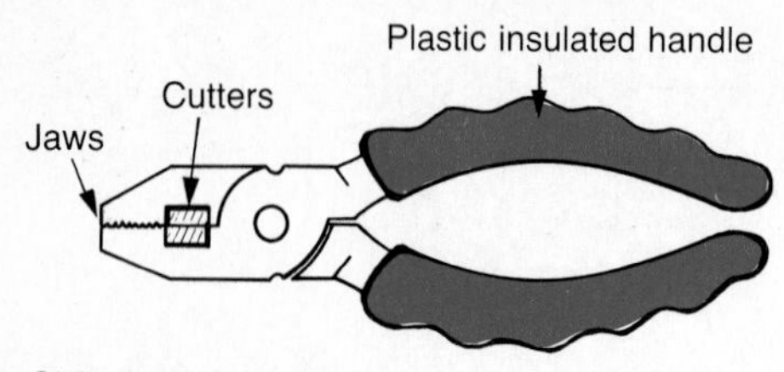

A. Side-cutting pliers

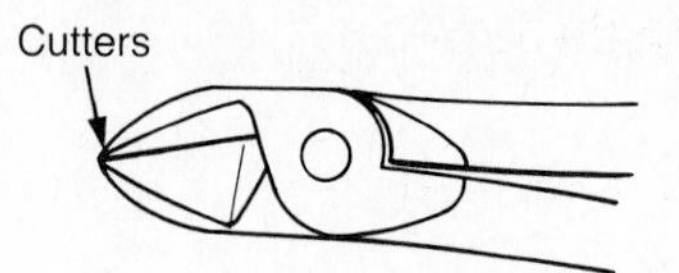

B. Diagonal-cutting pliers

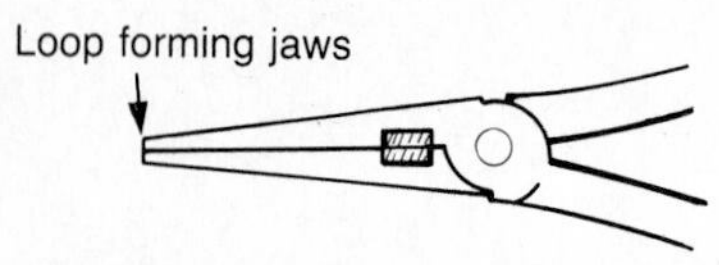

C. Needle-nose pliers

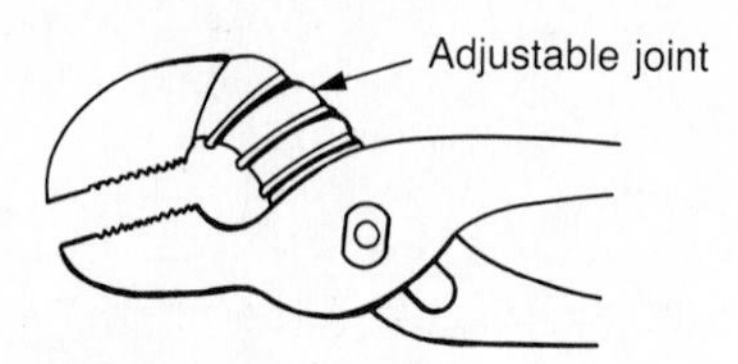

D. Curved-jaw pliers

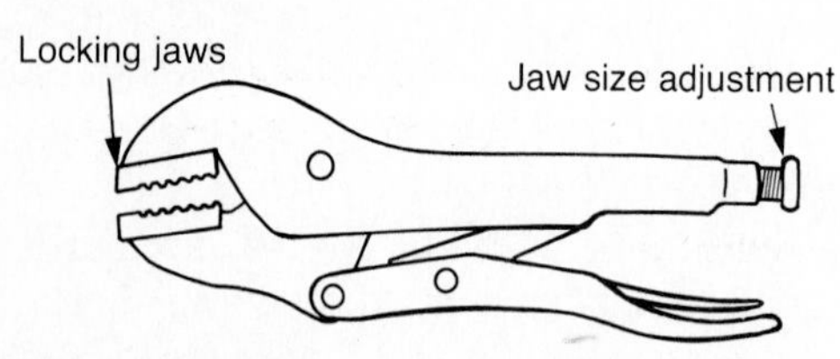

E. Vice-grip pliers

The *Phillips* screwdriver (Figure 2-6B) is designed for use on screws with an X-shaped insert in their heads. This type of screw is often used on the outside of electrical appliances because there is less likelihood of the screwdriver head slipping out of the slot and damaging the metal finish of the appliance.

The *Robertson* screwdriver (Figure 2-6C) is designed for use on screws with square-shaped inserts in their heads. This type of screw fits snugly into the screwdriver head, allowing the screw to be easily driven into wooden material. Such screws are sometimes used to secure outlet boxes to joists.

Pliers Electricians make use of several different types of pliers. The *side-cutting* pliers (Figure 2-7A) are used for gripping, twisting and cutting wires.

Diagonal-cutting pliers (Figure 2-7B) are designed specifically for cutting wire. They are used for close cutting jobs such as trimming the ends of wire on terminal board connections.

Needle-nose pliers (Figure 2-7C) are used to make loop ends on wire for connection to terminal screws.

Curved-jaw pliers (Figure 2-7D) are designed with an adjustable joint for gripping work of various sizes.

Vice-grip pliers (Figure 2-7E) are designed with jaws that can be locked onto the object.

Hammers The *claw* hammer (Figure 2-8) is the one most commonly used by the residential electrician. It is a good general purpose hammer. The face of the hammer is used to drive nails and staples. The claw is used for removing nails.

FIGURE 2-8 CLAW HAMMER

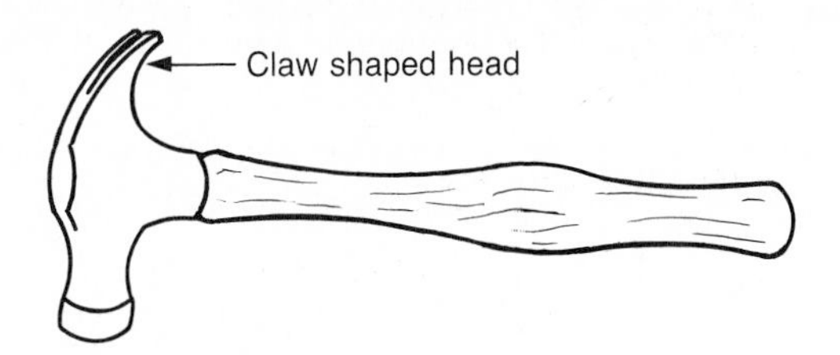

Saws A *cross-cut* type of wood saw (Figure 2-9A) is usually selected for sawing wood. The standard *hacksaw* (Figure 2-9B) is required for any metal cutting work.

A *keyhole* or *compass* saw (Figure 2-9C) is a narrow saw that is used for cutting holes into finished surfaces in order to mount outlet boxes.

Twist Drills and Auger Bits *Twist drills* are used in electric drills for drilling holes in metal chassis (Figure 2-10A). They are available in carbon tool steel and in high-speed steel. The more expensive high-speed drills are used for the drilling of hard materials because they can withstand a greater heat. The *auger bit* is used along with a brace for drilling holes in wood required for electrical cable runs (Figure 2-10B).

FIGURE 2-9 TYPES OF SAWS

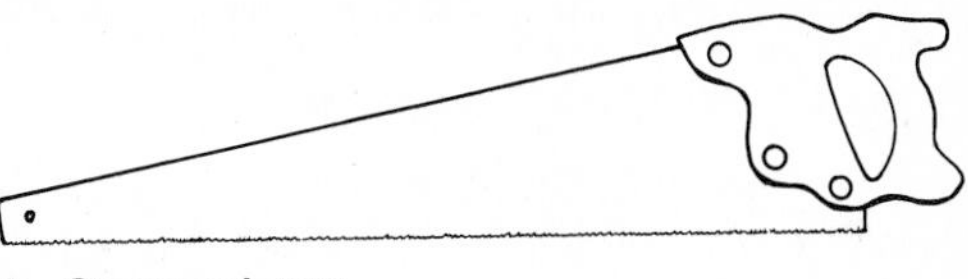

A. Cross-cut saw

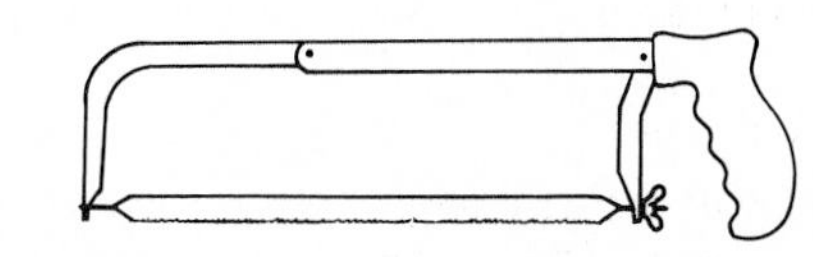

B. Hacksaw

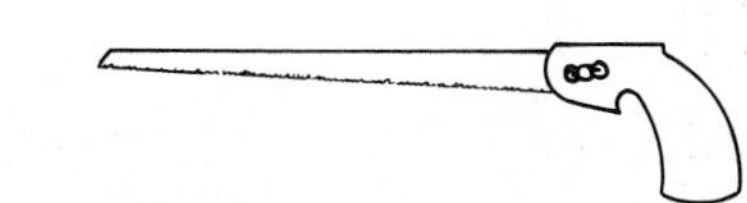

C. Keyhole saw

FIGURE 2-10 DRILLS

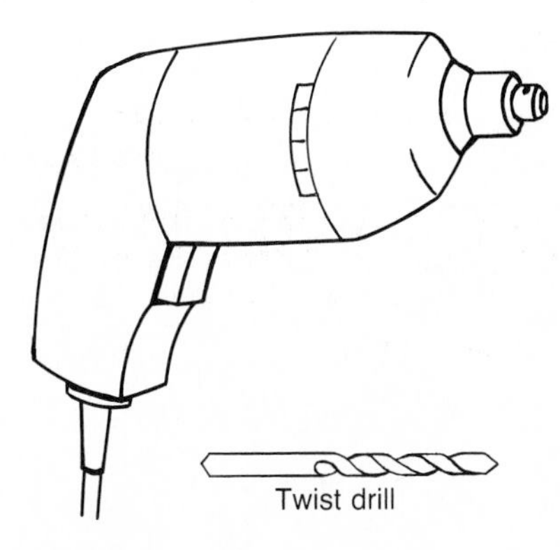

A. Electric drill

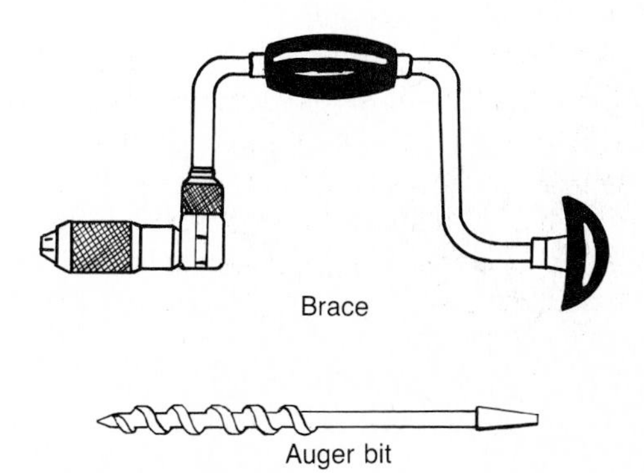

B. Brace and bit drill

CENTER PUNCH

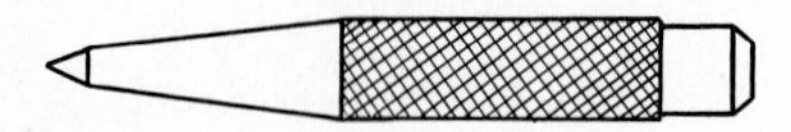

Punches A *center* punch (Figure 2-11) is used to mark the location of a hole to be drilled and is used most often by electricians. This assists a drill in entering exactly at the proper point.

Wrenches Commonly used wrenches include open-end wrenches, box-end wrenches, socket wrenches and adjustable wrenches. The wrench used must correctly fit the nut under pressure; otherwise the nut and the wrench will be damaged.

The *open-end* wrench (Figure 2-12A) is designed for use in close quarters. After each short stroke, the wrench can be turned over to fit other flats of the nut.

Box-end wrenches (Figure 2-12B) completely surround or "box in" the nut or bolt head when they are being used.

Socket wrenches (Figure 2-12C) can be positioned on a nut more quickly. These wrenches use an assortment of handles which make the electrician's or technician's work faster and easier.

Adjustable wrenches (Figure 2-12D) are especially convenient at times when an odd size of nut is encountered. When using an adjustable wrench the pulling force should always be applied to the stationary-jaw side of the handle.

FIGURE 2-12 WRENCHES

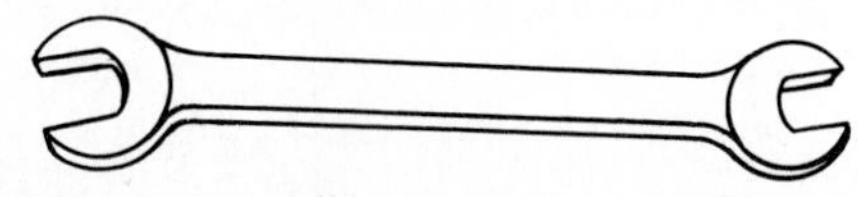

A. Open-end wrench

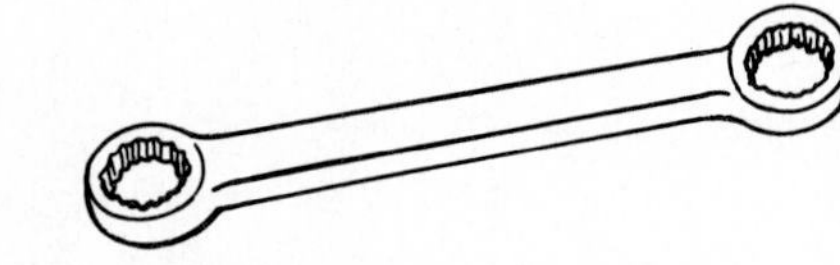

B. Box-end wrench

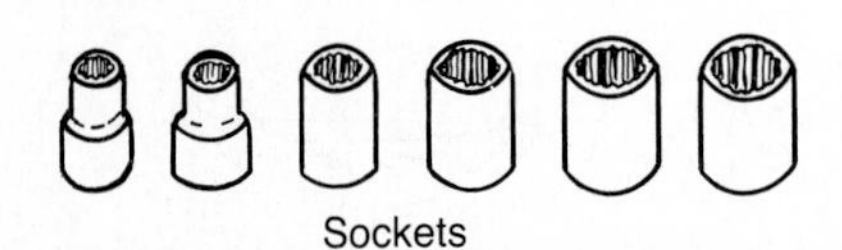

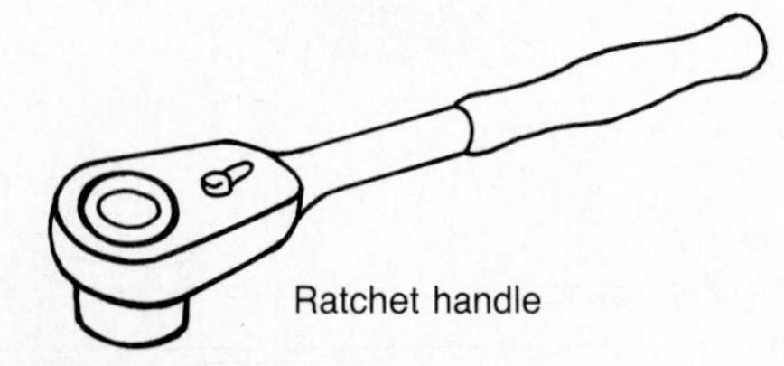

C. Ratchet handle

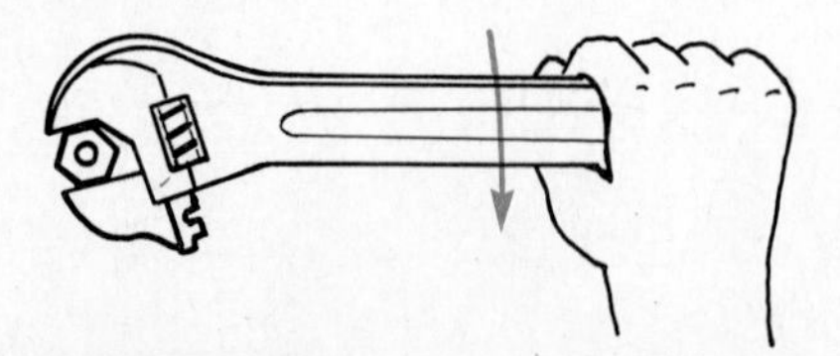

D. Adjustable wrench

Nut Drivers *Nut drivers* work like socket wrenches except that they use a straight handle similar to that of a screw driver (Figure 2-13). The sockets are used to tighten or loosen machine nuts on electrical and electronic equipment. Most nut driver shafts are hollow. This allows them to be used with nuts that are threaded on long bolts.

FIGURE 2-13 NUT DRIVER

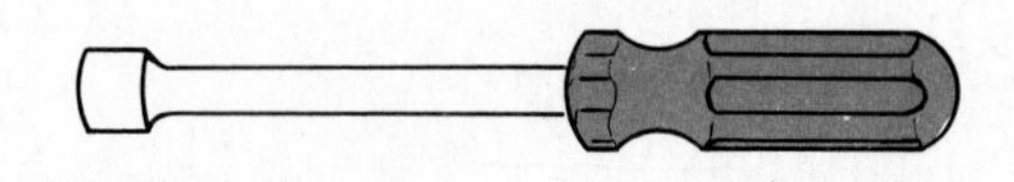

FIGURE 2-14 ALLEN KEYS

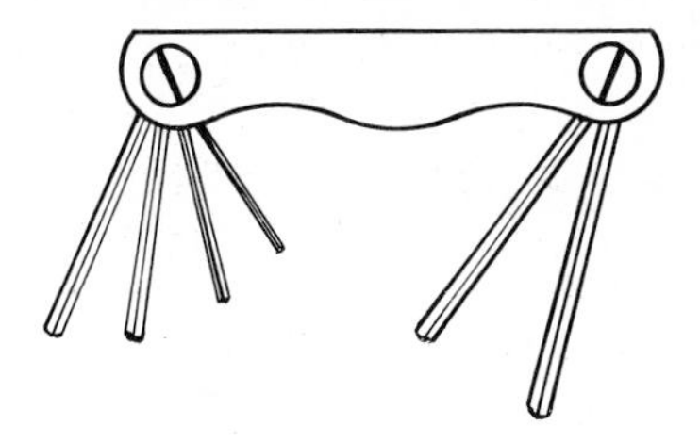

Allen Keys Some electronic control knobs are fastened in place by means of a small set screw with a hexagonal socket in the top of it called an Allen head. *Allen Keys* (Figure 2-14) are used for loosening and tightening this type of set screw.

Insulation Removing Devices A *wire stripper* or *knife* (Figure 2-15A, B) is used for removing insulation from wires and scraping the wire clean of oxides or wire enamels.

FIGURE 2-16 WOOD FILE

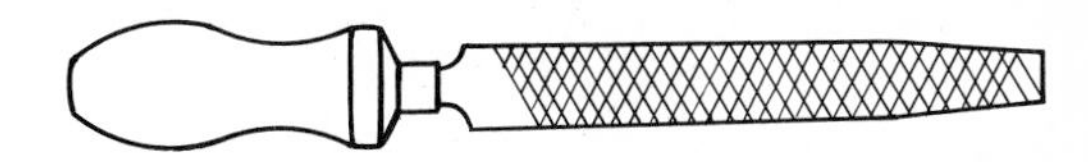

A *cable insulation stripper* (Figure 2-15C) is used to remove the insulating sheath from non-metallic sheath cable.

Files Both metal and wood files are commonly used. *Metal* files are used to remove sharp metal burrs produced as a result of cutting or drilling. *Wood* files (Figure 2-16) are used for fitting electrical outlet boxes into finished walls.

Chisels Two types of *chisels* are available. They are the *cold* chisel (Figure 2-17A) for metal work and the *wood* chisel (Figure 2-17B) for working with wood.

FIGURE 2-15 INSULATION REMOVING DEVICES

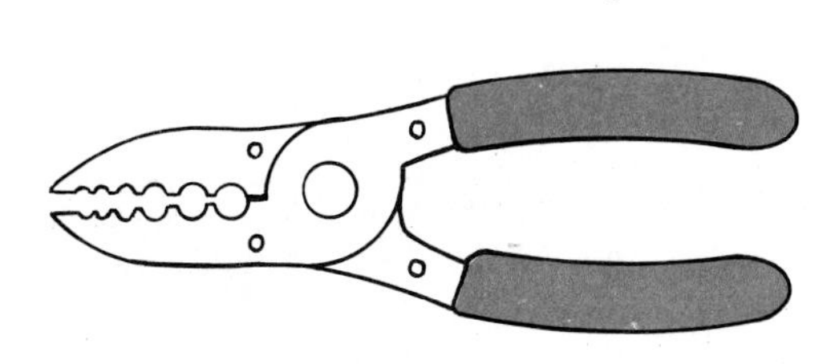

A. Wire stripper

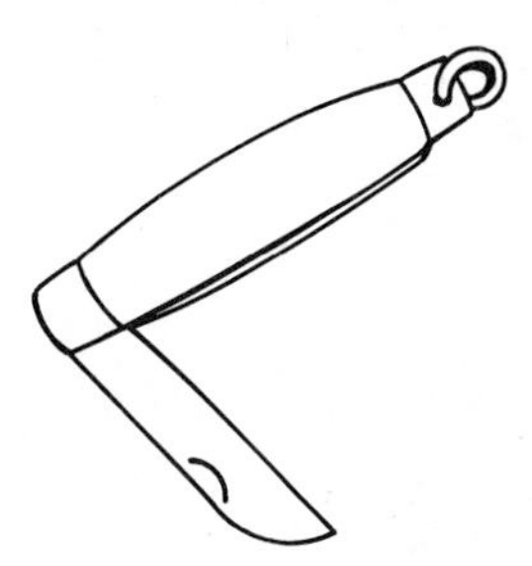

B. Knife

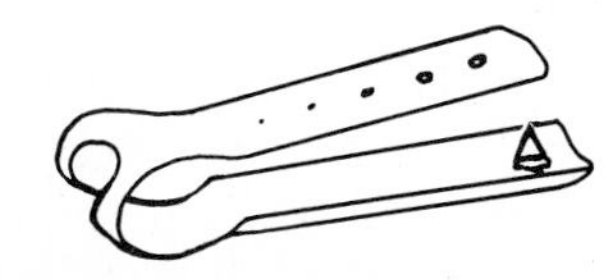

C. Sheathed cable insulation stripper

FIGURE 2-17 CHISELS

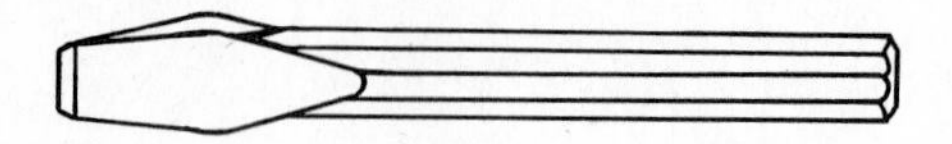

A. Cold chisel

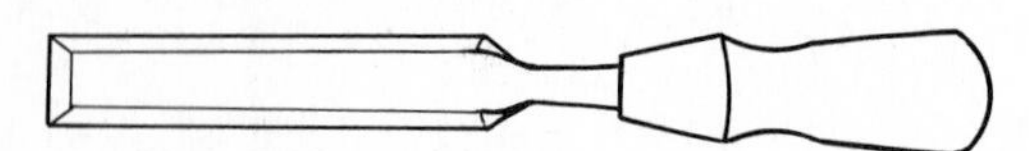

B. Wood chisel

Fish Tape A *fish tape and reel* (Figure 2-18) is a tool designed to fish and pull wire through wall partitions and electrical conduit.

FIGURE 2-18 A FISH TAPE AND REEL

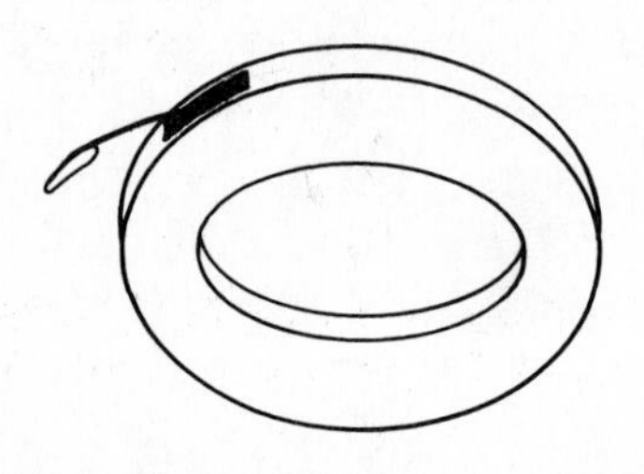

Soldering Equipment The *soldering gun* (Figure 2-19A) is a common soldering tool for general hand wired circuits. The *soldering pencil* (Figure 2-19B) is a common soldering tool used for printed circuit board soldering.

2-3 Proper Care and Use of Tools

A skilled worker is often judged by the quality and condition of his or her tools. Quality tools that are handled properly will last indefinitely. Some of the ways which will help keep your tools in good working condition are:

FIGURE 2-19 SOLDERING EQUIPMENT

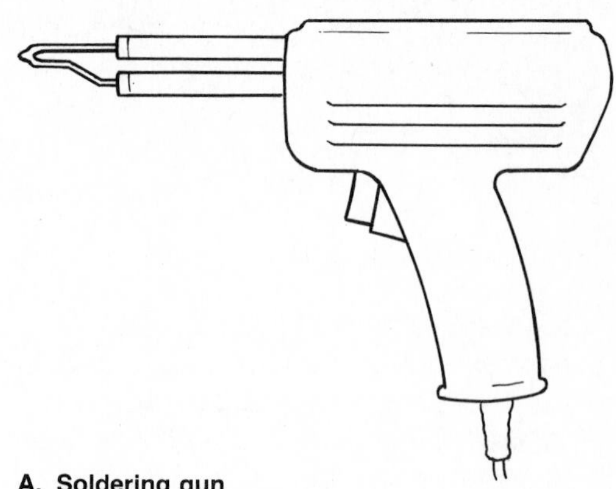

A. Soldering gun

B. Soldering pencil

1. Keep tools cleaned and well oiled.
2. Provide for proper storage of tools.
3. Use the right tool for the job.
4. Use the right size of tool for the job.
5. Keep drills, auger bits and saw blades sharp.
6. Replace dull hacksaw blades.
7. Never use a file without a firmly fitted handle.
8. Hammers with loose heads should be replaced.
9. Needle-nose pliers must be used on light wires only. The tips will break or bend if abused.
10. Pliers should not be used on nuts, as both the pliers and the nuts will become damaged.
11. Never use a screwdriver that has too large or too small a tip for the screw.

12. Never use a screwdriver as a pry bar or cold chisel.
13. Keep soldering gun or iron tips clean.
14. Whenever possible, pull rather than push on a wrench.
15. Do not use the hammer handle as a driver.
16. Always use an adjustable wrench large enough to handle the job. Using too small a wrench can cause the moveable jaw to break.
17. When replacing a hacksaw blade, be sure to mount the blade with its teeth slanting *away* from the handle.

2-4 Practical Assignments

The teacher will display each type of measuring device and tool mentioned in this unit. Each will be identified by a number. Record, in chart form, the number with the corresponding name of the device and list one specific application next to the name.

2-5 Self Evaluation Test

1. Name the tool or measuring device best suited for each of the following jobs:
   (a) measuring the current flow to a motor without making direct contact with the circuit.
   (b) checking for the presence of a voltage at a wall receptacle.
   (c) measuring voltage, current or resistance of an electronic circuit.
   (d) tighten or loosen a screw with an X-shaped insert in its head.
   (e) trimming the ends of wire on a terminal board connection.
   (f) making loop ends on wire for connection to a terminal screw.
   (g) tighten or loosen a screw with a *square-shaped* insert in its head.
   (h) cutting a piece of metal or electrical conduit.
   (i) twisting wires together.
   (j) cutting a hole in a finished plastered wall in order to mount an outlet box.
   (k) removing sheathed cable insulation.
   (l) removing metal burrs from a drilled hole.
   (m) soldering a resistor on a printed circuit board.
   (n) marking the location of a hole to be drilled in a metal chassis.
   (o) tighten or loosen a small machine nut threaded on a long bolt.
   (p) tighten or loosen a square head machine nut of an odd size.
   (q) remove a large number of standard size machine nuts with speed and ease.
   (r) notching a wooden wall partition to fit an electrical outlet box.
   (s) removing insulation from wires.
   (t) drilling a hole with an auger bit.
   (u) removing nails.
   (v) removing a volume control knob fastened by a set screw with a hexagonal socket.
   (w) pulling wires through electrical conduit.
   (x) drill a hole in a 2 cm hard steel plate.
2. What two general rules apply to the selection of a tool for a particular job?
3. What two rules apply to the storage of tools?
4. List five other rules that apply to the proper care and use of tools.

3

CONDUCTORS, SEMICONDUCTORS AND INSULATORS

OBJECTIVES

Upon completion of this unit you will be able to:

- Outline the basic structure of matter.
- Compare the particles within the atom with regard to type of electrical charge and mass.
- Draw the structure of an atom according to the Bohr model of an atom.
- Compare the atomic structure and type of charge of an atom and ion.
- Define electricity, conductor, semiconductor and insulator.
- Compare the atomic structure of a conductor, semiconductor and insulator.
- Classify common materials as being conductors, semiconductors or insulators.
- Draw the schematic circuit diagram for a simple continuity tester and explain its use as an out-of-circuit electrical tester.

3-1 Structure of Matter

All matter is made up of tiny particles which are called *molecules*. Molecules are so small that they are invisible to the unaided eye. In fact, millions of molecules can be found on the head of a pin. A molecule is defined as the smallest particle of matter which can exist by itself and still retain all the properties of the original substance.

Molecules are made up of even smaller particles, each of which is called an *atom*. Atoms, in turn, can be broken down into even smaller particles. These smaller sub-atomic particles are known as *electrons*, *protons* and *neutrons* (Figure 3-1).

FIGURE 3-1 A MODEL OF THE STRUCTURE OF MATTER

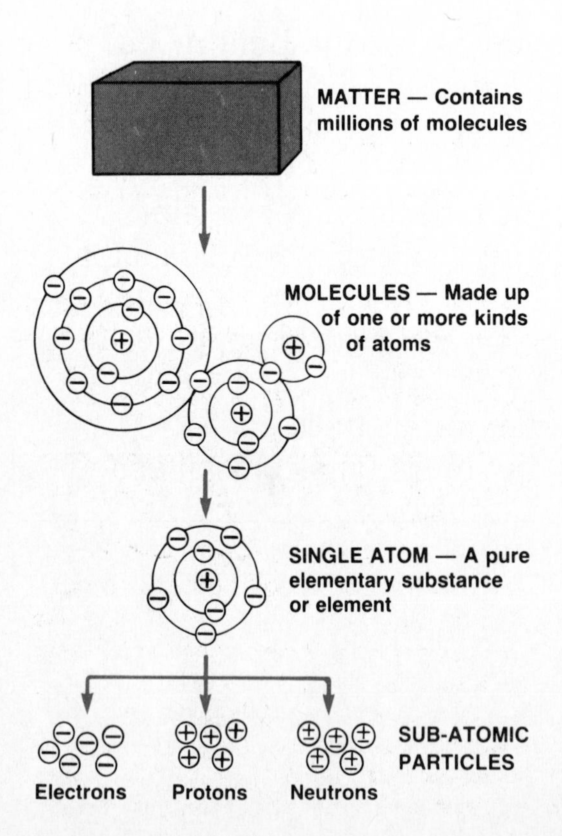

FIGURE 3-2 ROTATING ELECTRONS

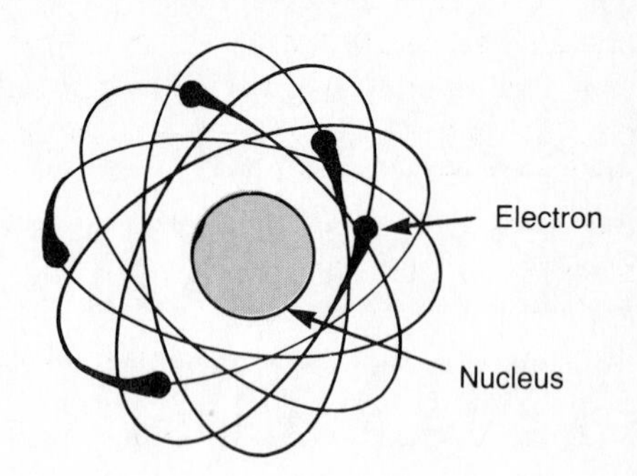

3-2 Bohr Model of Atomic Structure

The model of the atom as proposed by the physicist Niels Bohr gives a concept of its structure which is helpful in understanding the fundamentals of electricity. According to Bohr, the atom is similar to a miniature solar system. As with the sun in the solar system, the nucleus is located in the center of the atom. Tiny particles called electrons rotate in orbit around the nucleus just as the planets rotate around the sun (Figure 3-2).

The electrons are prevented from being pulled into the nucleus by the force of their momentum. They are prevented from flying off into space by an attraction between the electron and the nucleus. This attraction is due to the electrical charge on the electron and the nucleus. The electron is said to have a negative (−) charge and the nucleus, a positive (+) charge. These unlike charges attract each other.

Most of the mass of an atom is found in its nucleus. The particles that can be found in the nucleus are called protons and neutrons. A proton has a positive (+) electrical charge which is exactly equal in strength to that of the negative (−) charge of an electron. The proton is much heavier than the electron. The mass of the neutron is slightly more than that of the proton but it has no electrical charge; hence its name neutron.

Neutrons, as far as is known, do not enter into ordinary electrical activity. Every atom contains an equal number of electrons and protons, making its combined electrical charge zero or neutral. The total number of protons in the nucleus of an atom is called the *atomic number* of the atom. The total number of both protons and neutrons is known as the *atomic mass* of the atom (see the Aluminum atom example in Figure 3-3).

FIGURE 3-3 THE BOHR MODEL OF THE ATOM

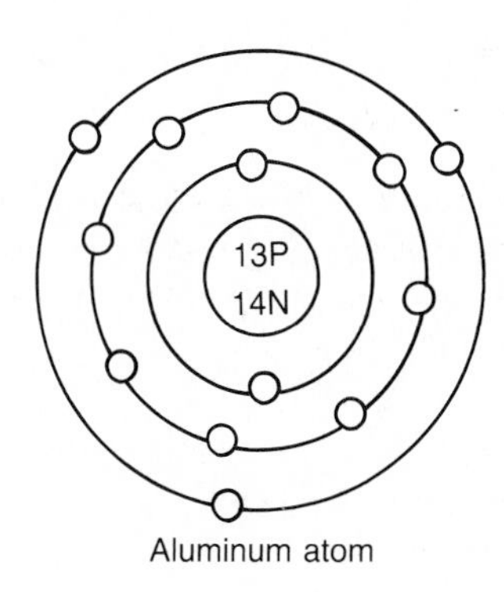

NOTE:

1. **No. Protons = No. Electrons**
   13 = 13
2. **Net electrical charge is neutral or zero.**
3. **Atomic Number = No. Protons**
   = 13
4. **Atomic Mass = No. Protons + No. Neutrons**
   = 13 + 14
   = 27

Energy Shells According to Bohr's model of the atom, electrons are arranged in *shells* around the nucleus. A shell is an orbiting layer or energy level of one or more electrons. The major shell layers are identified by numbers or by letters starting with K nearest the nucleus and continuing alphabetically outwards. There is a maximum number of electrons that can be contained in each shell. Figure 3-4 illustrates the relationship between the energy shell level and the **maximum** number of electrons it can contain.

FIGURE 3-4 ELECTRON SHELLS

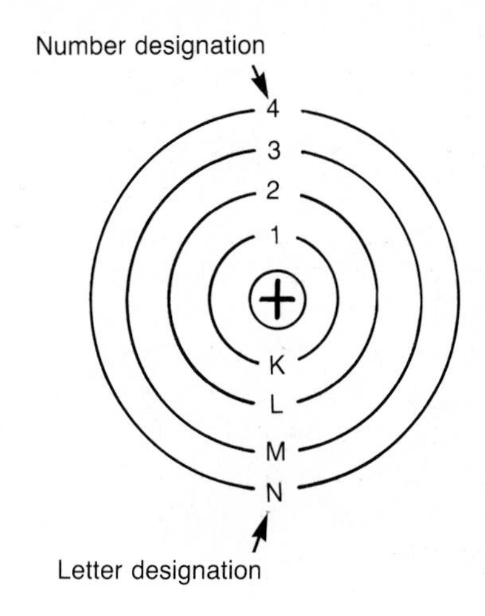

SHELL LETTER	MAX. ELECTRONS IT CAN HOLD
K	2
L	8
M	18
N	32

If the total number of electrons for a given atom is known, the placement of electrons in each shell can be easily determined. Each shell layer, beginning with the first, is filled with the maximum number of electrons in sequence. For example, a normal copper atom which has 29 electrons would have (Figure 3-5):

Shell K (or #1) = 2 (full)
Shell L (or #2) = 8 (full)
Shell M (or #3) = 18 (full)
Shell N (or #4) = 1 (incomplete)

FIGURE 3-5 PLACEMENT OF ELECTRONS IN A COPPER ATOM

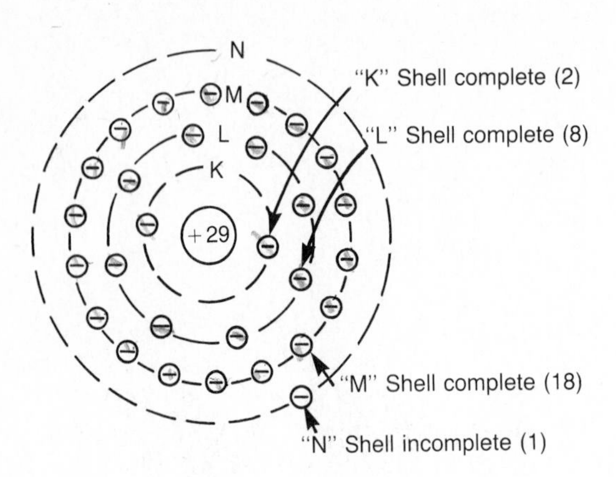

FIGURE 3-6 HOW ATOMS FORM IONS

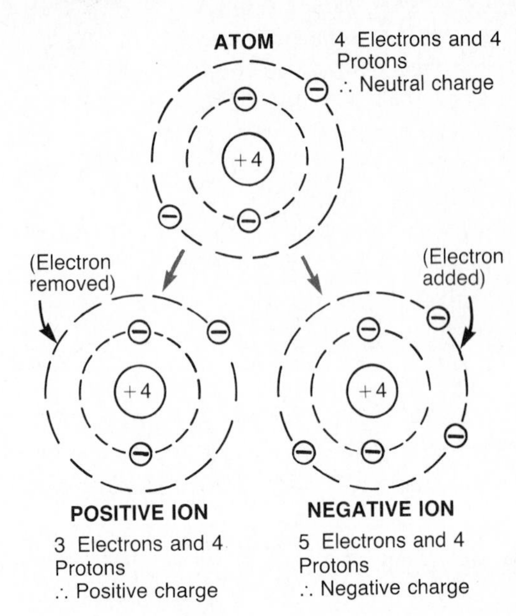

3-3 Ions

Normally the atom has just as many electrons as it has protons. Therefore, the total negative charges equal the total positive charges. The atom in this state is said to be electrically neutral. It is possible, however, through the action of some outside force, for the atom to lose or acquire electrons (Figure 3-6). This unbalanced atom is called an *ion*. A *negative ion* is an atom which has acquired electrons. It now has more electrons than protons and is said to be negatively (−) charged.

A *positive ion* is an atom that has lost electrons. It now has fewer electrons than protons and is said to be positively (+) charged.

3-4 Electricity Defined

The outer shell of the atom is called the *valence shell* and its electrons are called *valence electrons*. Because of their greater distance from the nucleus, and because of partial blocking of the electric field by electrons in the inner shells, the attracting force on the valence electrons is less. Therefore, valence electrons can be set free most easily. Whenever a valence electron is removed from its orbit it becomes known as a *free electron*. Electricity is commonly defined as **the flow of these free electrons through a conductor** (Figure 3-7).

3-5 Electrical Conductors, Insulators and Semiconductors

Conductors A *conductor* is a material that has many free electrons permitting electrons to move through it easily. Generally, conductors have incomplete valence shells of one, two or three electrons. Most metals are good conductors. Copper is the most common metal used as an electrical conductor because of its relatively low cost and good conducting ability. The following is a

FIGURE 3-7 ELECTRICITY — THE FLOW OF FREE ELECTRONS

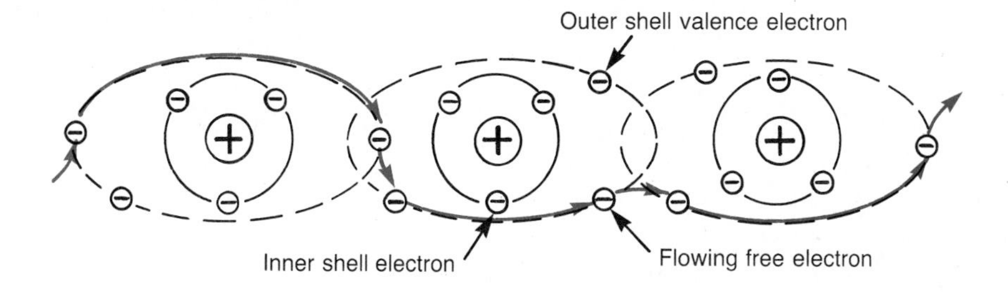

list of fairly good metal conductors, listed in order of their conductivity from good to poor:

silver	brass
copper	platinum
gold	iron (pure)
aluminum	tin
tungsten	lead
zinc	

Insulators An *insulator* is a material that has few, if any, free electrons and resists the flow of electrons. Generally, insulators have full valence shells of five, six or seven electrons. Some common insulators are air, glass, rubber, plastic, paper and porcelain.

No material has been found to be a perfect insulator. Every material can be forced to permit a flow of electrons from atom to atom if enough external force is applied. Whenever a material classified as an insulator is forced to pass an electric current, the insulator is said to have been broken down or to have become ruptured.

Semiconductors A *semiconductor* is a material that has some of the characteristics of both a conductor and an insulator. Semiconductors have valence shells containing four electrons. A pure semiconductor may either act as a conductor or insulator depending upon the temperature at which it is operated. Operated at low temperatures it is a fairly good insulator. Operated at high temperatures it is a fairly good conductor.

Common examples of pure semiconductor materials are silicon and germanium. Specially treated semiconductors are used to produce modern electronic components such as diodes, transistors and integrated circuit chips.

3-6 The Continuity Tester

A simple circuit made up of a lamp, battery and test leads can be constructed to test materials for their ability to conduct electricity. Connecting the two test leads across good conductors produces a flow of electrons through it causing the light to come on at full brightness (Figure 3-8A). Poorer conductors connected across the test leads produce varying brightness of light. Insulators produce little or no flow of electrons or light (Figure 3-8B).

This same circuit can also be used as a continuity tester to test electrical parts out-of-circuit (Figure 3-9). In this case a continuous conducting path across the test leads turns the lamp on fully and an open path produces no light at all. Practical applications for this type of testing include: checking extension cords for open leads, checking for blown fuses and checking for defective switches and pushbuttons. As mentioned,

FIGURE 3-8 SIMPLE TEST CIRCUIT FOR CONDUCTORS AND INSULATORS

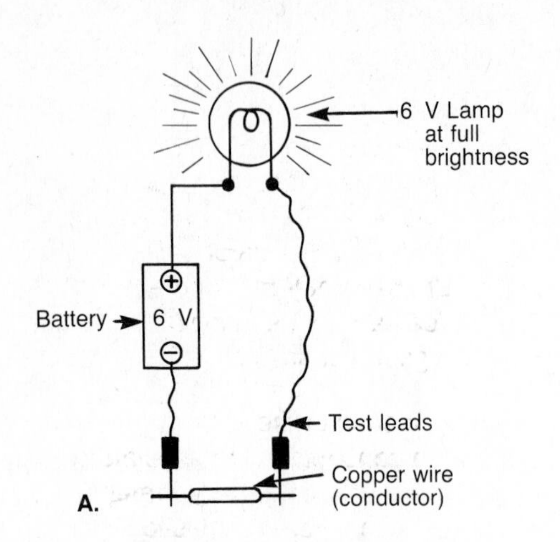

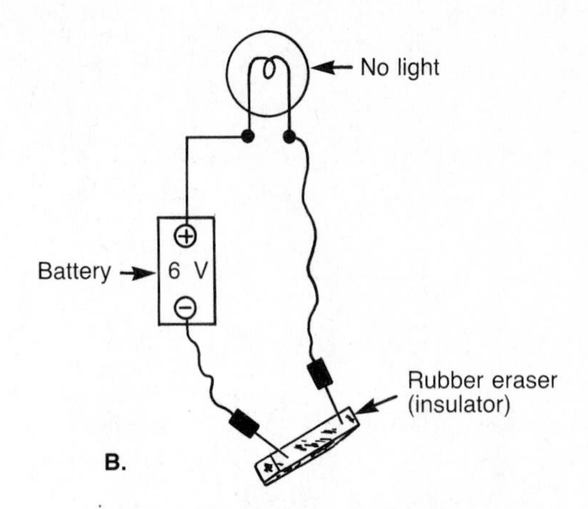

FIGURE 3-9 CONTINUITY TESTER

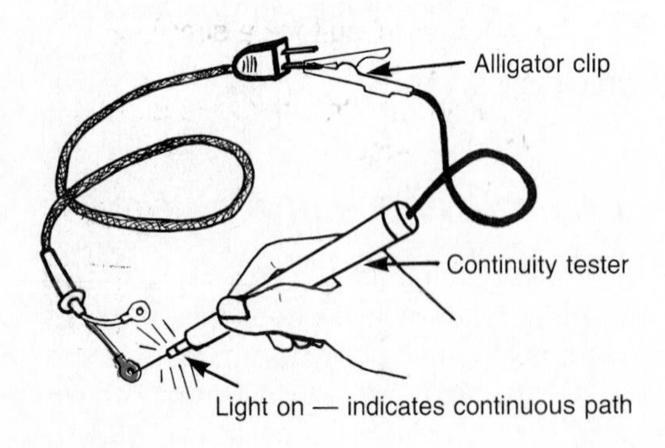

this tester is designed for checking electrical components out of their normal circuits. Under no circumstances should this circuit be connected to a circuit where other sources of voltage are present as a serious safety hazard would be created.

3-7 Practical Assignments

1. (a) Construct a simple continuity test circuit. Use this circuit to test and classify each of the following as being a conductor or an insulator:

paper	tap water
copper	plastic
wood	rubber
lead	steel

   (b) Repeat the above, using an ohmmeter.

2. (a) Use the continuity test circuit to classify each of the following sets of electrical components as being good or open circuited. The teacher will supply a color code set of one good and one open circuited component for each device listed.
   (i) plug fuse
   (ii) cartridge fuse
   (iii) single pole switch
   (iv) length of insulated copper wire
   (v) extension cord
   (vi) light bulb

   (b) Repeat the above, using an ohmmeter.

3-8 Self Evaluation Test

1. Draw a simple chart showing the basic structure of matter.

2. (a) State the type of electrical charge associated with an electron, proton and neutron.
   (b) Compare the mass of the proton and neutron with that of the electron.
3. A pure copper atom contains 29 electrons. Draw the structure of this atom, according to the Bohr model, showing the proper placement of these electrons.
4. (a) What type of electrical charge is associated with a single atom? Why?
   (b) What type of electrical charge is associated with a negative ion? Why?
   (c) What type of electrical charge is associated with a positive ion? Why?
5. According to the atomic theory of matter, what is electricity?
6. Define each of the following:
   (a) electrical conductor
   (b) electrical insulator
   (c) electrical semiconductor
7. (a) State the number of outer shell valence electrons usually associated with a conductor, semiconductor and insulator.
   (b) Under what temperature condition does a pure semiconductor act as a good conductor?
8. Classify each of the following as being a conductor or an insulator:

paper	tap water
copper	plastic
wood	rubber
lead	steel.

9. (a) Draw the circuit for a simple continuity tester.
   (b) Explain how this tester is used to check a fuse.

4

SOURCES AND CHARACTERISTICS OF ELECTRICITY

OBJECTIVES

Upon completion of this unit, you will be able to:

- Define static and current electricity.
- Explain how static positive and negative charges are produced.
- State the Law of Electric Charges.
- Explain the difference between *direct current* and *alternating current* electricity.
- List the basic sources of electricity and electrical devices used to convert the various energy forms.

4-1 Static Electricity

The term static means standing still or at rest. Static electricity is an electrical charge at rest. You may generate static electricity when you walk across a carpet or run a plastic comb through your hair.

For experimental purposes, a static charge can be produced by rubbing an ebonite or hard rubber rod with a piece of fur (Figure 4-1). Normally the atoms of both materials have the same number of electrons and protons and are therefore electrically balanced or neutral. When they are rubbed together, electrons move from one material to the other. In this case the ebonite rod takes on electrons giving it an excess of electrons or what is called a *negative charge*. At the same time, the fur, which has lost electrons, is considered to have an electron shortage or a *positive charge*.

A glass rod and silk cloth can also become charged when rubbed together (Figure 4-2). In this case the glass rod loses some of its electrons to the silk cloth. This leaves the

FIGURE 4-1 CHARGING AN EBONITE ROD

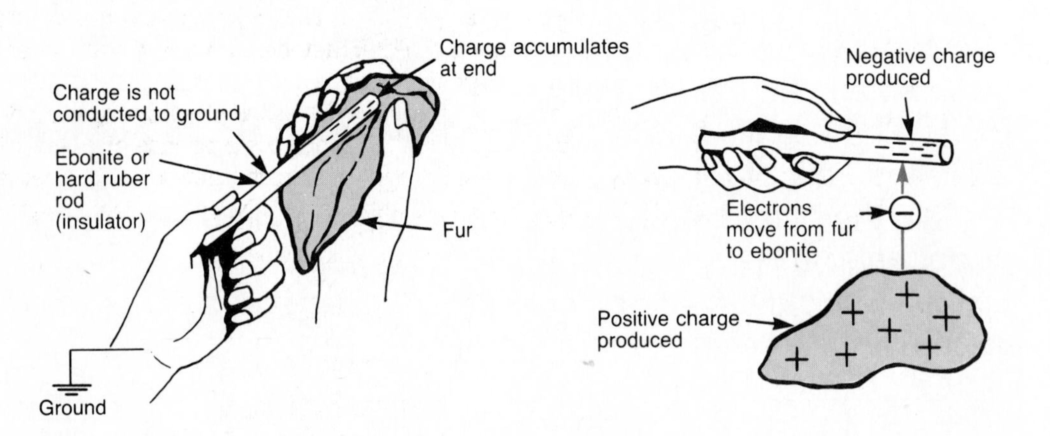

FIGURE 4-2 CHARGING AND DISCHARGING A GLASS ROD

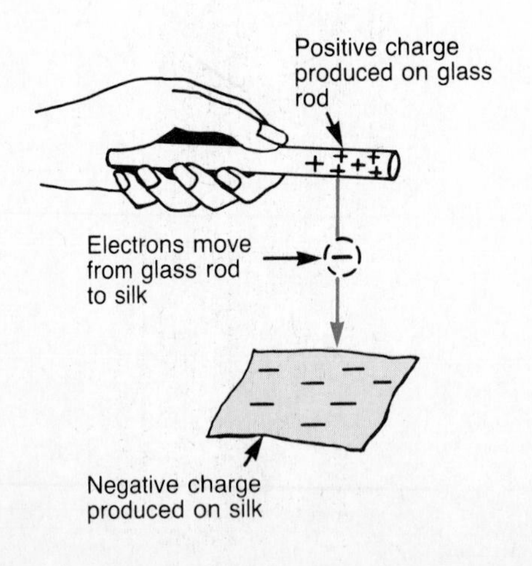

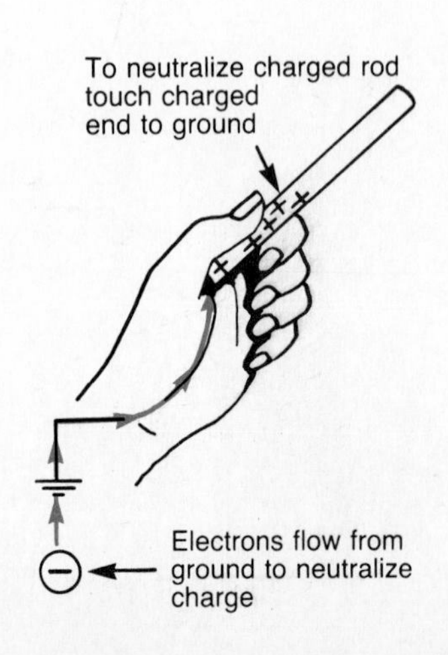

glass rod with a positive charge and the silk cloth with a negative charge. Note that the charges are produced by removing loosely held electrons from atoms. Since protons are located at the centers of atoms, they are so tightly bound that they do not normally move. Touching the charged end of the glass rod will immediately discharge or neutralize it. In this instance, electrons flow from your hand to neutralize the charged rod end.

4-2 Law of Electric Charges

When two charged objects are placed near each other, a force will exist in the space between them. This region is known as an *electric field of force* or *electrostatic field*. The force between the two charged objects varies directly with the quantity of charge on the objects and inversely with the distance between them. The force exerted by these charges can be either one of attraction or repulsion. Two **like** charges repel one another while two **unlike** charges attract one another. Stated simply, the *Law of Electric Charges* is: **like charges repel and unlike charges attract** (Figure 4-3).

4-3 Testing for a Static Charge

The charge on an object is found by seeing how it affects an object with a known charge. If the two repel, the charges are alike. If they attract, the charges are opposite. If you think the object is neutral, test it with another neutral object. A neutral object does not attract another neutral object. However, a neutral object is attracted by an object with either a positive or negative charge.

The **aluminum leaf electroscope** is a device for detecting the presence of an electric charge, and also for determining whether this charge is positive or negative (Figure

FIGURE 4-3 LAW OF ELECTRIC CHARGES

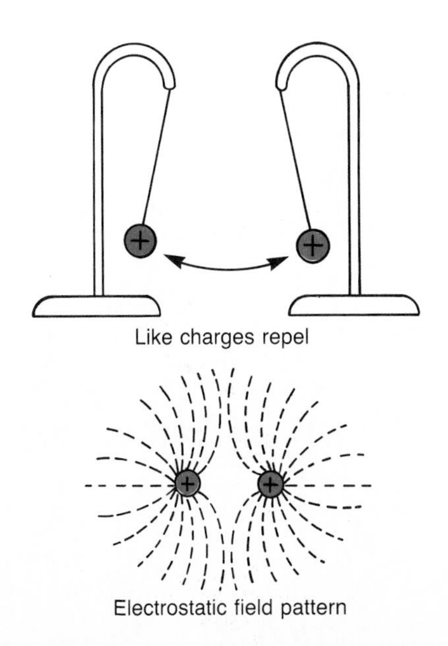

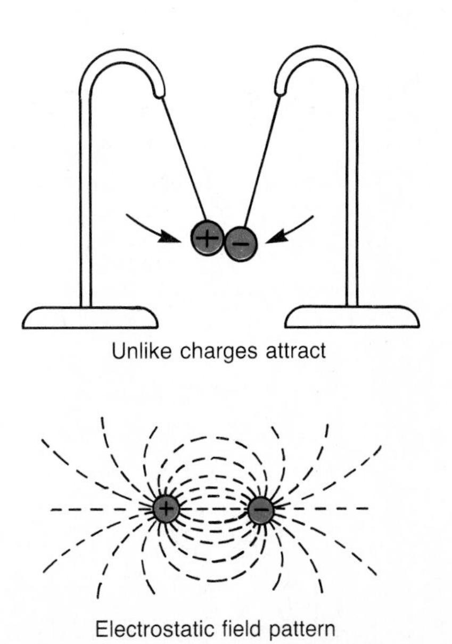

FIGURE 4-4 ALUMINUM LEAF ELECTROSCOPE

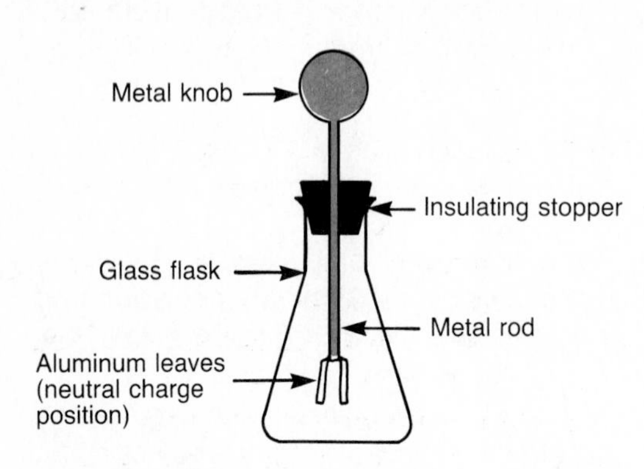

FIGURE 4-5 CHARGING THE ELECTROSCOPE

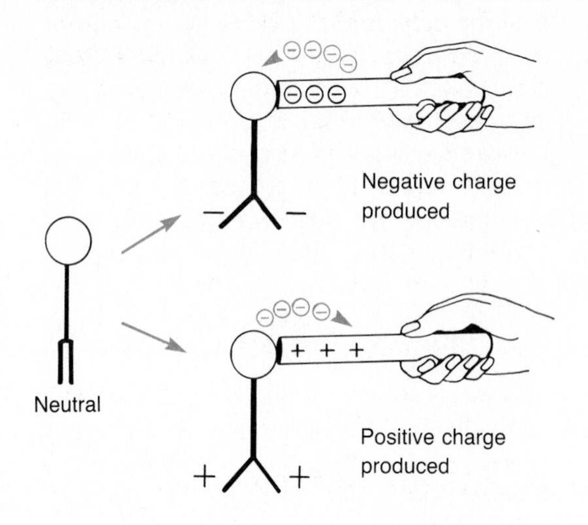

4-4). This device is made up of a glass flask with an insulating stopper. A metal rod passes through the center of the stopper. A metal knob is fastened to the outer end of the rod. Two very thin leaves of aluminum are fastened to the end of the rod inside the flask.

Normally the positive and negative charges within the electroscope balance each other leaving it neutral. When a negatively charged body touches the knob of the electroscope, electrons flow from the charged body into the knob and down to the aluminum leaves (Figure 4-5). Each leaf then becomes negatively charged. Since like charges repel, the leaves, both negative, will diverge indicating that the object contained a static charge. The extent to which the leaves diverge is an indication of the quantity or amount of the charge on the object. When the charged body is removed, the electroscope is left with an excess of electrons. The aluminum leaves remain diverged and the electroscope is said to be negatively charged.

4-4 Current Electricity

Current or dynamic electricity is defined as an electrical charge in motion (Figure 4-6). It consists of a flow of negative electron charges from atom to atom through a conductor. The external force that causes the electron flow is called the electromotive force (emf) or **voltage**.

The electromotive force or voltage of a battery consists of one positive and one negative terminal. The negative terminal has an excess of electrons while the positive terminal has a deficiency of electrons. When

FIGURE 4-6 CURRENT OR DYNAMIC ELECTRICITY

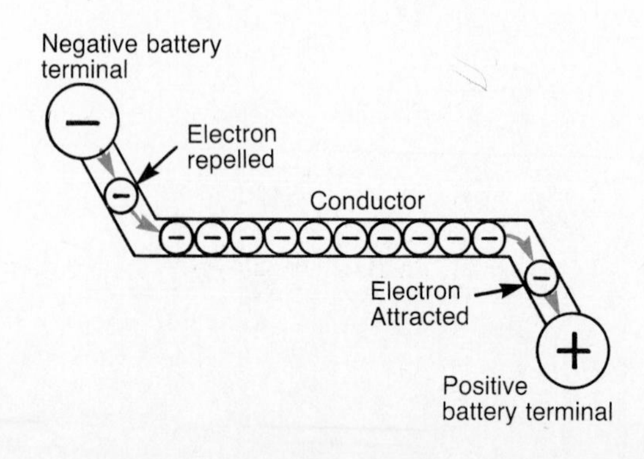

a conductor is placed across the two terminals of the battery, a flow of electrons occurs. The positive terminal of the battery has a shortage of electrons and thus attracts electrons from the conductor. The negative terminal has an excess of electrons which repels electrons into the conductor.

Although "static" and "current" electricity may seem to be different, they are really the same. Both consist of electric charges. Static electricity consists of electrons at rest on an insulated object and does little work. Current electricity moves and does useful work. When static electricity is discharged it is no longer static but current electricity.

Current electricity may also be classified as being ***direct current*** (**DC**) or ***alternating current*** (**AC**). Direct current produces a flow of electrons in one direction only. Alternating current produces a flow of electrons that constantly changes in direction. A battery is a common DC current source while an electrical wall outlet or receptacle is the most common AC current source (Figure 4-7).

4-5 Sources of Electromotive Force

As mentioned, for electrons to flow there must be a source of electromotive force (emf) or voltage. This voltage source can be produced from a variety of different primary **energy sources**. These primary sources supply energy in one form, which is then converted to electrical energy.

Light **Light energy** is converted directly into electrical energy by solar or photovoltaic cells. These are made from a semiconductor light-sensitive material that makes electrons available when struck by the light energy (Figure 4-8). A direct current output voltage is produced by the cell. The output voltage is directly proportional to the light energy striking the surface of the cell.

FIGURE 4-7 DC AND AC CURRENT ELECTRICITY

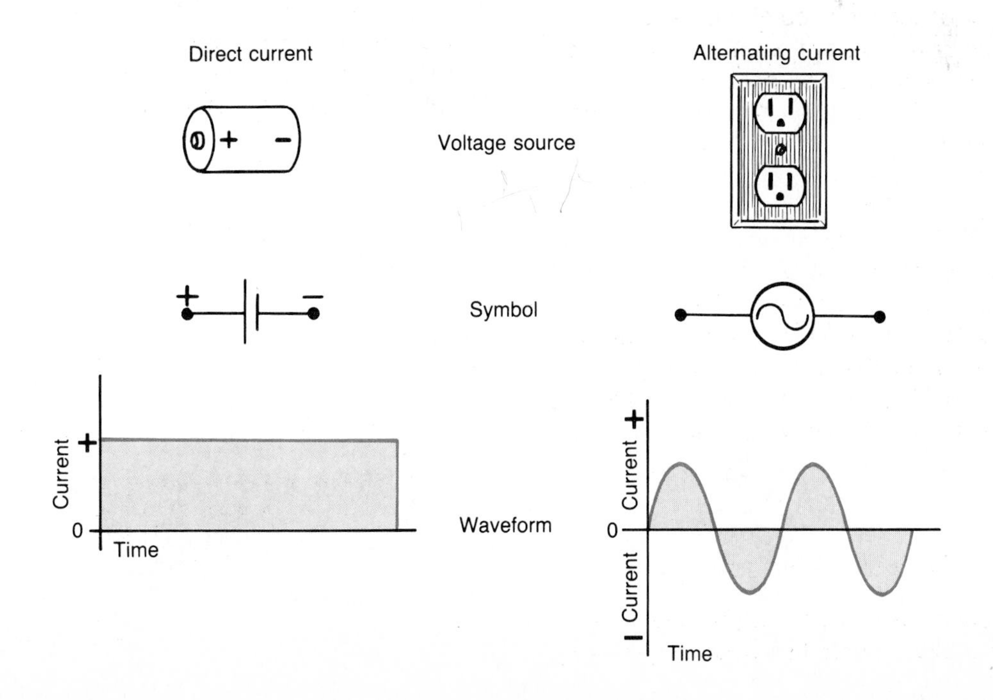

FIGURE 4-8 PRODUCING ELECTRICITY WITH LIGHT

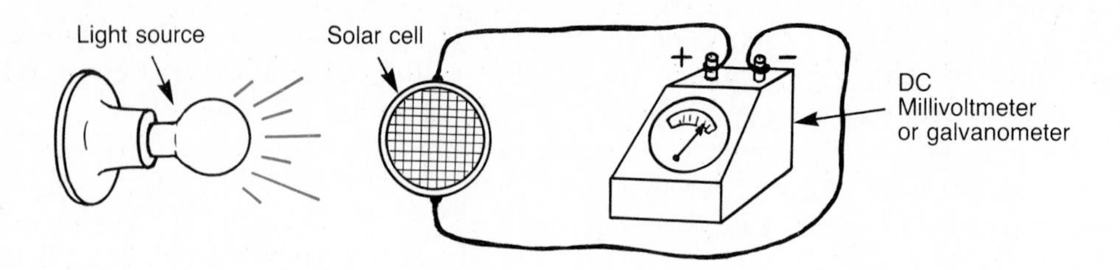

One of the best solar cells is the silicon cell. A single cell can produce up to 400 mV (millivolts) with current in the milliampere range. Solar cells are often used as sensing devices in light meters and automatic street lighting circuits. Banks of solar cells are used as electrical energy sources in satellites, digital watches and electronic calculators.

Chemical Reaction The **battery** or voltaic cell converts chemical energy directly into electrical energy (Figure 4-9). Basically, it is made up of two electrodes and an electrolyte solution. The chemical action within the cell causes the electrolyte to react with the two electrodes. As a result, electrons are transferred from one electrode to the other. This produces a positive and negative charge at the electrode terminals of the cell.

The battery is a popular low voltage, portable, DC voltage source. However, it is a relatively high cost electrical energy source and this limits its applications.

Heat Heat energy can be converted directly into electrical energy by a device called a thermocouple (Figure 4-10). A thermocouple is made up of two different types of metals joined at a junction. When heat is applied to the junction, electrons move from one metal to the other. The metal that loses electrons becomes positively charged while the metal that gains electrons takes on a negative charge. If an external circuit is connected to the thermocouple, a small amount

FIGURE 4-9 PRODUCING ELECTRICITY USING CHEMICAL ACTION

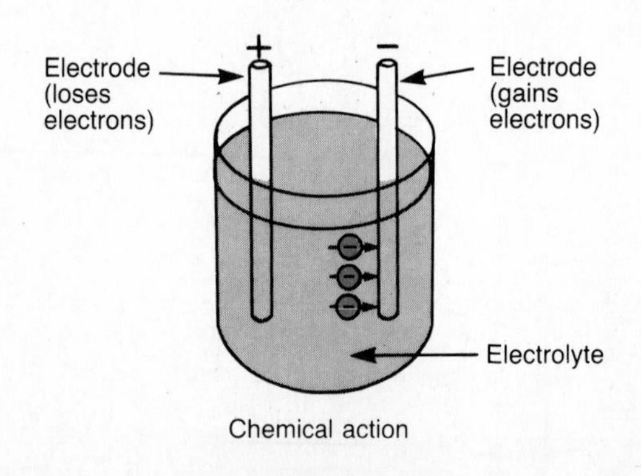

Chemical action

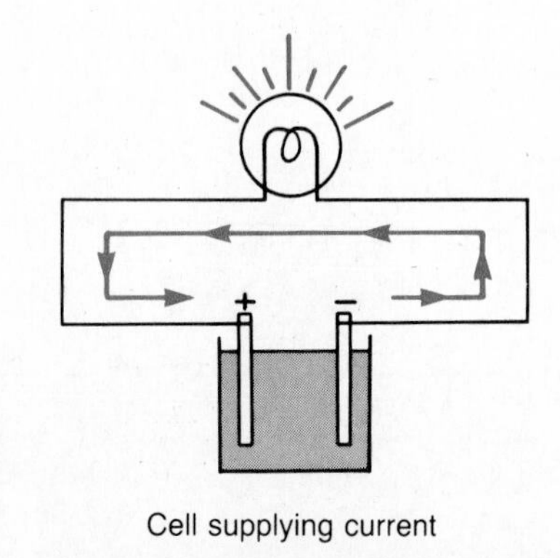

Cell supplying current

FIGURE 4-10 PRODUCING ELECTRICITY WITH HEAT

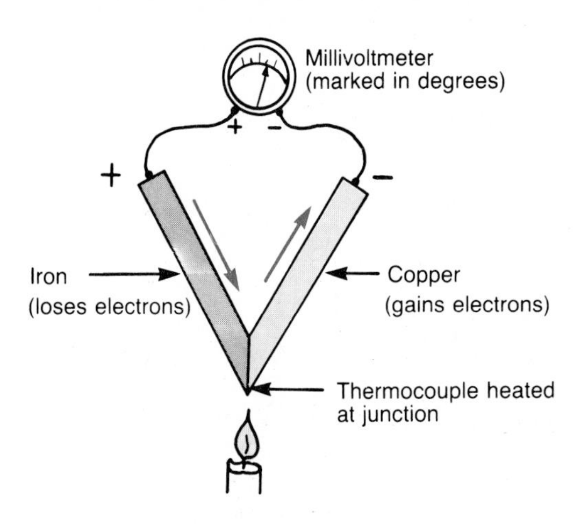

of DC current will flow as a result of the voltage between the two different metals.

The thermocouple can be used to measure high temperatures. Placed inside an industrial furnace it will produce a voltage that is directly proportional to the furnace temperature. A millivoltmeter, marked in degrees, is connected across the external thermocouple leads to indicate the temperature. Thermocouples are also used as part of electrical control systems to automatically maintain set temperature values.

Piezoelectric Effect A small voltage can be produced when certain types of crystals are put under pressure. This ability to change mechanical pressure into electricity is called the **piezoelectric effect**. This method of producing electricity is used in record players. The grooves in a disc record produce varying amounts of pressure on a crystal cartridge which in turn produces a minute electrical signal. These small signals are amplified and made to operate a speaker (Figure 4-11). The same principle applies to a crystal microphone, only in this case the pressure is produced by sound waves acting upon a diaphragm.

Friction As stated previously, static electricity is commonly produced by friction. The voltages produced by friction are difficult to harness for practical applications.

Mechanical-Magnetic Most of the electricity used is produced by converting mechanical and magnetic energy into electrical energy. If a conductor is moved through a magnetic field, a voltage is developed in the conductor (Figure 4-12). This is the principle used in generators. The magnetism can be provided by permanent magnets or electromagnets. Large generators are mechanically driven by water or steam turbines, electric motors or by gas or diesel engines. The generator may be designed to produce AC or DC current electricity.

4-6 Practical Assignments

1. (a) Using an electroscope, demonstrate the fact that a hard rubber

FIGURE 4-11 PRODUCING ELECTRICITY USING PRESSURE

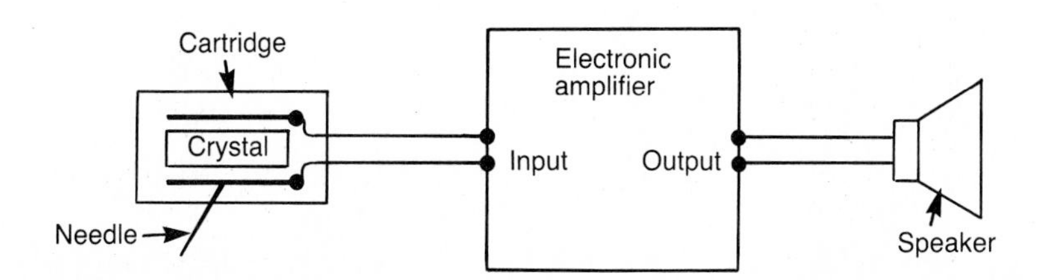

FIGURE 4-12 PRODUCING ELECTRICITY WITH MAGNETISM AND MOVEMENT

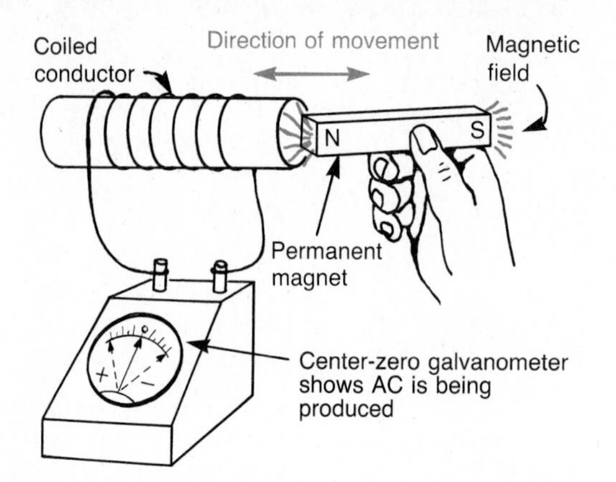

rod takes on an electrical static charge when rubbed with a piece of fur. With the aid of a simple sketch, explain how the metal leaves of the electroscope react to this static charge.
(b) With the electroscope charged negatively, touch a charged positive glass rod to the knob and make note of the reaction to the metal leaves. Explain this reaction.

2. Rub an ordinary blown-up balloon with a woolen or flannel cloth. Take the balloon and place it on a wall surface. Explain what happens.
3. Connect the output of a solar cell to a millivoltmeter or galvanometer. Record the output, as indicated by the meter, for three different light levels.
4. Connect the output from a thermocouple to a millivoltmeter or galvanometer. Heat the junction of the thermocouple with a soldering iron and record the output, as indicated by the meter, for three different heat levels.
5. Build a voltaic cell using a lemon, lime, or grapefruit as the electrolyte. Use a penny and nickel coin as the two electrodes. Record the output from the cell as indicated by a galvanometer or millivoltmeter.
6. Connect a crystal microphone to the input of an oscilloscope. Talk into the microphone and draw the shape of the waveform seen. Hum a constant note into the microphone and draw the shape of the waveform seen.
7. Using a coil of wire and a permanent magnet, demonstrate the generator principle of producing electricity. Connect the coil to a galvanometer or millivoltmeter and record the output as indicated by the meter.
8. (a) Connect a 6 V AC voltage source to the input of an oscilloscope. Draw the size and shape of the wave pattern.
   (b) Repeat for a 12 V AC voltage source.
   (c) Repeat for a 9 V DC voltage source.

4-7 Self Evaluation Test

1. What is the difference between static and current electricity?

2. The end of a rubber rod is rubbed with a piece of fur. Explain how and what charges are produced.

3. State the Law of Electrostatic Charges.

4. A positively charged glass rod is placed in contact with the knob of an electroscope. Explain the reaction which takes place.

5. Explain the difference between *direct current* and *alternating current* electricity.

6. List the six basic sources of electricity.

7. What is the basic energy source used in each of the following electrical devices:
   (a) generator
   (b) thermocouple
   (c) solar cell
   (d) crystal cartridge
   (e) battery

5

BASIC ELECTRICAL UNITS

OBJECTIVES

Upon completion of this unit you will be able to:

- Define each of the following basic electrical terms:
  - (a) current
  - (b) voltage
  - (c) resistance
  - (d) power
  - (e) energy
- State the basic unit used to measure:
  - (a) current
  - (b) voltage
  - (c) resistance
  - (d) power
  - (e) energy
- Identify the essential parts of a basic circuit and the purpose of each.
- Explain the relationship between current, voltage and resistance in a simple circuit.
- State the direction of current flow in a circuit according to:
  - (a) *electron* flow
  - (b) *conventional current* flow
- Measure the current, voltage and resistance of a simple circuit.

Electric Charge

As noted earlier, there is a tiny negative charge on each electron. For practical purposes, scientists decided to combine many of these charges to be able to measure them. The practical unit of electrical charge is the **coulomb**. One coulomb of charge is the total charge on 6.24×10^{18} electrons.

5-1 Current

The rate of flow of electrons through a conductor is called **current** (*I*). Current is measured in **amperes** (A). The term ampere refers to the number of electrons passing a given point in one second. This number is unbelievably large. If one could count the individual electrons, one would see approximately 6.24×10^{18} electrons go by during one second.

One ampere is equal to one coulomb of charge moving past a given point in *one* second. An instrument called an *ammeter* will measure electron flow in coulombs per second (Figure 5-1). The ammeter is calibrated or marked in amperes. For most practical applications, the term amperes is used instead of coulombs per second when referring to the amount of current flow.

5-2 Voltage

Voltage (*V*) is the difference of potential energy that forces electrons (current) to flow in a circuit. The voltage level or value is proportional to the difference in the electrical potential energy between two points. Voltage is measured in **volts** (V). A voltage of one volt is required to force one ampere of current through one ohm of resistance.

An instrument that will measure voltage is known as a *voltmeter* (Figure 5-2). The voltmeter is connected across the voltage source to read the applied or source voltage. It is connected across the load device to measure load voltage or potential energy difference.

5-3 Resistance

Resistance (*R*) is the opposition to the flow of electrons (current). Resistance is measured in **ohms**. The Greek symbol Ω (omega) is often used to represent ohms.

FIGURE 5-1 AN AMMETER CONNECTED TO MEASURE CURRENT FLOW

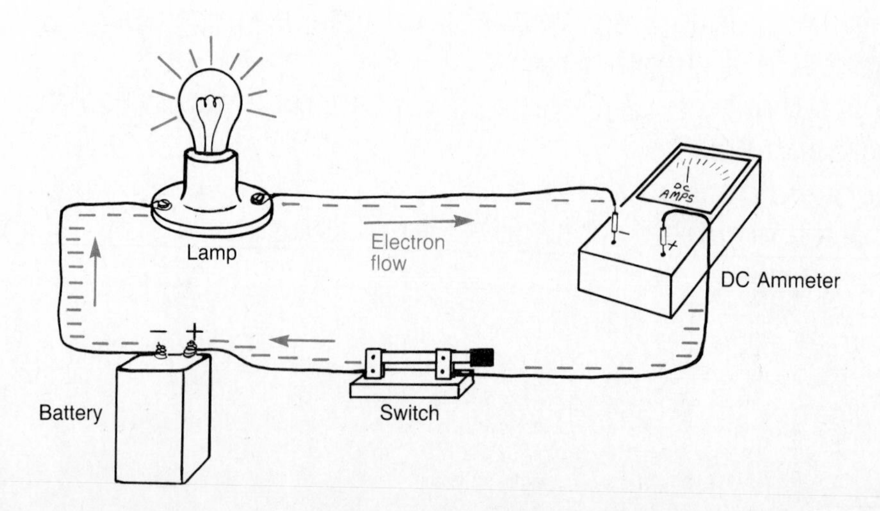

FIGURE 5-2 VOLTMETERS CONNECTED TO MEASURE DIFFERENT VOLTAGES

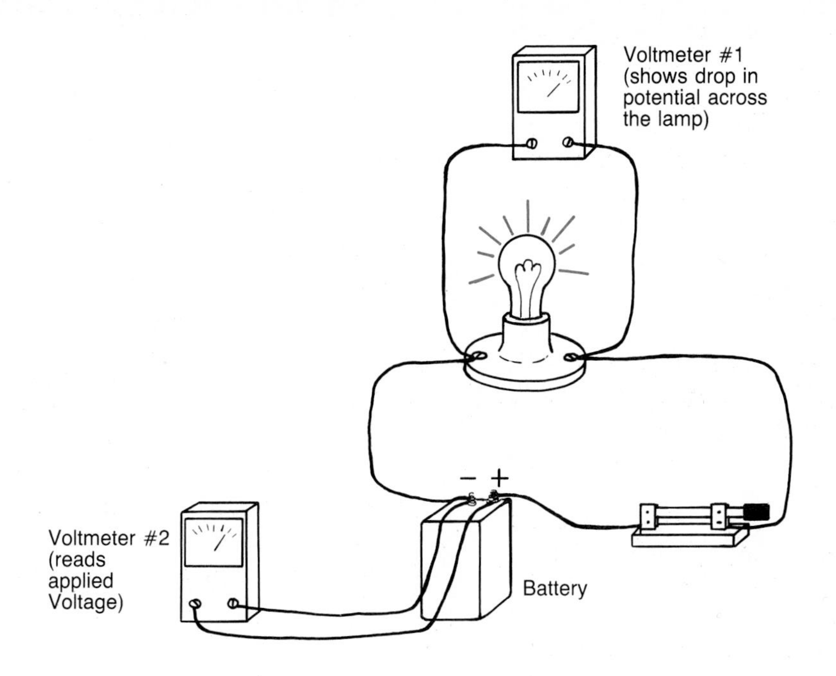

FIGURE 5-3 AN OHMMETER CONNECTED TO MEASURE RESISTANCE

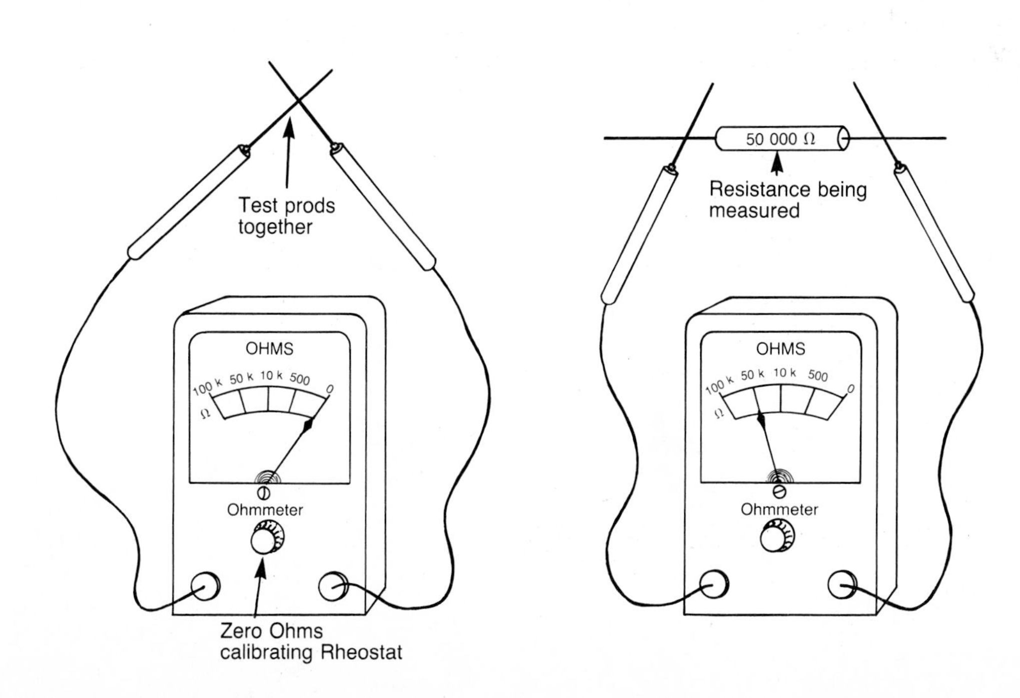

Resistance is due in part to each atom resisting the removal of an electron by its attraction to the positive nucleus. Collisions of countless electrons and atoms, as the electrons move through the conductor, also create additional resistance. The resistance created causes heat in the conductor when current flows through it.

An instrument that will measure ohms of resistance is known as an **ohmmeter** (Figure 5-3). The resistance of any material depends on the type, size and temperature of the material. Even the best conductor offers some opposition to the flow of electrons.

5-4 Power

Electrical power (P) refers to the amount of electrical energy converted to another form of energy in a given length of time. Power is measured in **watts** (W). Power in an electric circuit is equal to:

$$P = VI$$

where:

P is the power in watts
V is the voltage in volts
I is the current in amperes

All electrical loads have a specific power rating. This rating is not always specified simply in terms of watts. Ratings may give voltage and current or voltage and watts (Figure 5-4).

5-5 Energy

In a complete electrical circuit, the voltage or electromotive force pushes and pulls electrons into motion. When electrons are forced into motion, they have kinetic energy or the energy of motion. **Electrical energy** (E) is measured in **joules** (J). Energy in an electric circuit is equal to:

$$E = VIt$$

where:

E is the energy in joules
V is the voltage in volts
I is the current in amperes
t is the time in seconds

The kilowatt-hour (kW·h) is the non-SI metric unit of measurement of electrical energy. Energy measurements are used in calculating the cost of electrical energy. A kilowatt-hour meter connected to the home

FIGURE 5-4 POWER RATING OF ELECTRICAL DEVICES

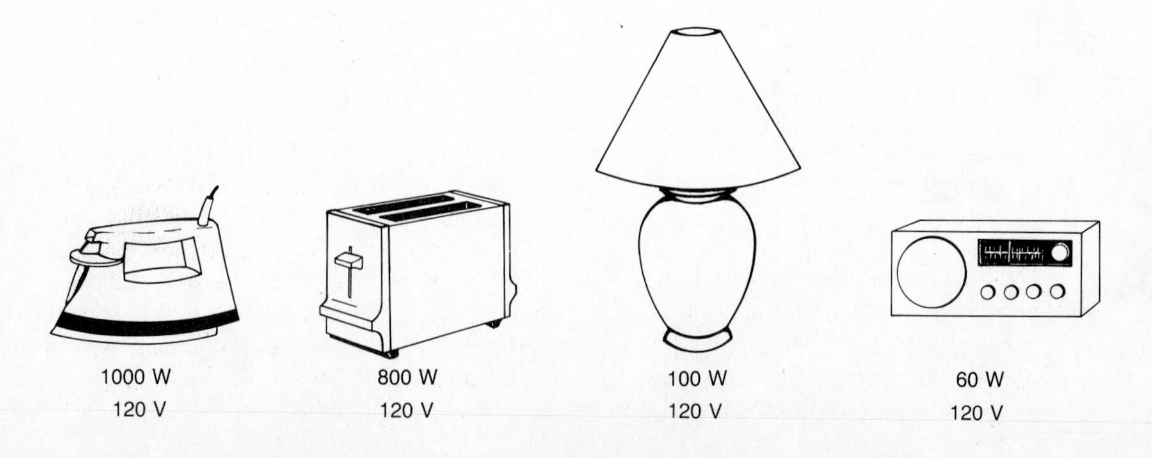

FIGURE 5-5 KILOWATT-HOUR ENERGY METER

Courtesy Ferranti-Packard Ltd.

electrical system is used to measure the amount of energy used (Figure 5-5). Energy as measured in kilowatt-hours is equal to:

$$E = \frac{Pt}{1\,000}$$

where:

E is the energy in kilowatt-hours
P is the power in watts
t is the time in hours

5-6 Basic Electrical Circuit

A closed electrical circuit can be defined as a complete electric path from one side of a voltage source to the other. Its essential parts consist of a source, load and conductors (Figure 5-6).

A **battery** or cell is an example of a **source**. The battery creates a potential energy difference or voltage across its two terminals. The electrical energy from the battery is transported through the circuit by moving electrons. These moving electrons drift from

FIGURE 5-6 BASIC ELECTRICAL CIRCUIT

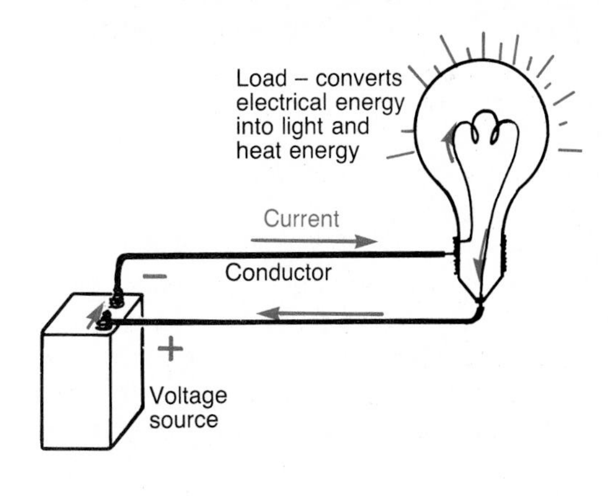

atom to atom in the direction of the positive terminal of the battery.

A **load** is any device that stores electrical energy or changes it into other forms. A lamp is an examle of a load in a circuit. Electrons flow through the lamp coverting the electrical energy of the source into light and heat energy.

Conductors provide a low resistance path from the source to the load. In an ideal circuit, electrons lose all of their available energy going through the load. In reality, a slight energy loss occurs in the conductors as electrons flow through the circuit.

5-7 Relationship Between Current, Voltage and Resistance

The amount of current (electron) flow in a circuit is determined by the voltage and resistance (Figure 5-7). The amount of current flow is directly proportional to the source voltage. In other words, when the voltage increases, the current increases; when the voltage decreases the current decreases.

If the voltage is held constant, the current will change as the resistance changes, but

FIGURE 5-7 CURRENT, VOLTAGE AND RESISTANCE RELATIONSHIPS

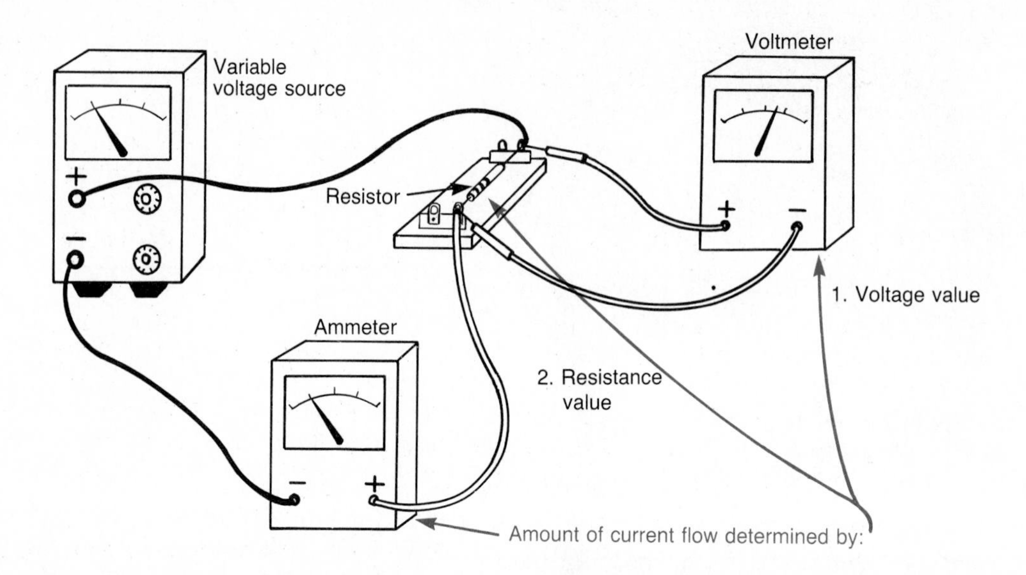

in the opposite direction. The current will decrease as the resistance increases and will increase as the resistance decreases.

5-8 Direction of Current Flow

The direction of current flow in a circuit can be marked according to "electron" flow or "conventional current" flow (Figure 5-8A, B). Electron flow is based on the electron theory of matter and, therefore, indicates the flow of current from negative to positive. Conventional current flow is based on an older fluid theory of electricity and assumes a current flow direction from positive to negative.

Both conventional current flow and electron flow are acceptable and used for different applications. It is important to

FIGURE 5-8 DIRECTION OF CURRENT FLOW

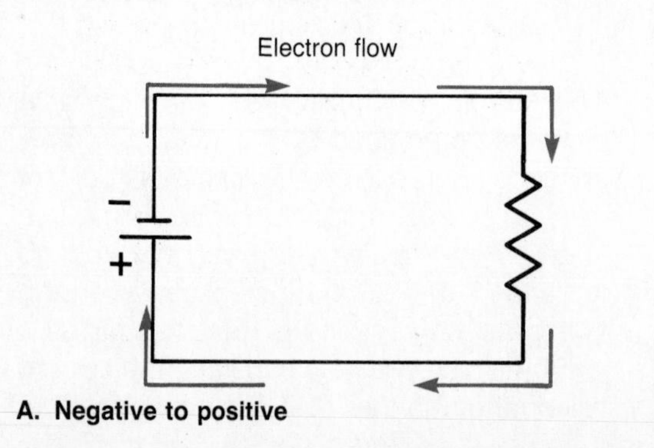

A. Negative to positive

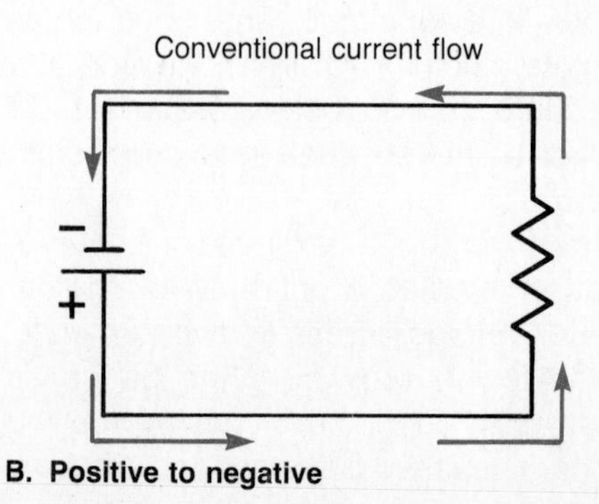

B. Positive to negative

understand and be able to think both in terms of conventional current direction and in terms of electron flow. Unless otherwise specified, the direction of current flow assumed in this text will be according to the **electron flow** (negative to positive).

5-9 Practical Assignments

1. (a) Using an ohmmeter, measure and record the resistance of any two electrical load devices (for example: lamp or resistor).
   (b) Connect each load to a battery voltage source. Using an ammeter, measure and record the current drawn by each device.
   (c) Using a voltmeter, measure and record the voltage across each load as well as that of the voltage source.
   (d) Explain the transfer of energy that took place for each of the loads tested.

2. (a) Connect a center-zero galvanometer in series with a DC source and lamp. Demonstrate how the direction of the current flow can be changed.
   (b) Draw a sketch of the circuit and indicate the direction of current according to electron flow.
   (c) On this sketch, label the source, conductors and load.

3. (a) Apply 5 V to a 100 Ω, 1 W carbon resistor. Record the current flow.
   (b) Apply 5 V to a 1 000 Ω, 1 W carbon resistor. Record the current flow.
   (c) Apply 10 V to a 100 Ω, 1 W carbon resistor. Record the current flow.
   (d) Apply 10 V to a 1 000 Ω, 1 W carbon resistor. Record the current flow.
   (e) Which circuit produced the most current? Why?

5-10 Self Evaluation Test

1. Define each of the following:
   (a) current (d) power
   (b) voltage (e) energy
   (c) resistance

2. State the basic SI unit used to measure each of the following:
   (a) current (d) power
   (b) voltage (e) energy
   (c) resistance

3. (a) Draw a basic electrical circuit and correctly label the different parts.
   (b) Explain the purpose of each of the circuit parts.

4. Express the direction of current in terms of electron flow and conventional current flow.

5. Explain how the amount of current flowing in a circuit is related to the voltage of the circuit.

6. Explain how the amount of current flowing in a circuit is related to the resistance of the circuit.

6

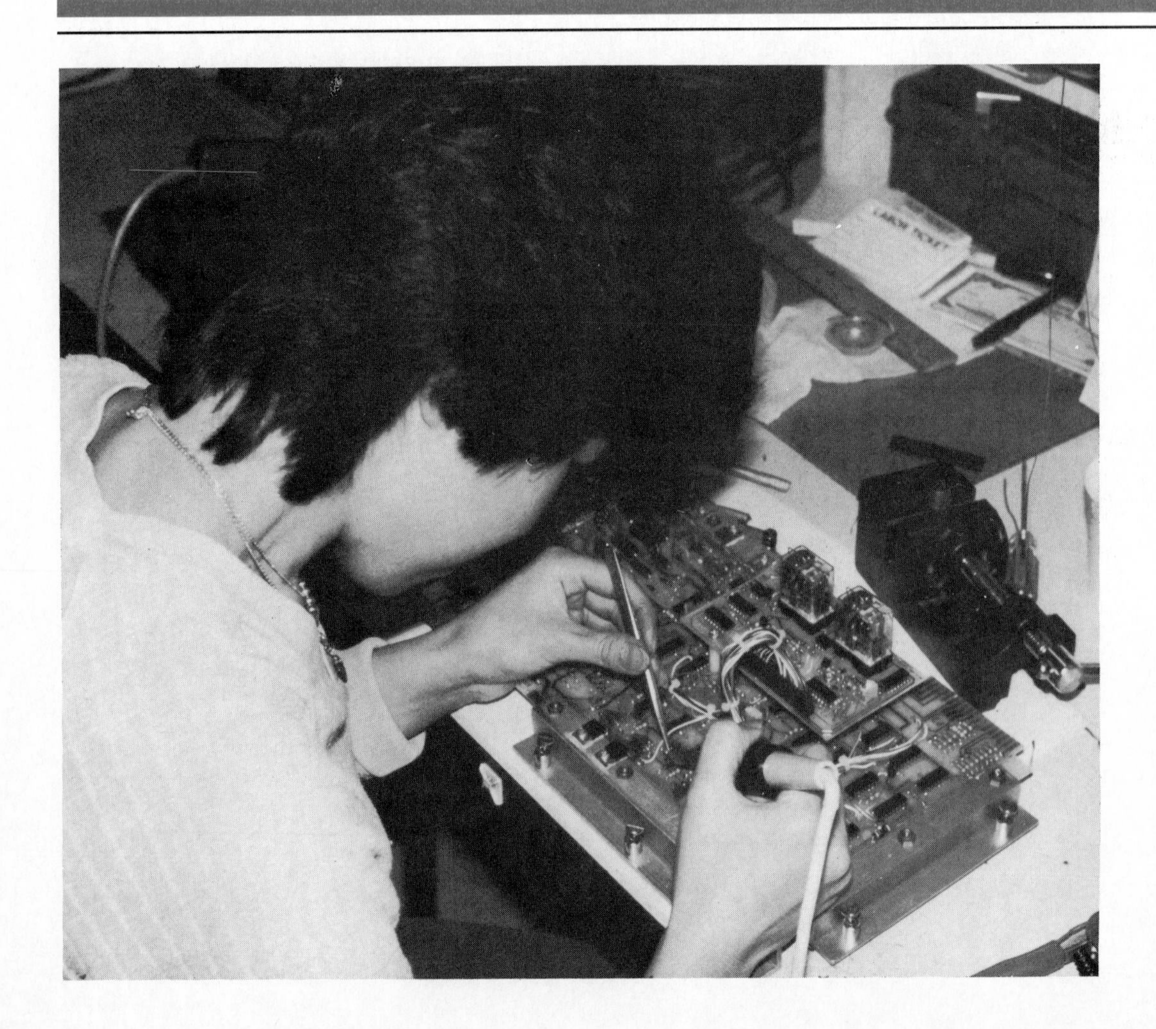

ELECTRICAL CONNECTIONS

OBJECTIVES

Upon completion of this unit you will be able to:

- Describe the negative effects of a poorly formed electrical connection.
- Properly remove insulation from a wire.
- Make a proper terminal screw connection.
- Outline the National Electrical Code® requirements for soldered electrical splices.
- Outline the proper soldering techniques used to solder a splice and terminal connection.
- Make a proper Western Union and pigtail splice.
- Correctly solder wires to a soldered terminal connection.
- Correctly install a twist-on, set-screw and compression connector.

6-1 The Necessity for Proper Electrical Connections

Almost all electrical installations and repairs consist of connecting wires to terminals or to other wires. Cutting, splicing and connecting electrical wires must be done well or problems will result. A poorly made electrical connection will have a much higher than normal resistance. This results in an excessive amount of heat being produced at the connection when normal current flows through the circuit (Figure 6-1). A poor electrical connection will also reduce the total energy normally available for the load. This is due to the fact that a portion of the energy supplied is used to produce unwanted heat at the faulty connection point.

FIGURE 6-1 HIGH RESISTANCE CONNECTIONS

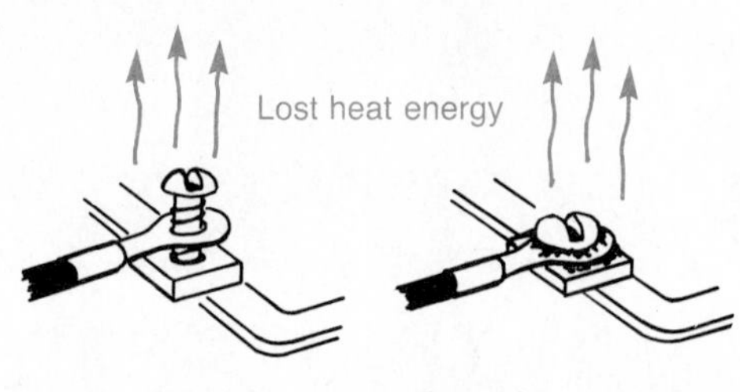

6-2 Preparing Wire for Connection

Electrical wire is always fully insulated with a covering of rubber or plastic. This covering must be removed (stripped) before the wire can be properly connected to anything. The amount of insulation to be stripped from the end of the wire depends on the kind of connection you are going to make.

The most efficient, as well as the least damaging, method of stripping insulation from any wire is done with the use of a special wire stripper tool (Figure 6-2). This tool has a series of sharp openings in its scissor blade to allow stripping of wire of different gauge sizes or diameters. The gauge size of the wire must be matched to the opening in the wire stripper to prevent cutting into the wire and weakening it.

If you are stripping wire with a knife, you must be careful not to nick the wire and weaken it. Do not cut straight into the insulation. Slant your knife blade at an angle towards the end of the wire as you would in

FIGURE 6-2 USING A WIRE STRIPPER

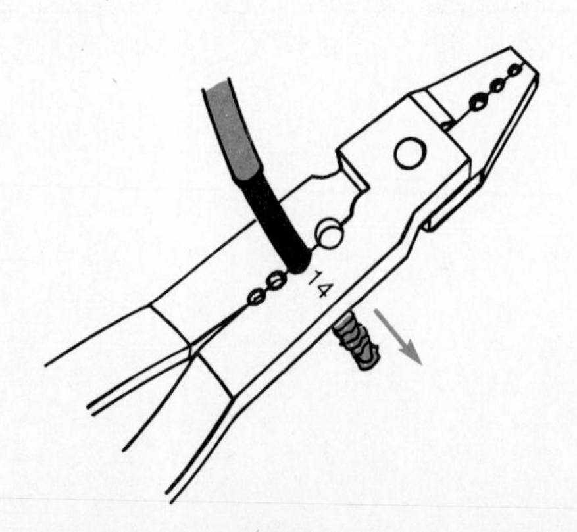

How to use a wire stripper

1. Insert the wire end into the proper diameter hole.
2. Close the stripper around the wire. The tool should cut through the insulation but not the wire.
3. Pull the insulation from the end of the wire.

FIGURE 6-3 USING A KNIFE TO REMOVE INSULATION

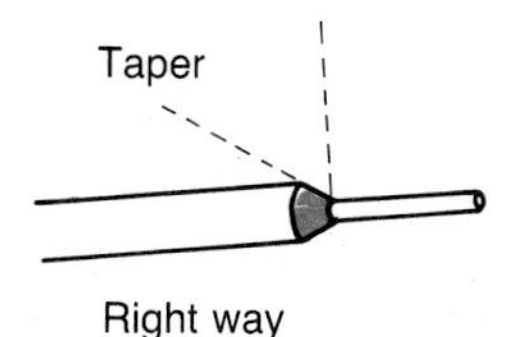

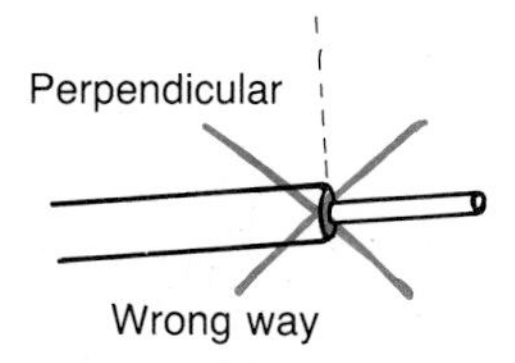

sharpening a pencil (Figure 6-3). Once cut, pull the insulation off the wire, using your fingers or a pair of pliers.

6-3 Terminal Screw Connections

The simplest and most common type of electrical connection is the terminal screw type. When you are wiring electrical devices such as switches or lampholders, this is the type of connection most often used. Connections to screw type terminals are made by forming the end of the wire into a loop that is fitted around the head of the screw (Figure 6-4).

FIGURE 6-4 CONNECTING TO TERMINAL SCREWS

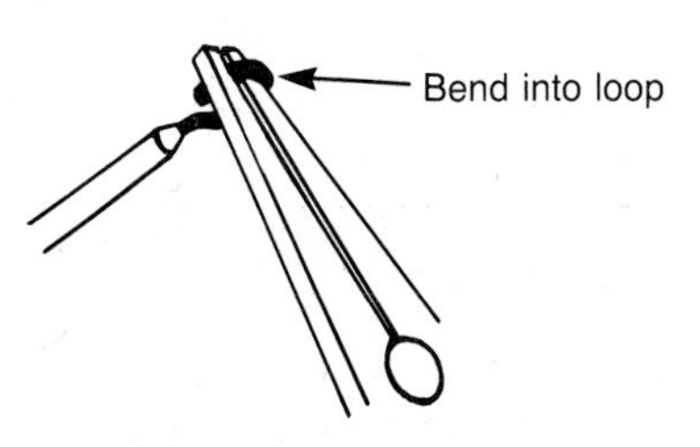

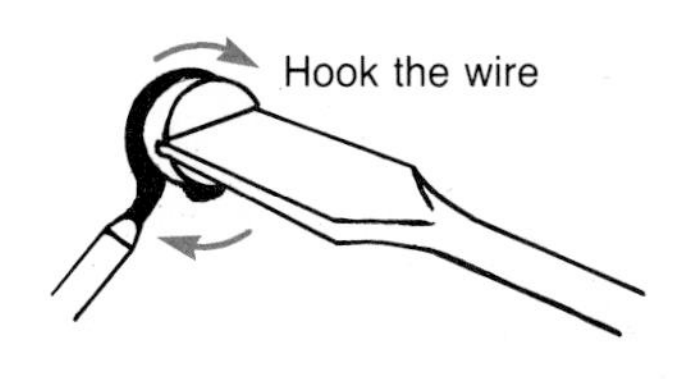

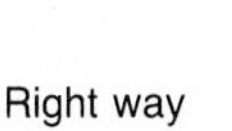

How to connect to terminal screws

1. Remove only enough insulation to make a loop of bare wire around the screw terminal (about ¾ in.). If the wire is stranded, the bare strands should be tightly twisted together, and then tinned.
2. Use needle-nose pliers to bend the bared wire into a loop.
3. Loosen the screw terminal with a screwdriver, but do not remove it from its hole.
4. Hook the wire over the top of the screw and tighten in a **clockwise** direction. In this way the wire will be drawn around the screw and not pushed away from it when the screw is tightened.
5. Close the loop around the screw with the pliers and tighten the screw.
6. Almost no bare wire should extend beyond the screw head. If it does, you have stripped away too much insulation and the bare wire should be shortened.

6-4 Splicing Wires

Electrical connections made by means of wire splicing involve twisting two wires together, then securing and taping the splice. Soldering, once an acceptable method for joining wires, has been replaced by solderless connectors that are easier and safer to use.

The National Electrical Code requires that all electrical splices be mechanically and electrically secure before soldering. **Mechanically secure** means that the joint must be as strong as if the wire were unbroken. **Electrically secure** means that the contact points between the two conductors must be such that current will pass from one wire to the other as if the wire were in one piece. Scraping the conductor clean with the back of a knife before forming your splice helps to improve the electrical contact between the two conductors.

One method of splicing a wire without the use of a connector is known as the **Western Union splice**, named a century ago by telegraph linemen (Figure 6-5).

A **pigtail splice** can be used to join two wires together in an electrical junction box. The method used to make this splice is illustrated in Figure 6-6.

6-5 Soldering Splices

Some types of splices should be soldered. Soldering involves applying a thin coat of solder to the splice. Soldering improves the mechanical strength of the splice. It also helps to lower the electrical resistance of the splice by increasing the contact area between the conductors and preventing any corrosion or oxidation of the copper.

FIGURE 6-5 MAKING A WESTERN UNION SPLICE

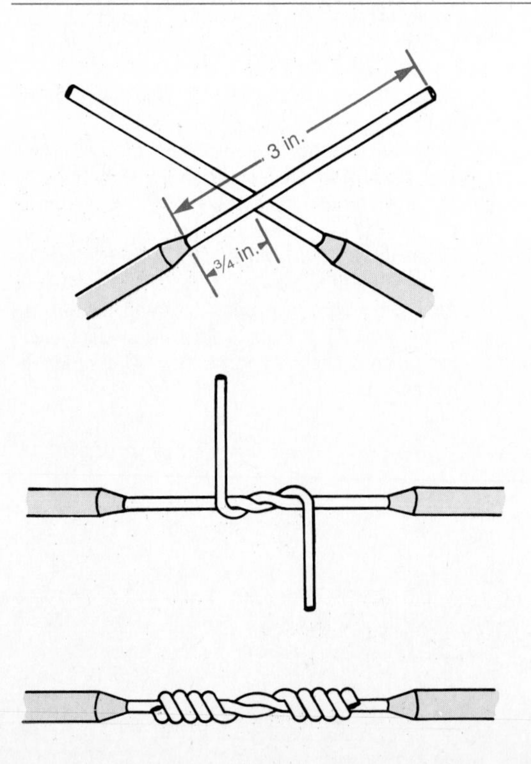

How to make a Western Union splice

1. Remove about 3 in. of insulation from each conductor.
2. Cross the conductors about ¾ in. from the insulation and bend their ends so that one wire wraps around the other.
3. Wrap one wire in a clockwise direction until it reaches the insulation. Wrap the other in a counter-clockwise direction for the same number of turns.
4. Cut off any excess wire and bend the ends down.
5. Solder the entire splice and tape.

FIGURE 6-6 MAKING A PIGTAIL SPLICE

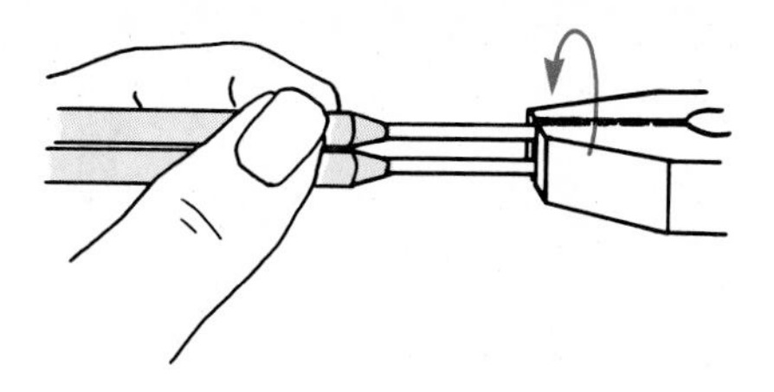

How to make a pigtail splice

1. Remove about 1½ in. of insulation from each conductor.
2. Hold the wires tightly parallel with your left hand and begin twisting the free ends.
3. Twist the free ends in a clockwise direction forming about 6 turns and cut off the wire ends.
4. Solder the entire splice and tape.

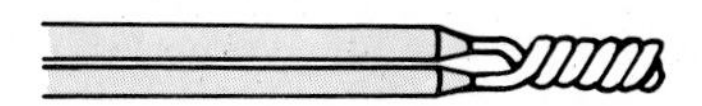

Solder is an alloy made up of tin and lead and has a low melting point. The tin/lead ratio determines the strength and melting point of the solder. For most electrical and electronic work, 60/40 wire type solder with a resin flux core is recommended (Figure 6-7).

When soldering, the copper surfaces being soldered must be free from dirt and oxide otherwise the solder will not adhere to the splice. Wires should be cleaned by a light scraping or sanding to remove any dirt before they are made into splices.

Heating the copper splice for application of the solder increases the oxidation process. For this reason a soldering flux or paste must be applied to the splice when soldering (Figure 6-8). The flux prevents oxidation of the copper surfaces by insulating the surface from the air. Both acid and resin based

FIGURE 6-7 WIRE TYPE RESIN CORE SOLDER

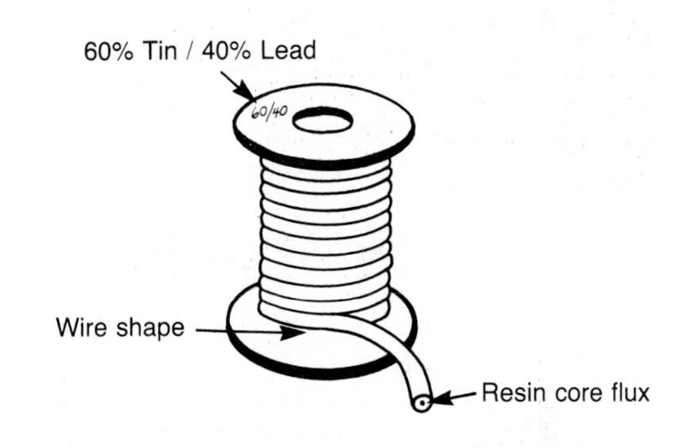

FIGURE 6-8 RESIN BASED FLUX

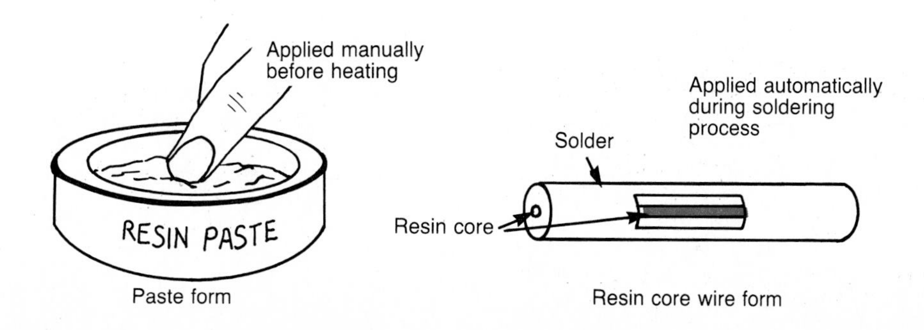

FIGURE 6-9 SOLDERING HEAT SOURCES

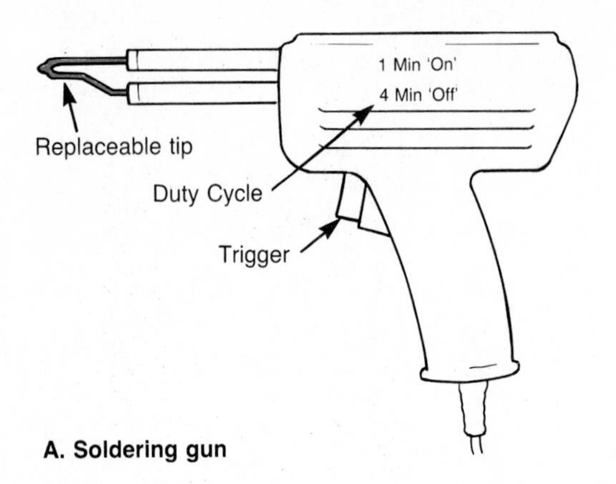

A. Soldering gun

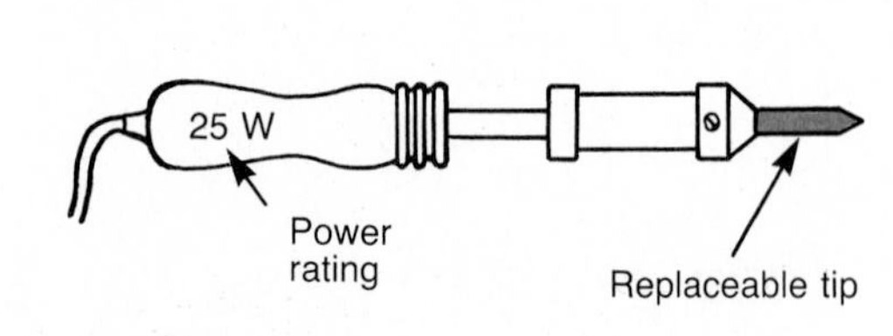

B. Soldering iron

fluxes are available. It is important to note that acid based fluxes should not be used in electrical work as they tend to corrode the copper wires. Resin flux is available in paste form or as a continuous core inside solder wire.

The most common method of applying heat to a splice is by means of a soldering gun or soldering iron (Figure 6-9). Heat in a soldering gun is produced by means of transformer action. The soldering gun heats up very quickly but has a short duty cycle. Heat in a soldering iron is produced by means of a heating element similar to that found in a toaster. The soldering iron heats up at a slower rate but has a longer duty cycle.

For the best soldering results, the copper heating tip of the soldering iron or gun must be kept clean or well tinned. New tips must be tinned before they can be used. This can be accomplished by applying solder to the heated tip and wiping it clean. A well tinned tip will conduct the maximum amount of heat from the tip to the surface being soldered. Safety glasses should be worn during both the tinning and soldering processes to provide adequate eye protection.

To do the actual soldering, position the heated copper tip below the splice and establish a good contact between the tip and splice (Figure 6-10). Wait until the splice is hotter than the melting point of the solder before beginning to solder. Then apply the solder from the top. Gravity and adhesion

FIGURE 6-10 SOLDERING A SPLICE

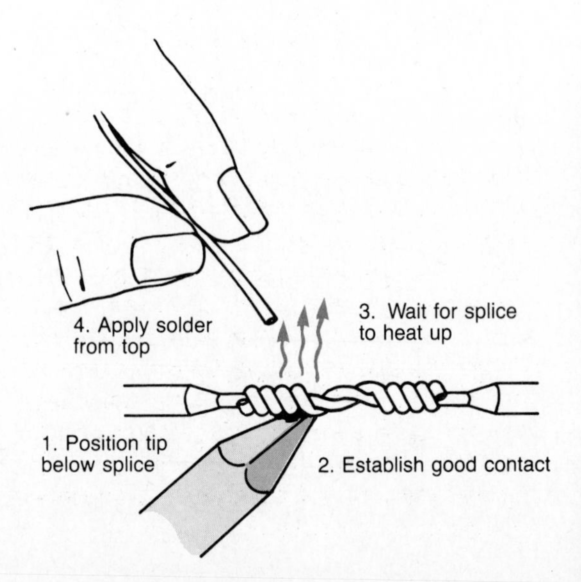

FIGURE 6-11 TAPING SPLICES

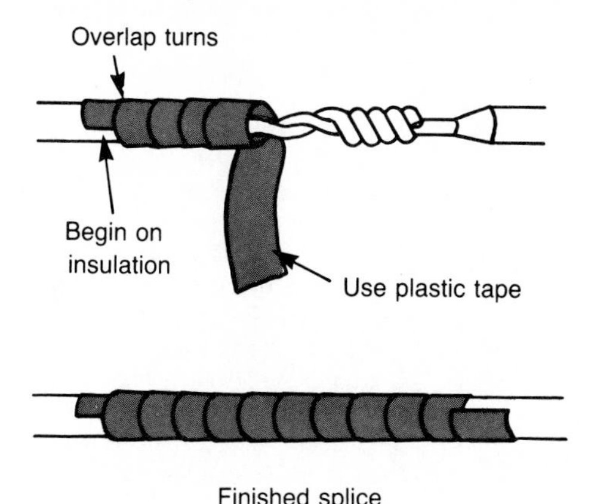

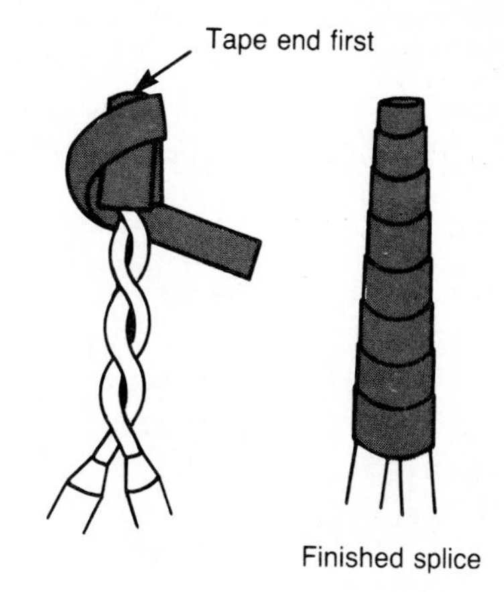

will cause the flux and solder to reach all parts of the splice. A good soldered splice should be covered entirely with a thin coat of glossy solder.

6-6 Taping Splices

The National Electrical code states that the insulation used to cover splices must be equivalent to the amount of insulation that was removed from the wire. When taping splices, vinyl thermoplastic or a layer of friction tape wrapped over a layer of rubber tape may be used. The plastic tape is preferred and is most often used because it has a high insulation value, is easier to handle and takes up less space in boxes (Figure 6-11). Friction tape should never be used by itself on any electrical splices because it has a very low electrical insulation value. To do the actual taping, begin on the insulation of the conductors and wrap the plastic tape tightly around the entire splice ending back on the insulation. The turns should overlap by half the width of the tape to produce a double layer of insulation.

6-7 Soldered Terminal Connections

Many electronic installations make use of soldered terminal connections for connecting component parts and wiring. Connections to soldered terminals are made by forming a tight mechanical connection that is soldered in place (Figure 6-12).

A good soldered connection has just enough solder to bond the wire to the terminal. It should have a smooth, semi-shiny appearance, and edges that seem to blend cleanly and smoothly into the terminal and wire. Any other appearance is a sign of improper soldering. A large glob on the terminal indicates too much solder was applied. If the soldered connection has a dull and pitted appearance, or has a ball of solder that does not blend into the metal, then the temperature of the metal was too low.

FIGURE 6-12 SOLDERING TO A TERMINAL

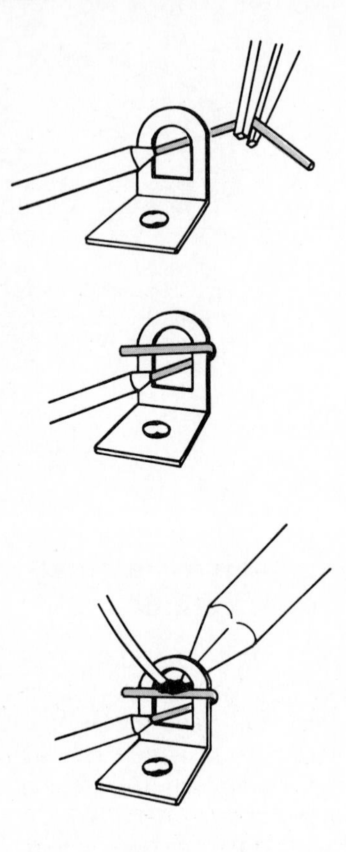

How to solder to a terminal

1. Remove only enough insulation to make a loop of bare wire around the terminal.
2. Use needle-nose pliers to bend the bare wire through and around the terminal hole.
3. Apply heat from a soldering iron or gun to the back side of the terminal. After a few seconds, touch the resin-core solder to the wire at the front of the terminal. If the solder melts, the connection is sufficiently heated and the solder will flow. (Do not touch the solder to the iron.)
4. When the connection has been coated with solder, remove the length of solder. Leave the iron on the connection to boil away the resin. If any resin remains under the solder, an insulation barrier could be formed. Now remove the iron and let the connection cool. Do not move the connection while it is cooling.

6-8 Solderless Mechanical Connectors

Approved **mechanical connectors** that require no soldering or taping are now used by electricians for making most electrical connections. These devices save both time and labor and, as a result, are used extensively. The three basic types of solderless connectors are: twist-on type, set-screw type and compression type.

The *twist-on connector* uses a metal spring that threads itself around the conductors. As the connector is rotated it holds the conductors in place (Figure 6-13). The internal spring design takes advantage of leverage and vise action to multiply the strength of a person's hand. It is available in a range of sizes for splicing conductors from No. 18 gage up to No. 8 gage. When using any type of solderless mechanical connector, it is very important to match the size of the connector with the gage size of the wire to make a proper connection.

The set-screw connector is a two-piece connector that uses a set-screw to hold the conductors in place (Figure 6-14). This design allows conductor connections to be interchanged easily. They are used mainly in commercial and industrial circuits where equipment must be changed frequently for maintenance purposes.

FIGURE 6-13 INSTALLING A TWIST-ON CONNECTOR

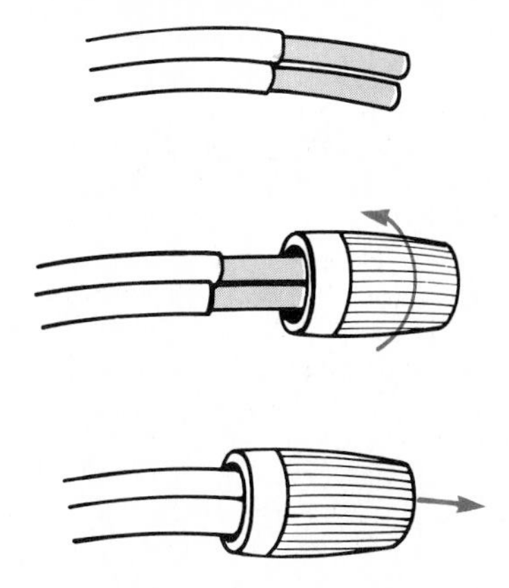

How to install a twist-on connector

1. Remove about ¾ in. of insulation from both conductors and scrape clean.
2. Hold the two conductor ends even and insert into connector shell.
3. Twist the connector clockwise onto the wires until it is tight.
4. Be certain no bare wire is visible when the cap is in place. Test the connection by trying to pull the connector away from the wires.

FIGURE 6-14 INSTALLING A SET-SCREW CONNECTOR

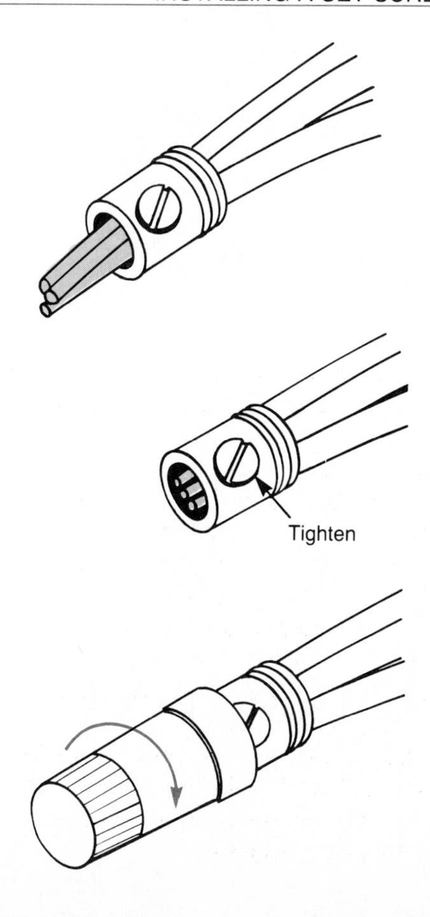

How to install a set-screw connector

1. Remove about ¾ in. of insulation from conductors and scrape clean.
2. Remove the brass fitting from the connector and loosen the set screw.
3. Hold the two conductor ends even and insert them into the brass fitting so that the threaded shoulder is next to the insulation.
4. Tighten the set screw and cut off any excess wire that extends beyond the end of the brass fitting.
5. Thread the plastic cap onto the brass fitting until snug.
6. Be certain no bare wire is visible when the cap is in place. Test the connection by trying to pull the connector away from the wires.

FIGURE 6-15 INSTALLING A COMPRESSION CONNECTOR

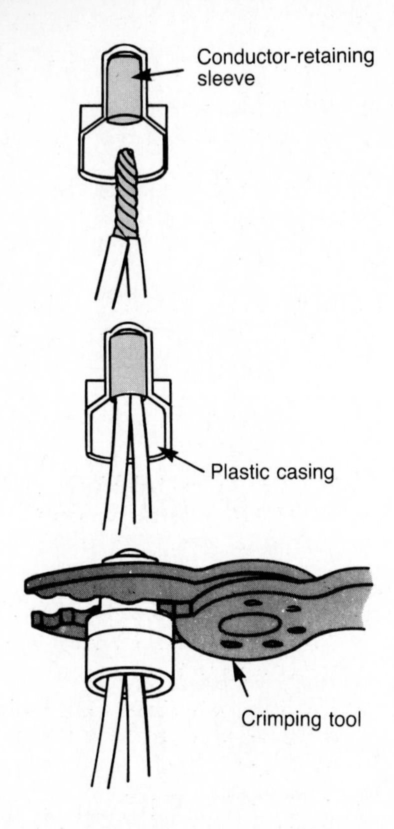

How to install a compression connector

1. Remove about ¾ in. of insulation from both conductors and twist the ends together.
2. Insert the conductor ends into the conductor-retaining sleeve. Be certain no bare wire is visible below the plastic casing.
3. Squeeze the top of the connector firmly with a crimping tool. The inner retaining-sleeve will collapse and anchor the conductors. Test the connection by trying to pull the connector away from the wires.

With the **compression** or **crimp-on connector** the conductors are held together by crimping or compressing them in a special conductor-retaining sleeve. A special crimping tool must be used to crimp the wires in the sleeve. This type of connector is used for *permanent* type installations because the conductors cannot be easily removed from the retaining sleeve. (Figure 6-15).

6-9 Practical Assignments

1. Make several connections to a terminal screw board.
2. Make a splice using each of the following mechanical connectors:
   (a) twist-on type
   (b) set-screw type
   (c) compression type
3. Properly tin a soldering gun or iron tip.
4. (a) Make a Western Union and pigtail splice.
   (b) Solder and insulate each of these splices.
5. Make several connections to a soldered terminal board.

6-10 Self Evaluation Test

1. Describe two negative effects of a poor electrical connection.

2. What is the best way of removing insulation from wire? Why?

3. State two important features of a properly made terminal screw connection.

4. What National Electrical Code rule applies to the forming of a wire splice by twisting two wires together?

5. State two positive effects of soldering a splice.

6. (a) What characteristic of solder is determined by its tin/lead ratio?
   (b) What tin/lead ratio is most often recommended for electrical work?

7. When soldering copper surfaces, what three things could cause the solder not to adhere to the copper surface?

8. What type of flux should not be used for electrical work? Why?

9. Compare the soldering gun and iron with regard to:

   (a) how heat is produced
   (b) duty cycle
   (c) heat-up time

10. What is the reason for keeping the soldering tip well cleaned or tinned?

11. What safety precaution should be observed when tinning or soldering?

12. What type of tape is most often used on electrical splices? Why?

13. State two important features of a properly taped splice.

14. Describe two signs of an improper soldered terminal connection.

15. State the main advantages that solderless connectors have over the soldered splice type of connection.

16. (a) When installing a twist-on connector, the bare wires can be twisted together before they are inserted into the connector shell (YES or NO)?
    (b) When installing a set-screw connector, into which end of the brass fitting are the bare wire ends inserted?
    (c) What special piece of equipment is required to install a compression connector?
    (d) What final check can be made on all solderless connectors to test the mechanical strength of the connection?

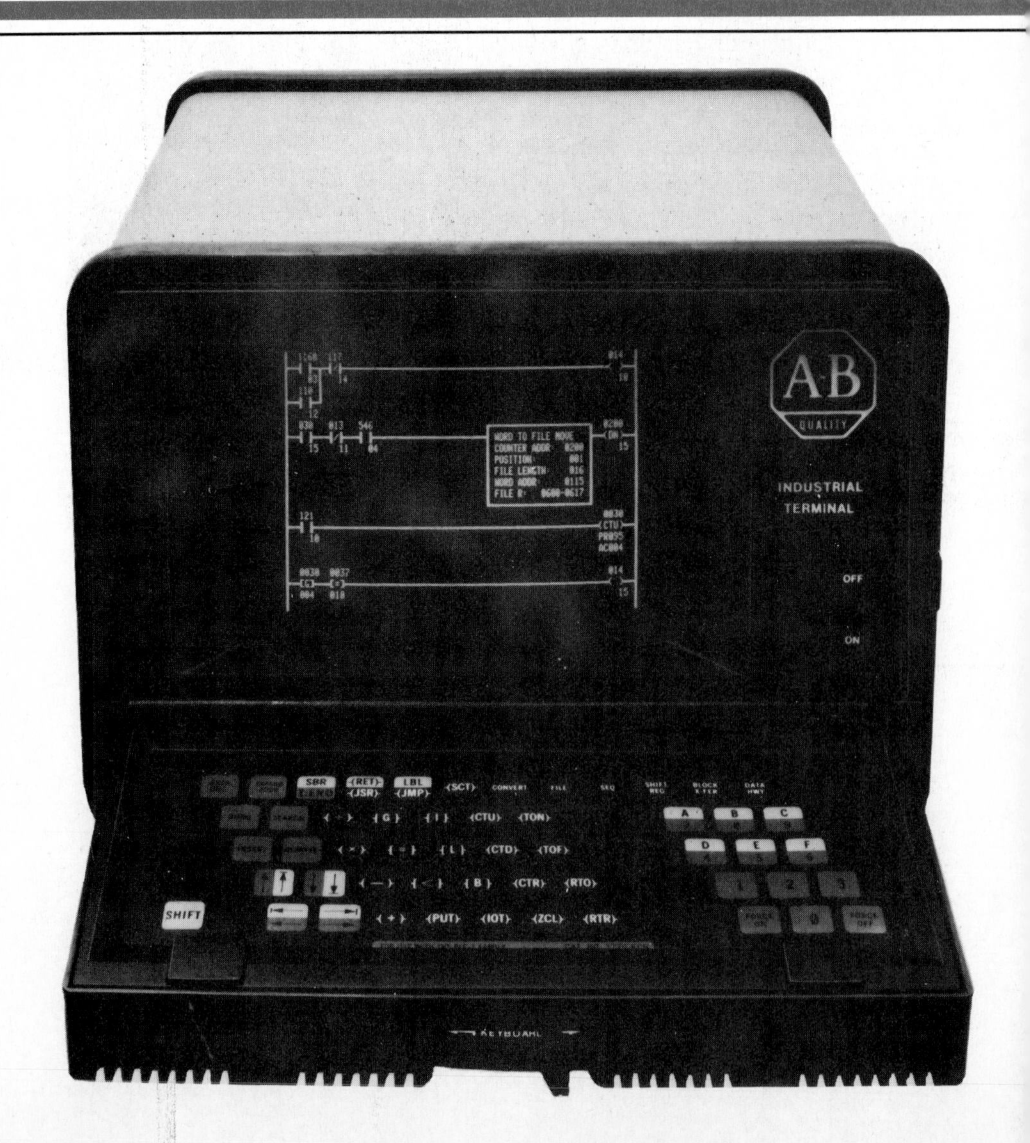
A-B
QUALITY
INDUSTRIAL
TERMINAL
OFF
ON
WORD TO FILE MOVE
SBR
(RET)
(JSR)
LBL
(JMP)
(SCT)
CONVERT
FILE
SEQ
(CTU)
(TON)
(CTD)
(TOF)
(CTR)
(RTO)
SHIFT
(PUT)
(IOT)
(ZCL)
(RTR)
A
B
C
D
E
F
1
2
3
0

SIMPLE SERIES AND PARALLEL CIRCUITS

OBJECTIVES

Upon completion of this unit you will be able to:

- Identify the basic components of a circuit.
- Identify common electrical symbols used in schematic diagrams.
- Compare electrical pictorial, schematic and wiring diagrams.
- Compare the operation of series and parallel connected load devices.
- Compare the operation of series and parallel connected control devices.
- Draw the schematic diagram of a circuit from a set of written instructions.
- Read a simple schematic electrical diagram.
- Complete a wiring number sequence chart and wiring diagram from a given schematic diagram.
- Satisfactorily wire circuits with the aid of a wiring diagram.

7-1 Circuit Components

Three fundamental invisible quantities, voltage, current and resistance, are present in every electrical circuit. These quantities are controlled and directed by the proper arrangement of component parts to produce the desired function of the circuit. The component parts that make up any circuit are: the *energy source*, the *protective device, conductors,* the *control device* and the *load device*.

The energy source supplies the voltage required to move the free electrons along the conducting path of the circuit. It is also referred to as the power supply. Two types of sources, direct current (DC) and alternating current (AC) are used.

The purpose of the *protective device* is to protect circuit wiring and equipment. It is designed to allow only currents within safe limits to flow. When a higher current flows, this device will automatically open the circuit. This effectively shuts off current until the problem is corrected. Two types of protective devices, namely fuses and circuit breakers are normally used. Often the protection device is a part of the voltage source or power supply device. In this unit we will assume the protective device to be a part of the power supply. As such, we will not include it in our diagrams.

Conductors or wires are used to complete the path from component to component. Conductors provide a low resistance path for electrons. They are usually insulated to protect against accidental contact with the circuit. The most common conductor type is rubber- or plastic-insulated copper wire.

A *control device* is usually included in the circuit to allow you to easily start, stop or vary the electron flow. The most common control devices are switches and pushbuttons.

The *load* is the part of the circuit that converts the electrical energy to produce the desired function of the circuit. Lamps, motors, heaters and resistors are just a few common load devices. For practical purposes, all the circuit resistance is considered to be contained in the load device.

7-2 Circuit Symbols

The use of symbols to represent electrical and electronic components can be considered to be a form of technical shorthand. The use of these symbols tends to make circuit diagrams less complicated and easier to read and understand. The symbols we will be using in this unit are listed in Figure 7-1.

7-3 Circuit Diagrams

Pictorial Diagrams There are three types of diagrams used to show the layout of the circuit: pictorial, schematic and wiring diagrams. A *pictorial diagram* (Figure 7-2) is used to show the physical details of the circuit as seen by the eye. The advantage here is that a person can simply take a group of parts, compare them with the pictures in the diagram and wire the circuit as shown. The main disadvantage is that many circuits are so complex that this method is impractical.

Schematic Diagram A *schematic diagram* (Figure 7-3) uses symbols to represent the various components and, as a result, is not as cluttered as a pictorial diagram. The components are arranged in a manner that makes it easier to read and understand the operation of the circuit. This type of diagram is most often used to explain the sequence of operation of a circuit. The "ladder" type schematic diagram is the one most often used in industry. In this type of schematic the two power lines connect to the power source and the various circuits connect across them like rungs in a ladder.

FIGURE 7-1 CIRCUIT SYMBOLS

A. Direct current source

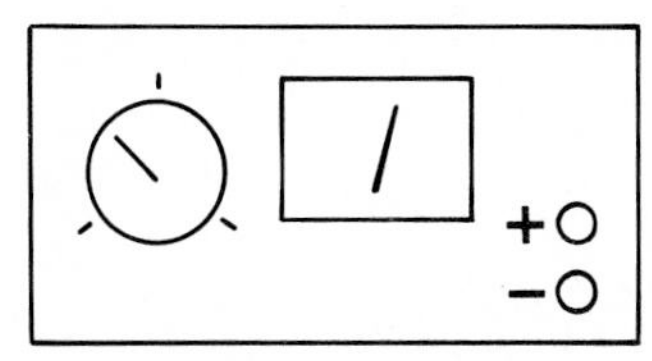
DC Power supply

Battery

Cell

\+ −

Symbol

B. Alternating current source

Receptacle

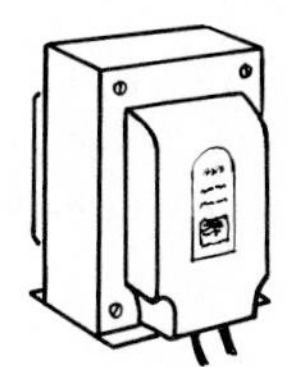
Transformer

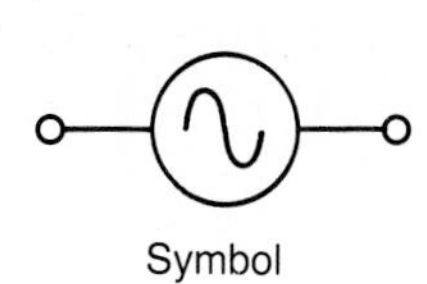
Symbol

C. Switch

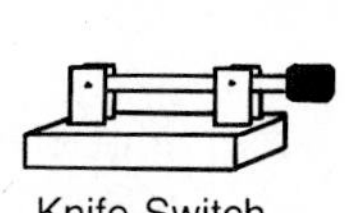
Knife Switch

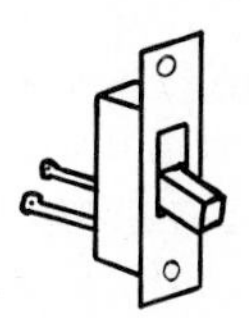
Slide Switch

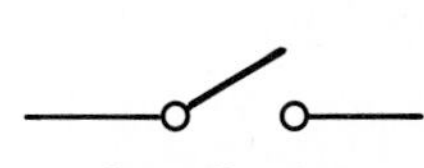
Open Symbol

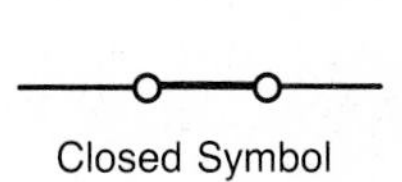
Closed Symbol

D. Pushbutton

Symbol

E. Buzzer

Symbol

F. Bell

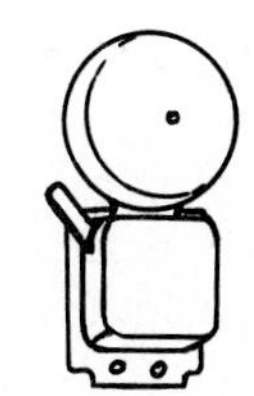

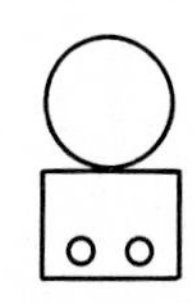
Symbol

G. Lamp

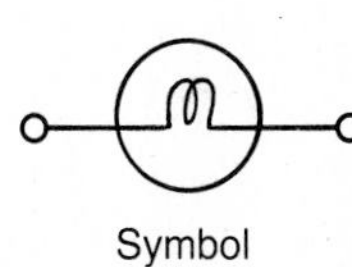
Symbol

H. Conductors

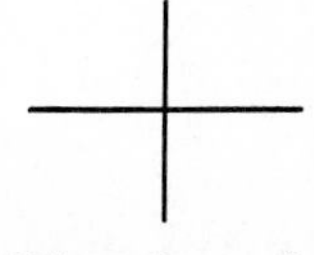
Wires, Crossed

Wires, Joined

I. Resistor

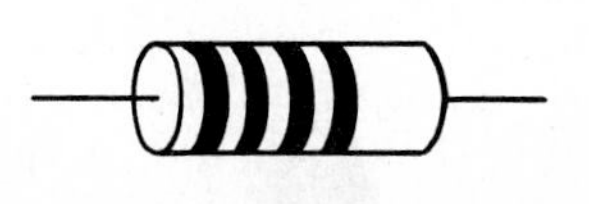

Symbol

FIGURE 7-2 PICTORIAL DIAGRAM

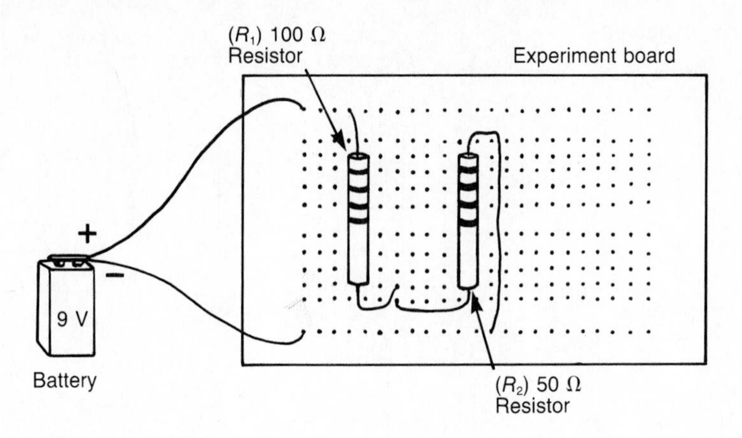

FIGURE 7-3 SCHEMATIC DIAGRAM (LADDER TYPE)

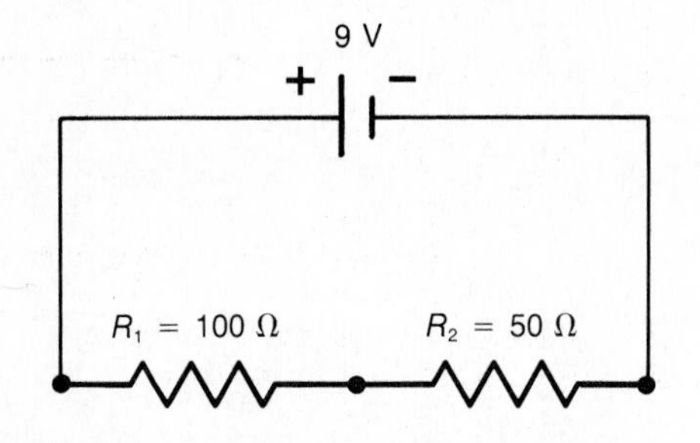

Wiring Diagram A *wiring diagram* (Figure 7-4) is similar to a pictorial diagram except that the components are shown in a symbolic form. Unlike the schematic, the components are shown in their relative physical position making it a useful diagram to use when doing the actual wiring for the circuit.

FIGURE 7-4 WIRING DIAGRAM

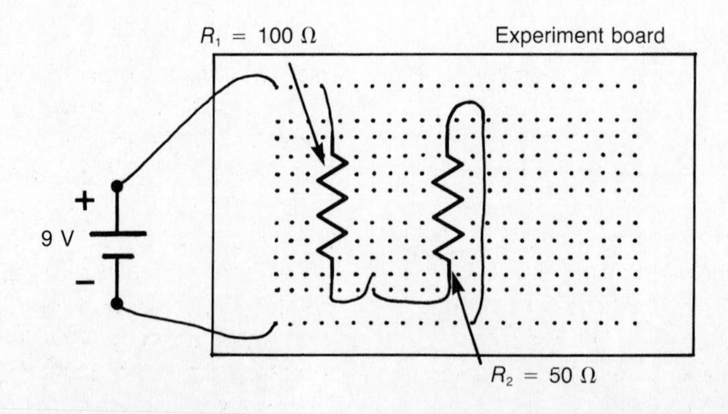

FIGURE 7-5 SIMPLE CIRCUIT

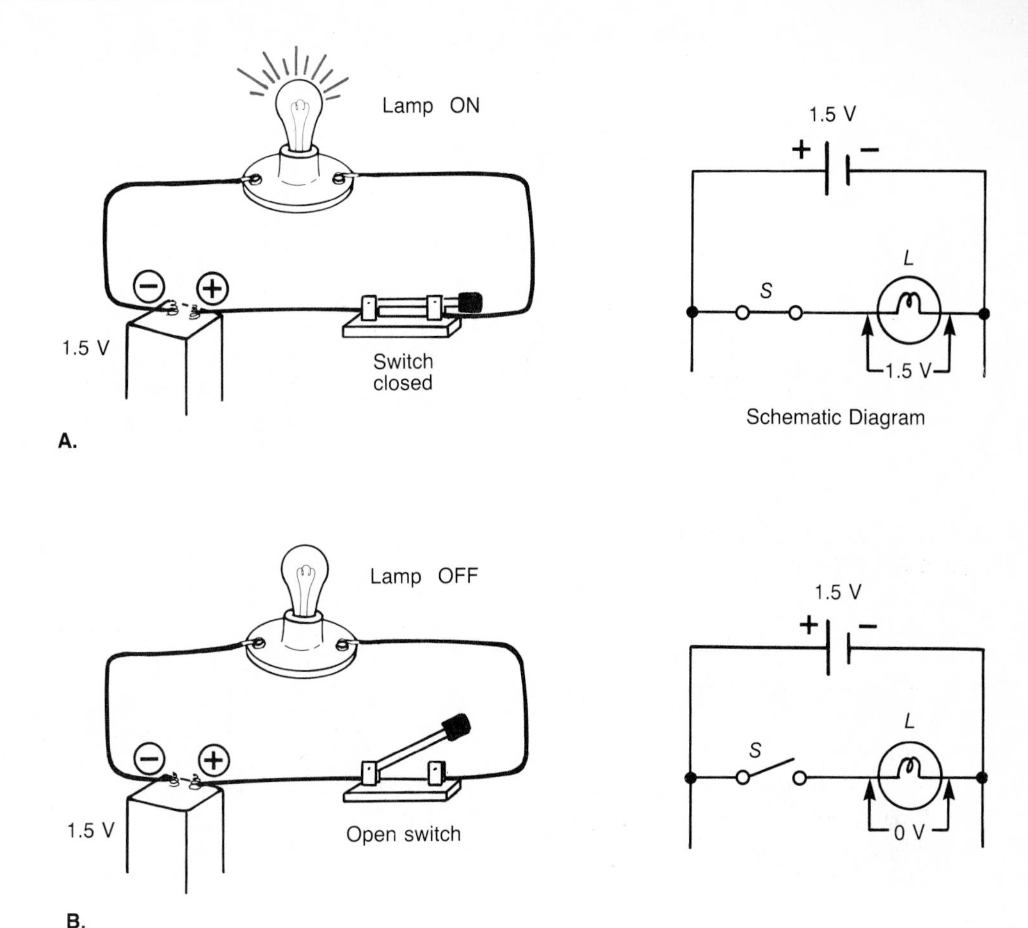

7-4 Simple Circuit

A *simple circuit* is one that has only one control and one load device. A single lamp controlled by a single switch is one example of a simple circuit. Each component is connected or wired "end-to-end." This simple circuit is controlled by opening and closing the switch. When the switch is closed, electrons flow to turn the lamp on (Figure 7-5A). When the switch is open the flow is interrupted and the lamp turns off (Figure 7-5B). When on, the voltage at the lamp is the same as the voltage of the source.

7-5 Series Circuit

Series Connected Loads If two or more loads are connected end-to-end they are said to be connected in series (Figure 7-6). The same current flows through each of them. Also, there is only one path for the current to flow. The current stops if the circuit is broken at any point. For example, if two lamps are connected in series and one burns out, *both* lamps will go out.

When loads are connected in series, each receives part of the applied source voltage. For example, if three *identical* lamps are

FIGURE 7-6 TWO LAMPS CONNECTED IN SERIES

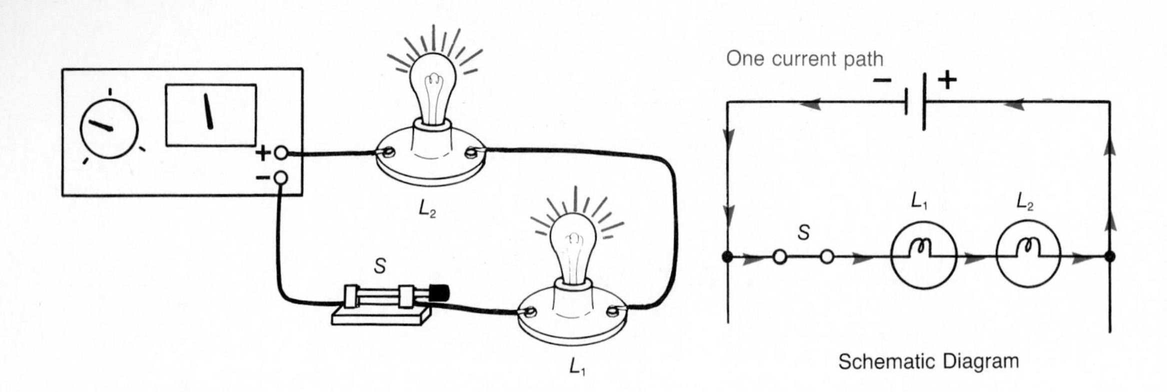

connected in series each will receive one-third of the applied source voltage (Figure 7-7). The amount of voltage each load receives is dependent upon its electrical resistance.

Series Connected Control Devices Two or more control devices may also be connected in series. The connection is the same as that used for loads, which is end-to-end. Connecting control devices in series results in what can be called an *AND* type of control circuit. Take the example of two switches, A and B, connected in series to a lamp (Figure 7-8). In order for the lamp to turn on both switch A *AND* switch B would have to be closed. Series connection of control devices is used in computer circuits and industrial control systems.

7-6 Parallel Circuit

Parallel Connected Loads If two or more loads are connected across the two voltage source leads they are said to be connected in parallel. The parallel connection of load devices is used where each device is designed to operate at the same voltage as the power supply (Figure 7-9). This is the

FIGURE 7-7 VOLTAGE DROP ACROSS IDENTICAL SERIES CONNECTED LAMPS

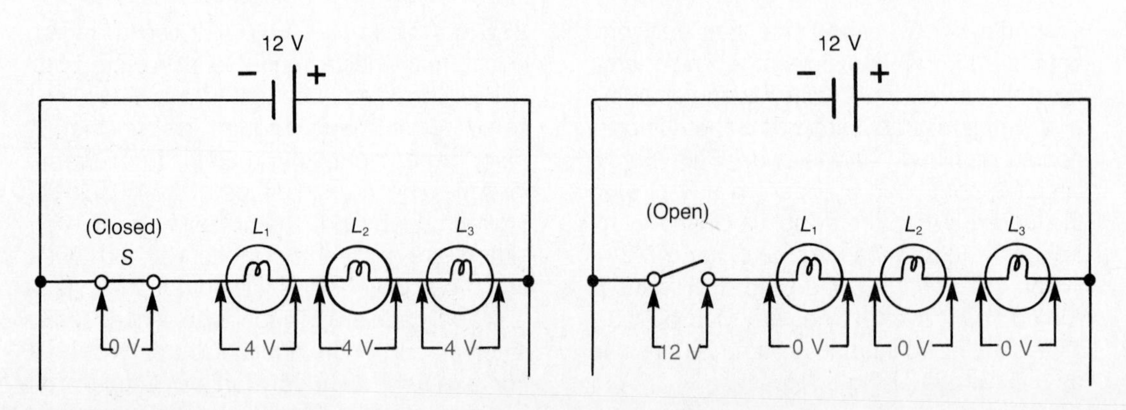

FIGURE 7-8 TWO SWITCHES CONNECTED IN SERIES

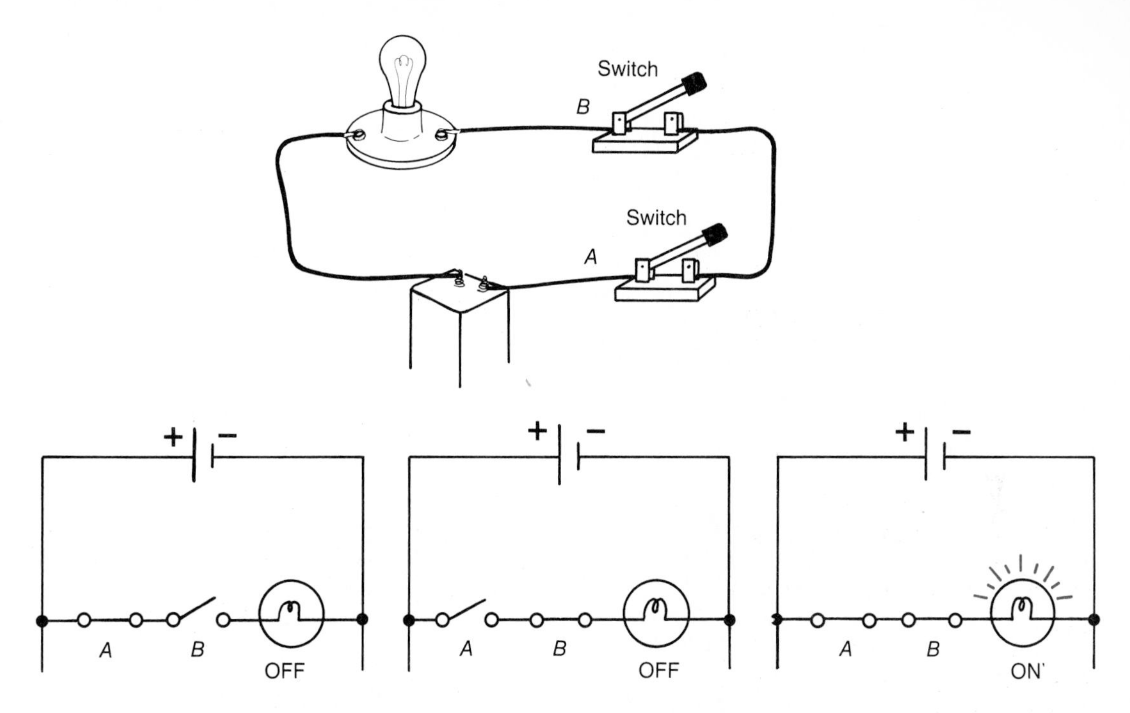

"*AND*" Type control circuit *A* "AND" *B* Are closed to switch lamp **ON**

case in wiring lights and small appliances in the home. Here, with the source voltage 120 V, all appliances and lights connected in parallel to it must be rated for 120 V. The use of lower voltage devices, such as 12 V devices would cause such units to burn out.

A parallel circuit has the same voltage applied to each device. However, the current

FIGURE 7-9 THREE LAMPS CONNECTED IN PARALLEL

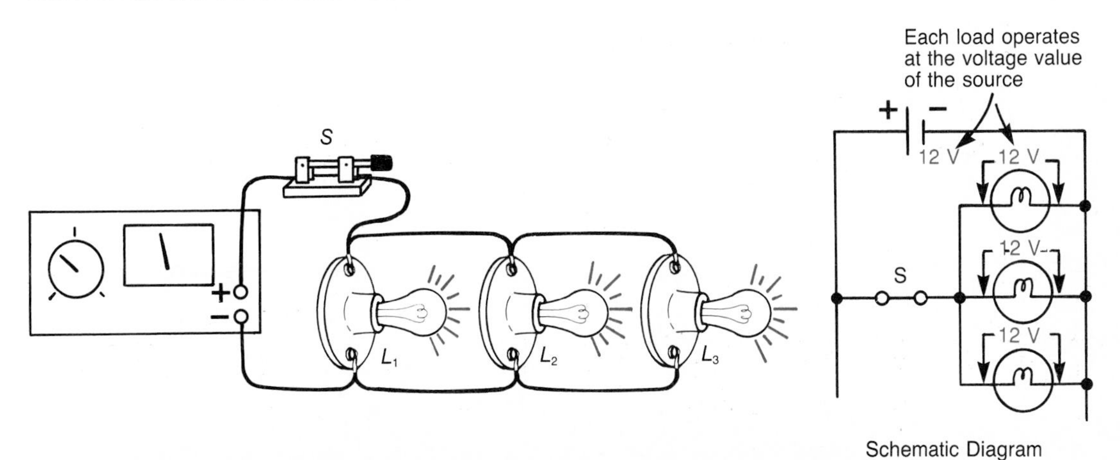

Schematic Diagram

FIGURE 7-10 CURRENT PATHS FOR TWO LAMPS CONNECTED IN PARALLEL

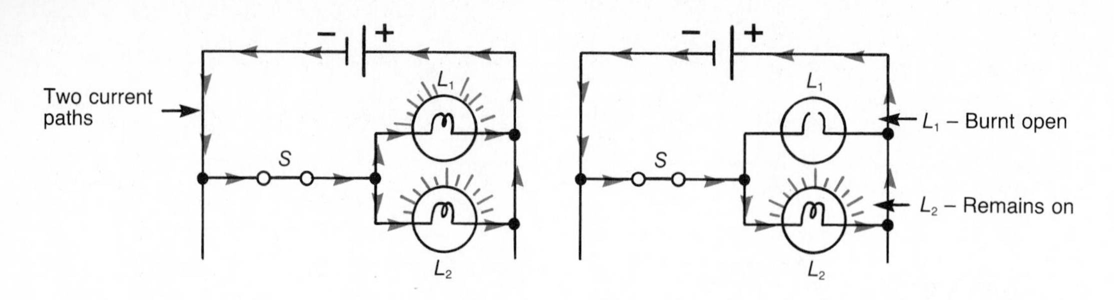

to each device will vary with the resistance of the device.

When load devices are operated in parallel, each load operates independently of the others. This is because there are as many current paths as there are loads. For example, if two lamps are connected in parallel there will be two paths created. If one lamp burns out, the other will not be affected (Figure 7-10).

Parallel Connected Control Devices When two or more control devices are connected across each other they are said to be connected in parallel. Connecting control devices in parallel results in what can be

FIGURE 7-11 TWO PUSHBUTTONS CONNECTED IN PARALLEL

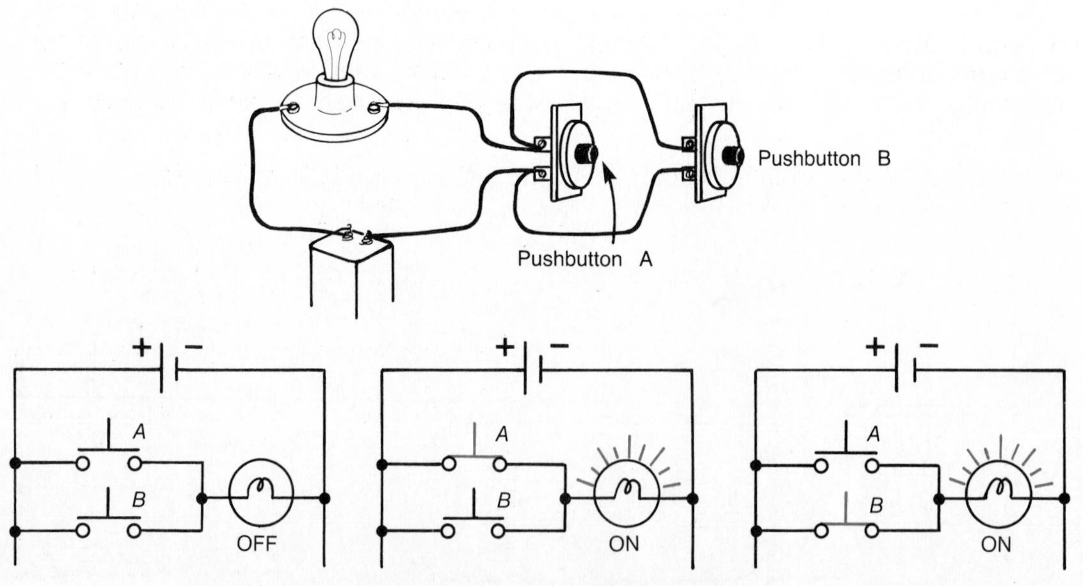

called an *OR* type of control circuit. Take the example of two pushbuttons, *A* and *B* connected in parallel to a lamp. In order for the lamp to turn on, either pushbutton A *OR* pushbutton B or both would have to be pressed (Figure 7-11). Parallel connection of control devices is also used in computer and industrial control circuits.

7-7 Constructing Wiring Projects Using Schematic Diagrams

Planning is an important part of an electrical wiring project. It involves *thinking through* the job to be done. What do you want the circuit to do? What type of electrical diagrams are required? What should be done first, second, and so on? All of these questions should be answered before you begin to construct the project.

The best way to begin is with a schematic diagram of the circuit. This may either be given to you or drawn from a set of instructions. Take for example a circuit consisting of two lamps connected in parallel controlled by a single pushbutton. Assume a 12 V DC source and two 12 V lamps are to be used (Figure 7-12).

Your next step is to make a wiring number sequence chart for the circuit. This will help you make the right wire connections. Begin by assigning a number (start with 1) to each terminal of each component part. Next, all common terminal numbers are grouped together. Common terminals represent common points in the circuit that are connected together. Record these common number groups in chart form as illustrated in Figure 7-13.

The final step is to complete a wiring diagram of the circuit. This diagram shows the different components and wires, as they would be located in the actual circuit. Note that the component parts in the wiring layout do not appear in the same position as they did in the schematic (Figure 7-14). The first step in completing the wiring diagram is to number the terminals of each component. The numbers assigned are the same as those on the schematic.

FIGURE 7-13 HOW TO USE A WIRING NUMBER SEQUENCE CHART

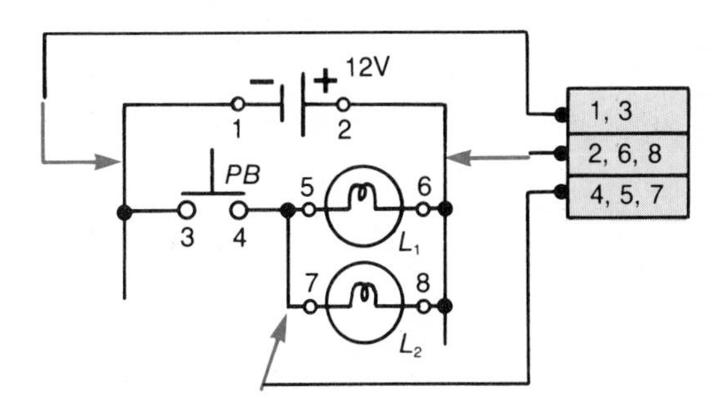

FIGURE 7-12 CIRCUIT SCHEMATIC DIAGRAM

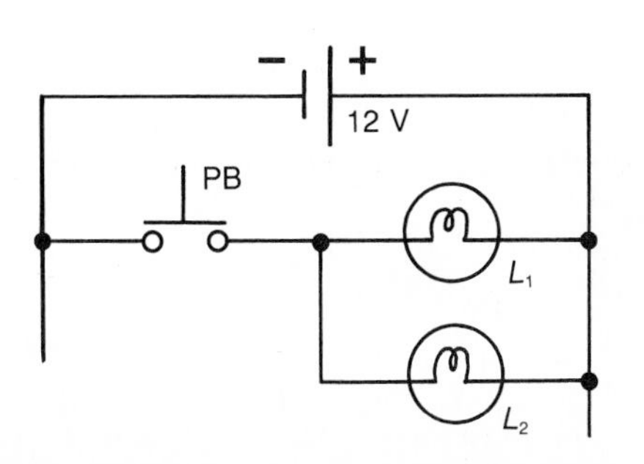

FIGURE 7-14 CIRCUIT WIRING LAYOUT

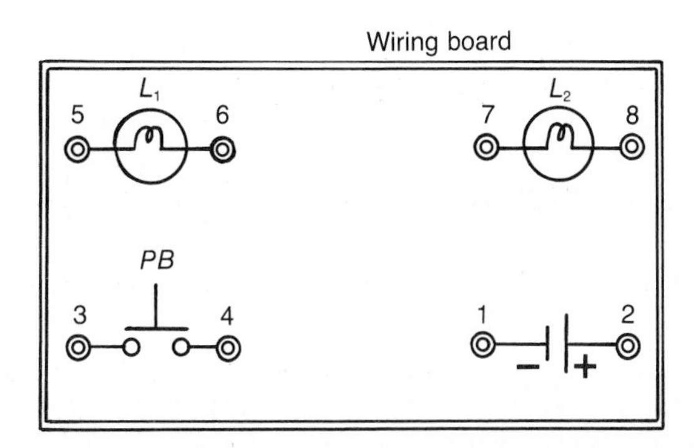

FIGURE 7-15 CIRCUIT WIRING DIAGRAM

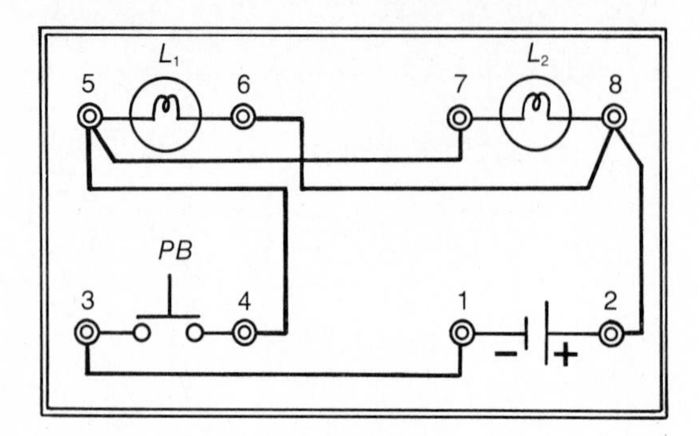

To complete the wiring diagram, all common sets of terminals are interconnected. The wiring number sequence chart is used to determine what terminals connect together. Draw wires neatly from point to point using straight lines with right angle turns. Make all connections at terminal screws. Plan the "wire runs" from point to point so that you use the minimum amount of wire (Figure 7-15).

The project is now ready to be wired. The type of wiring board used to wire the actual project varies from school to school. One type of board that can be used for teaching basic electrical circuits is shown in Figure 7-16. The actual components are fastened below the board while their schematic symbols appear on top. Terminal connections are made by way of binding posts. The actual picture of the top of the board can be reproduced for use by students in making wiring diagrams.

FIGURE 7-16 A WIRING BOARD

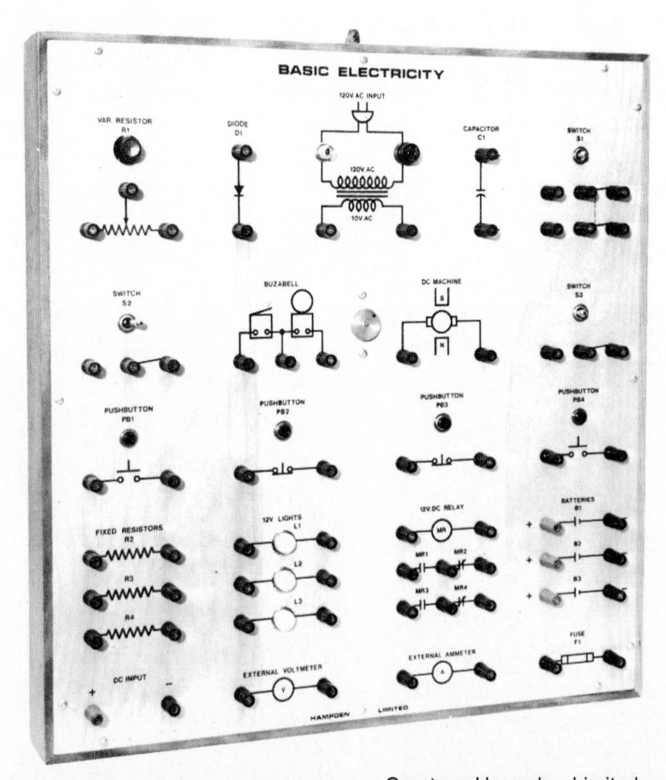

Courtesy Hampden Limited

7-8 Practical Assignments

1. (a) Complete a schematic diagram, wiring number sequence chart and wiring diagram for each of the following lighting circuits:
   - (i) One light controlled by one pushbutton.
   - (ii) Two lights in series controlled by one switch.
   - (iii) Two lights in parallel controlled by one switch.
   - (iv) One light controlled by two pushbuttons connected in series.
   - (v) One light controlled by two pushbuttons connected in parallel.

   (b) Wire each circuit and have it checked for layout and operation.

   (c) Compare the operation of the circuit when the two lights are connected in series and parallel.

   (d) Compare the operation of the circuit when the two pushbuttons are connected in series and parallel.

2. (a) Complete a wiring number sequence chart and wiring diagram for each of the following bell-buzzer signal circuits (Figure 7-17A–F).

   (b) Wire each circuit and have it checked for layout and operation.

FIGURE 7-17 CIRCUITS FOR PRACTICAL ASSIGNMENT 2

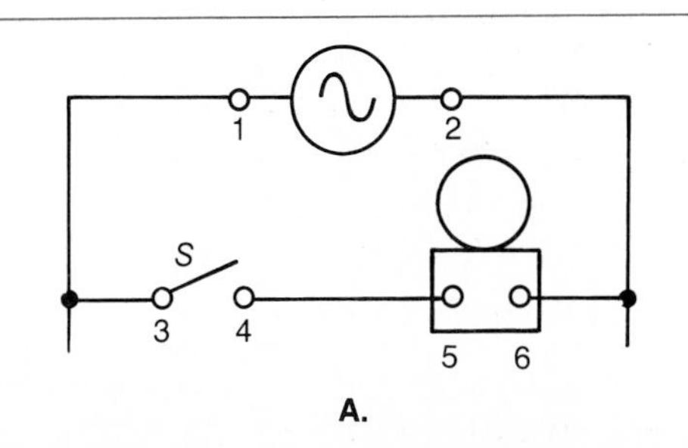

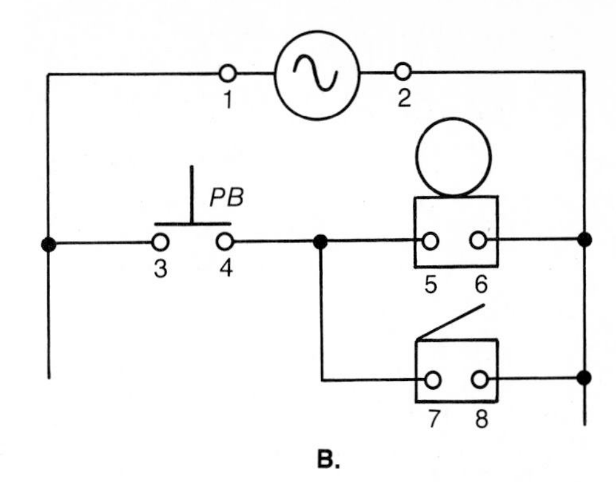

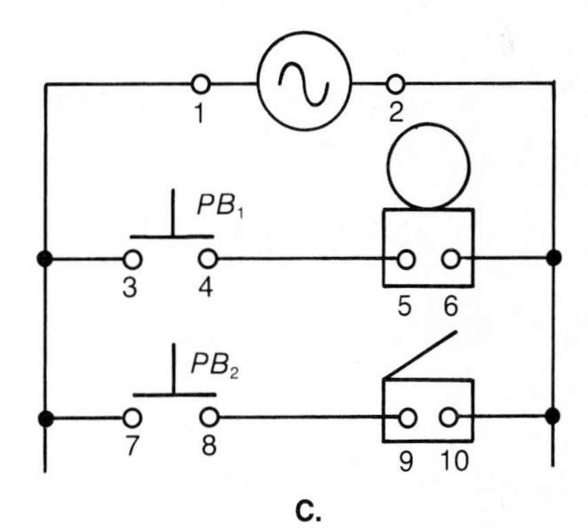

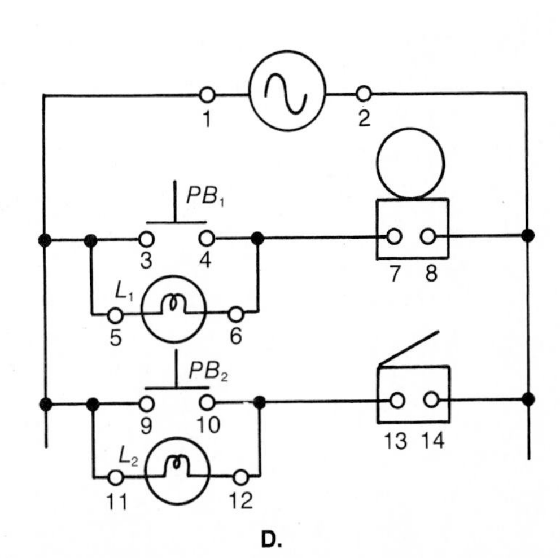

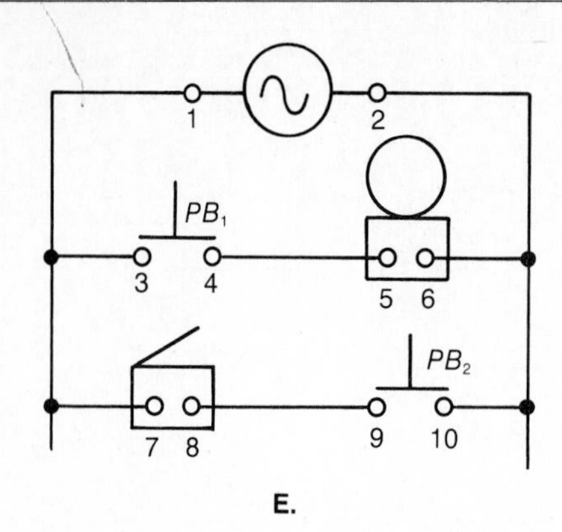

E.

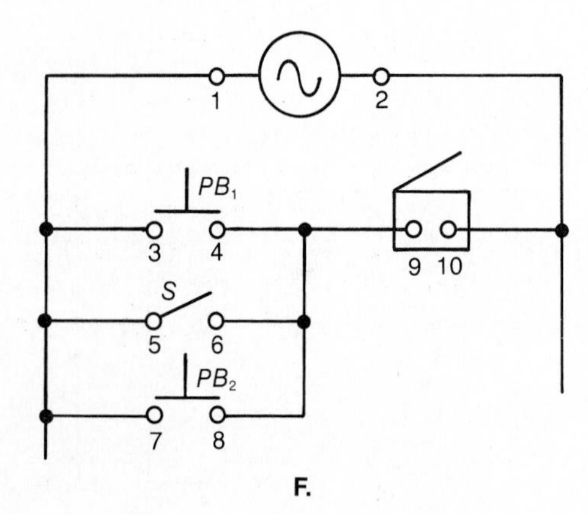

F.

7-9 Self Evaluation Test

1. Name five component parts of any basic circuit.
2. Draw an acceptable symbol that can be used on a schematic diagram to represent:
   (a) a DC power supply
   (b) an AC voltage source
   (c) a lamp
   (d) a resistor
   (e) a bell
   (f) a buzzer
   (g) a switch
   (h) a pushbutton
3. Describe one advantage and one disadvantage of the pictorial type of electrical diagram.
4. In what way is a wiring diagram different from a pictorial diagram?
5. Why is a schematic diagram easier to read?
6. Draw the schematic diagram for each of the following circuits:
   (a) one lamp controlled by one pushbutton and operated from an AC source
   (b) two lamps connected in series controlled by one switch and operated from a DC source
   (c) one lamp controlled by two pushbuttons connected in series and operated from an AC source
   (d) one lamp controlled by two switches connected in parallel and operated from a DC source
   (e) two lamps connected in parallel controlled by one pushbutton and operated from an AC source
7. Two identical 12 V lamps are connected in series to a 12 V DC source.
   (a) How many current paths are produced?
   (b) What is the value of the voltage drop across each lamp?
   (c) Comment on the brightness of each lamp (full or dim)?
   (d) Assume that one lamp burns open—what happens as a result of this?
8. Two identical 12 V lamps are connected in parallel to a 12 V DC source.
   (a) How many current paths are produced?
   (b) What is the value of the voltage across each lamp?
   (c) Comment on the brightness of each lamp (full or dim)?
   (d) Assume that one lamp burns open—what happens as a result of this?

9. A buzzer is to be operated when either one or the other of two pushbuttons is pressed. What type of electrical connection should be used to connect the two pushbuttons?

10. What connection of two or more pushbuttons can be considered to be an *AND* type of control circuit?

FIGURE 7-18 CIRCUIT FOR SELF EVALUATION TEST 11

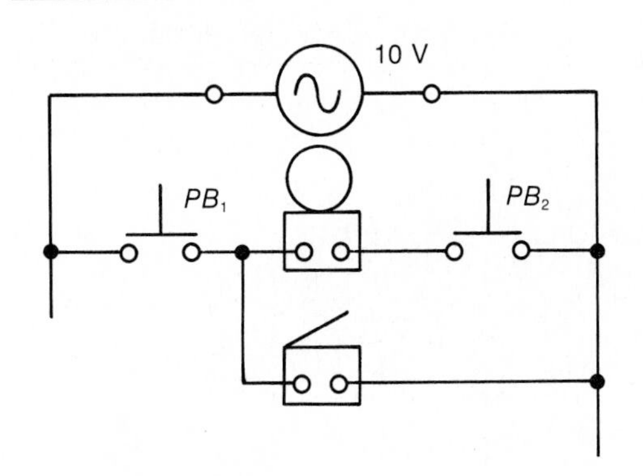

11. Answer each of the following questions with regard to the schematic circuit in Figure 7-18.
    (a) What type of voltage source is used?
    (b) What operates when PB_1 is pressed?
    (c) What operates when PB_2 is pressed?
    (d) What operates when both PB_1 and PB_2 are pressed?
    (e) How much voltage is across the bell and buzzer terminals when they are both operating?

12. (a) Re-draw the schematic circuit of Figure 7-18.
    (b) Number the terminals of each component on your schematic.
    (c) Complete a wiring number sequence chart for the circuit.
    (d) Complete a wiring diagram for the circuit.

8

1.980
HICKOK MX-333
PUSH ON
AUDIO
LOGIC
200Ω
2KΩ
20KΩ
200KΩ
2000KΩ
20MΩ
10A
200mV
1000V
2000mA
20Ω
SPECIAL FUNCTION
DC
AC
OHMS
POWER

INSTRUMENTS OF MEASUREMENT

OBJECTIVES

Upon completion of this unit you will be able to:

- Properly connect a voltmeter, ammeter and ohmmeter into a circuit.
- Correctly read an analog meter scale.
- State the precautions to be observed when using a voltmeter, ammeter and ohmmeter.
- Compare the *analog* with the *digital* type of meter.
- Explain the function of a multimeter.

8-1 Basic Electrical Test Instruments

The three basic electrical test instruments are the *voltmeter, ammeter* and *ohmmeter*. They are used to obtain accurate information about the voltage, current and resistance of a circuit. Both *analog* and *digital* types are available (Figure 8-1). Analog meters use a mechanical type of meter movement to indicate the measurement. Digital meters use an internal electronic circuit which indicates the measurement on a digital display. Analog meters usually cost less than the digital type but are not as accurate nor are they as easy to read as the digital type.

8-2 Reading Analog Meters

Reading a single range voltmeter or ammeter scale is similar to reading a ruler scale (Figure 8-2). Generally, the major divisions are marked and the value of the minor di-division. The scale in Figure 8-2 is read as follows:

Value of each major division = 1
Value of each minor division = 0.2
∴ Reading = 2.4

FIGURE 8-2 READING SINGLE RANGE METERS

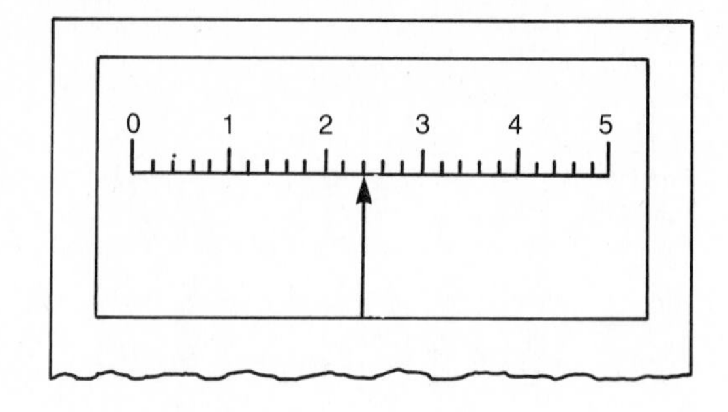

FIGURE 8-1 METER TYPES AND SYMBOLS

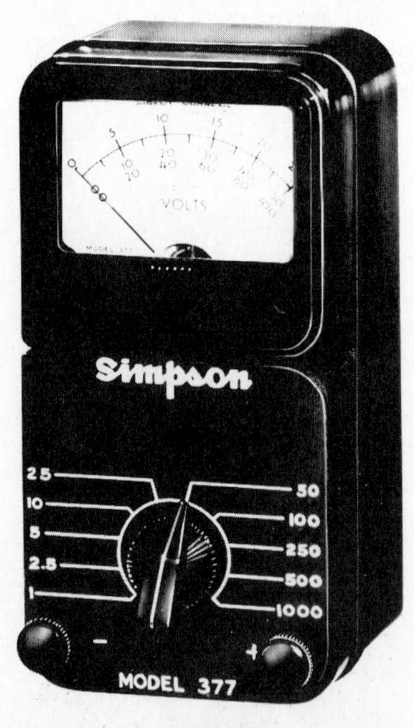

A. Analog

B. Digital

Symbols

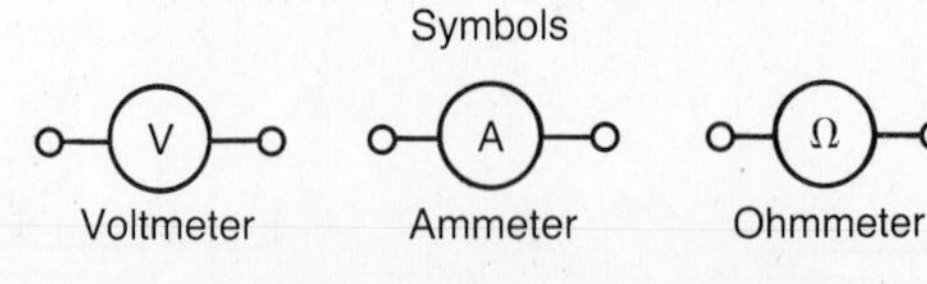

Courtesy Bach-Simpson Limited

FIGURE 8-3 READING A MULTIRANGE VOLTMETER OR AMMETER SCALE

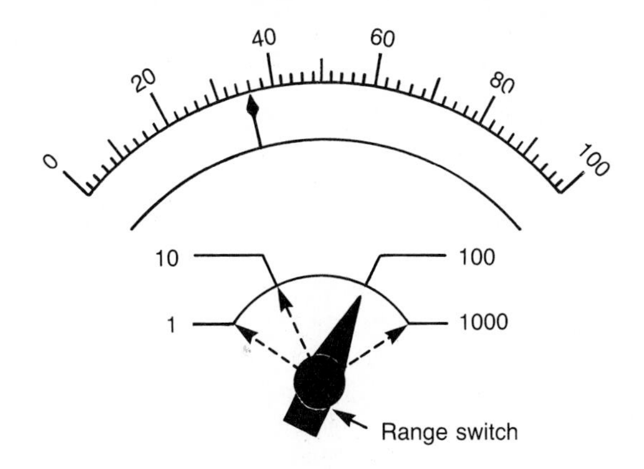

RANGE SWITCH	SCALE READING	X or ÷	CORRECT READING
100	36	X 1	36
1 000	36	X 10	360
10	36	÷ 10	3.6
1	36	÷ 100	0.36

Multirange voltmeters or ammeters are harder to read because one scale is often used with two or more ranges. To read this type of meter, first determine the reading on the scale and then apply the appropriate multiplier or divider as indicated by the selector switch (Figure 8-3).

Some voltmeters and ammeters have both multiranges and multiscales. These are the most difficult to read and can only be mastered with practice. If the setting of the range switch number is the same as the number at the end of one of the scales, that scale is then read directly to determine the correct reading (Figure 8-4).

Assume that the setting of the range switch number is 10 times more than the number at the end of one of the scales (Figure 8-5). The scale reading is then multiplied by 10 to determine the correct reading.

Assume that the setting of the range switch number is 10 times less than the number at

FIGURE 8-4 READING WHEN THE RANGE SWITCH NUMBER IS THE SAME AS THE SCALE NUMBER

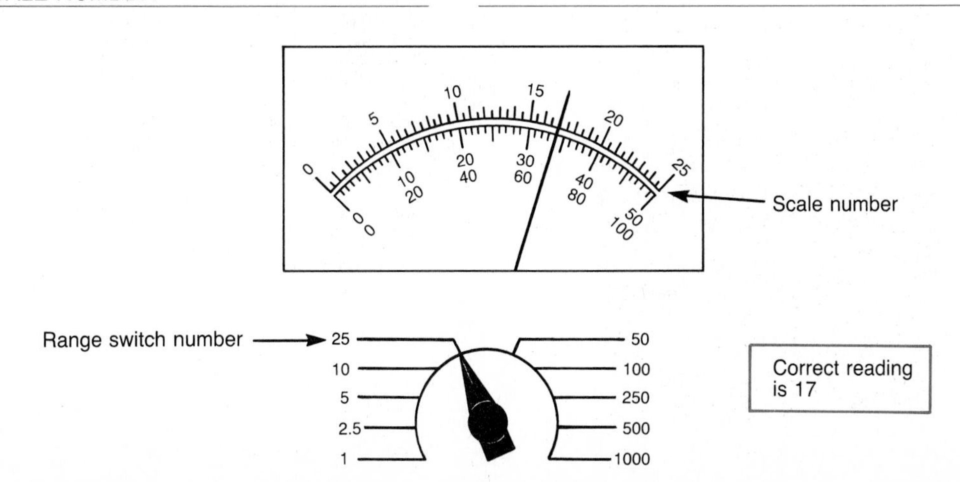

FIGURE 8-5 READING WHEN THE RANGE SWITCH NUMBER IS 10 TIMES GREATER THAN THE SCALE NUMBER

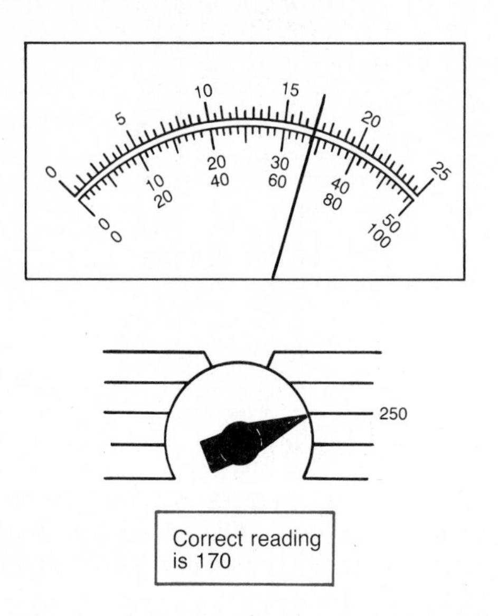

FIGURE 8-6 READING WHEN THE RANGE SWITCH NUMBER IS 10 TIMES LESS THAN THE SCALE NUMBER

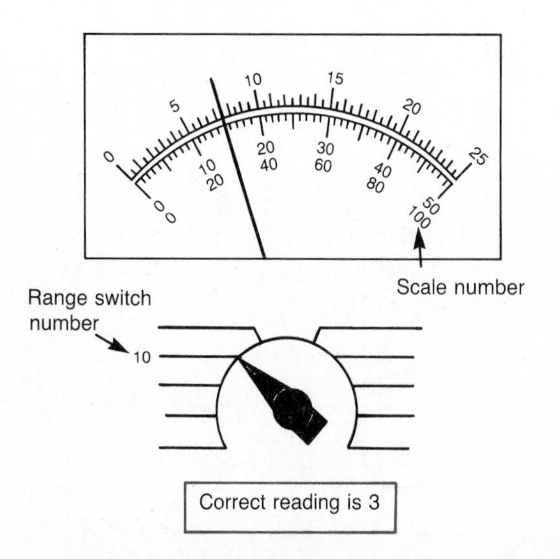

the end of one of the scales (Figure 8-6). The scale reading is then divided by 10 to determine the correct reading.

8-3 The Voltmeter

The *voltmeter* is used to measure the voltage or potential energy difference of a load or source. Voltage exists between two points and does not flow through a circuit like current. As a result, a voltmeter is connected across or in parallel with these points.

DC or AC types of voltmeters are selected according to the type of voltage to be measured. When connecting a DC voltmeter the polarity of the meter's test lead must match that of the circuit. In other words, the voltmeter's positive and negative leads must connect to the circuit's positive and negative points respectively (Figure 8-7). If the polarities are reversed, the DC meter will tend to read backwards.

FIGURE 8-7 DC VOLTMETER CONNECTION

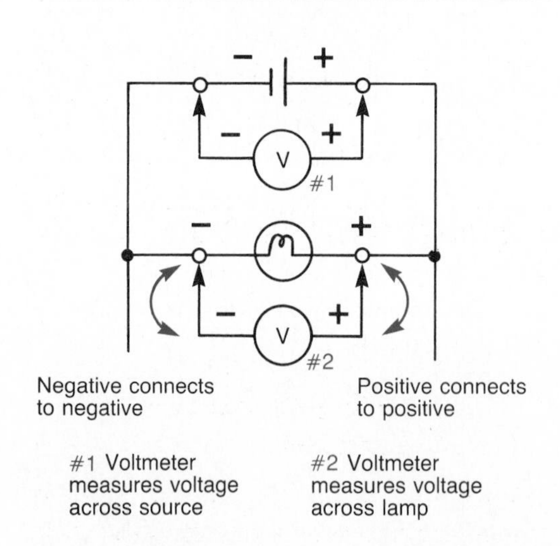

The voltmeter range switch should always be set higher than the voltage you expect to read. You cannot measure 12 V on a range that ends at 5 V. If you try to do so you may damage the meter. For unknown voltage levels, always set the range switch to the highest range and work down from there.

FIGURE 8-8 VOLTAGE TESTER

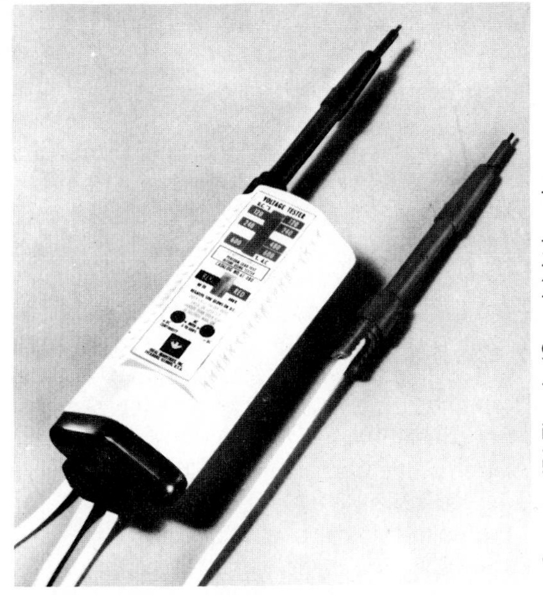

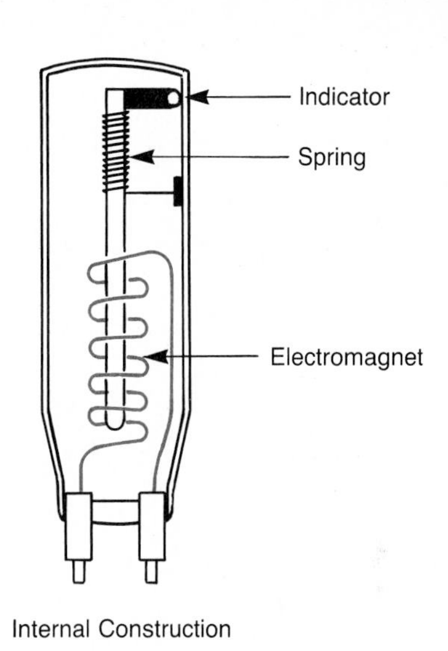

Internal Construction

The voltage tester is a special type of voltmeter commonly used by electricians (Figure 8-8). Its rugged construction makes it ideally suitable for on-the-job voltage measurements. The voltage tester indicates the approximate level of the voltage present and not the exact value. It is used primarily to test for the presence or absence of a voltage.

8-4 The Ammeter and Milliammeter

The *ammeter* and *milliammeter* are both used to measure electrical current flow in a circuit. Unlike voltage, current flows through a circuit. As a result, the ammeter or milliammeter must be connected into the circuit or in series to measure the current. A milliammeter is designed to measure much lower values of current than an ammeter. Both are connected into the circuit in series.

DC or AC types of ammeters are selected according to the type of current flow to be measured. DC ammeters must be connected into the circuit observing the correct polarity. Again the positive and negative leads of the meter must connect to the positive and negative leads of the circuit respectively (Figure 8-9).

Since the ammeter must be connected in series with the load, its resistance must be low to allow normal current to flow during the metering process. Accidentally connecting it in parallel with a load or voltage source may permanently damage it. The meter can also be overloaded by attempting to measure current in excess of the range it is set to measure.

FIGURE 8-9 DC AMMETER CONNECTION

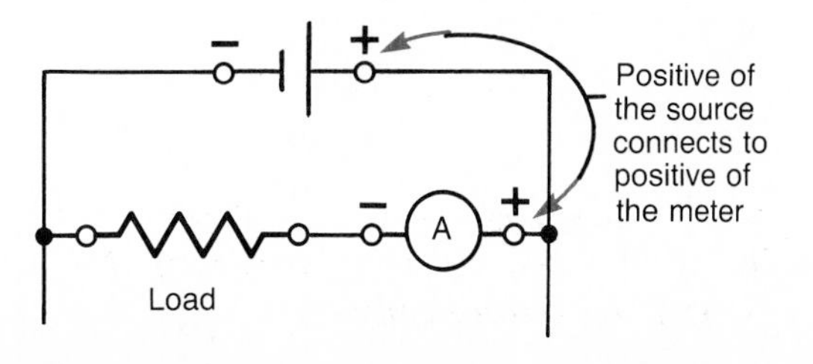

FIGURE 8-10 CLAMP-ON AMMETER

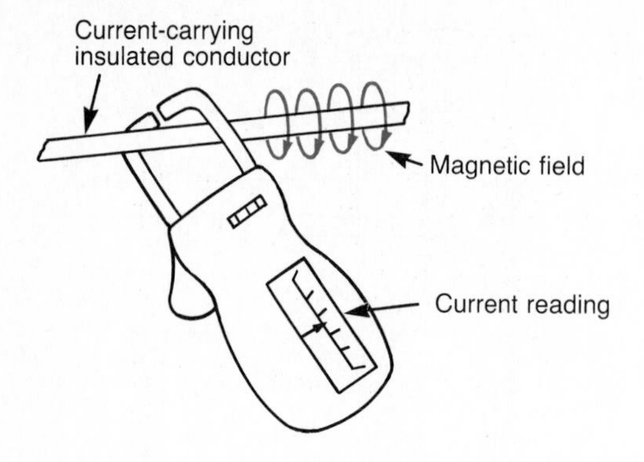

The clamp-on ammeter is a special type of ammeter commonly used by electricians to measure higher AC currents in the ampere range (Figure 8-10). It is more convenient to use than a standard ammeter in that the circuit does not have to be opened to take a current reading. This instrument clamps around the circuit conductor and indicates the current flow in the conductor by measuring the strength of the magnetic field about the current-carrying conductor.

8-5 The Ohmmeter

The *ohmmeter* is an instrument used to measure the amount of electrical resistance offered by a complete circuit or a circuit component (Figure 8-11). It has its own power supply in the form of batteries located within the meter itself and can be damaged if connected to a live circuit. Before taking a reading with an analog type of ohmmeter the meter scale must be set for zero. This is accomplished by connecting the two test leads of the ohmmeter together and adjusting the zero adjust knob either way to set the pointer for a reading of zero ohms.

In addition to measuring resistance, the ohmmeter can also be used as a continuity tester. A complete conducting path is indicated by a zero resistance reading on the meter. An open in the circuit is indicated by a reading of infinity on the meter scale.

FIGURE 8-11 OHMMETER CONNECTION

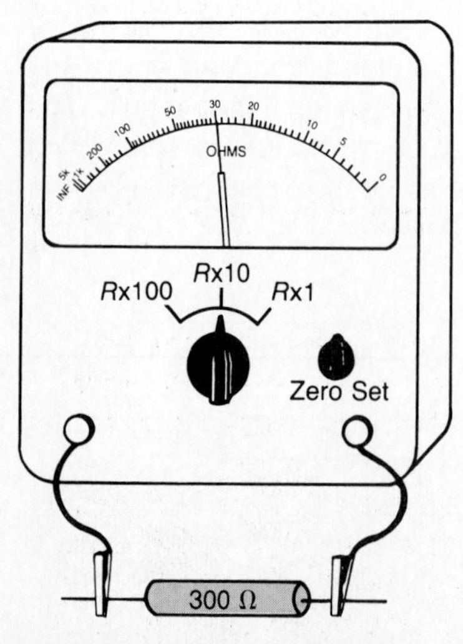

A. Out of circuit measurement

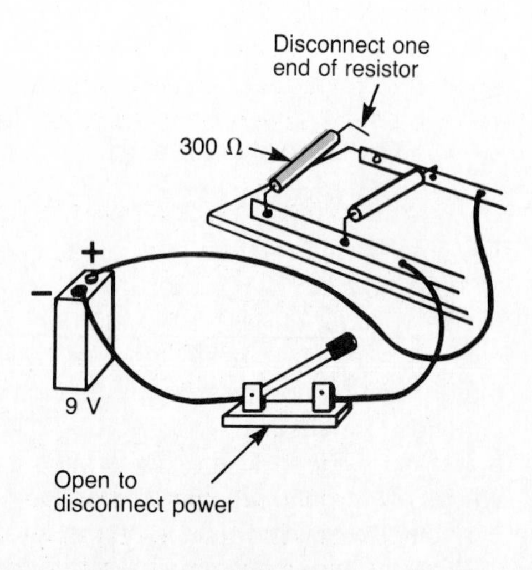

B. In circuit measurement

FIGURE 8-12 MULTIMETERS

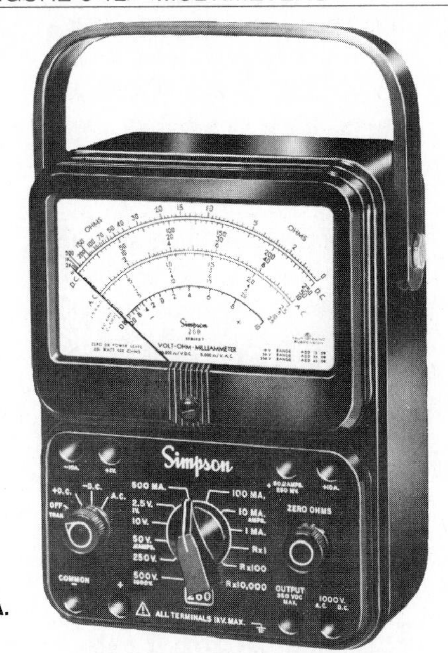

A.

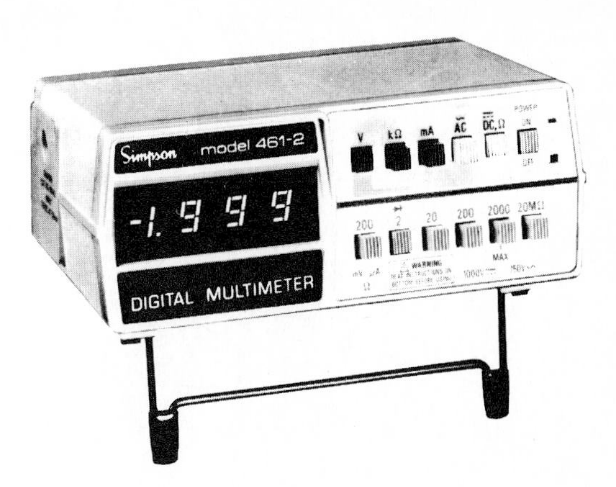

B. Courtesy Bach-Simpson Limited

8-6 The Multimeter

The *multimeter* is a combination ammeter, voltmeter and ohmmeter in one single, cased instrument. It is the most popular of all test instruments. Both analog and digital types are available (Figure 8-12A, B). Such instruments contain front panel switching arrangements that allow you to select the function and range desired. When connecting the multimeter, the same voltmeter, ammeter and ohmmeter connection rules apply.

8-7 Practical Assignments

1. Connect a multirange analog DC voltmeter to a variable DC voltage source. Have the teacher vary the voltage of the source. Record the voltage reading of the meter for ten different settings of the voltage source.

2. (a) Complete a wiring number sequence chart and wiring diagram for the schematic circuit in Figure 8-13.

FIGURE 8-13 CIRCUIT FOR PRACTICAL ASSIGNMENT 2

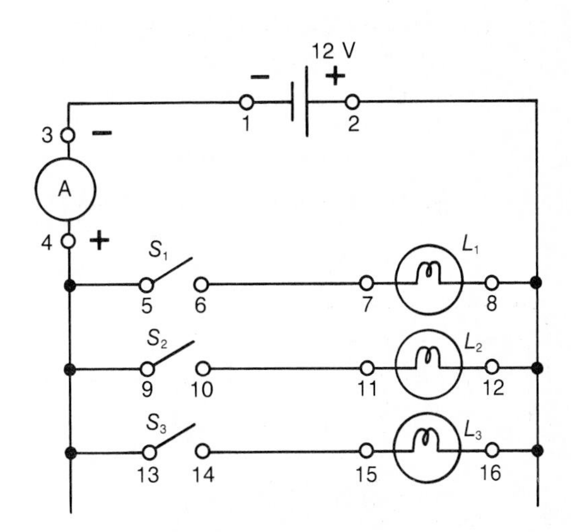

(b) Wire the circuit and record the reading on the ammeter for each of the following switching positions:

(i) S_1 closed – S_2 and S_3 open
(ii) S_2 closed – S_1 and S_3 open
(iii) S_3 closed – S_1 and S_2 open
(iv) S_1 and S_2 closed – S_3 open
(v) S_1 and S_3 closed - S_2 open

(vi) S_3 and S_2 closed – S_1 open
(vii) S_1, S_2, and S_3 closed

(c) What connection of lamps is used in this circuit?

3. Using an ohmmeter, measure and record the resistance of ten unmarked resistors.

FIGURE 8-14 CIRCUIT FOR PRACTICAL ASSIGNMENT 4

4.

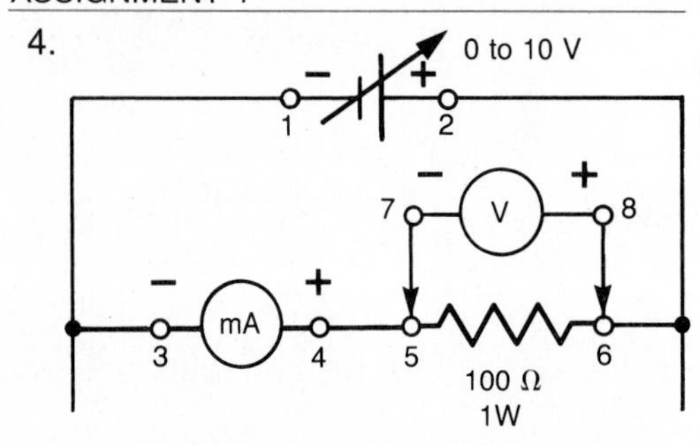

(a) Complete a wiring number sequence chart and wiring diagram for the schematic circuit in Figure 8-14.
(b) Wire the circuit. Increase the voltage across the resistor from 0 to 10 V, in ten 1 V steps. Record the voltage and corresponding current for each step.

FIGURE 8-15 CIRCUIT FOR PRACTICAL ASSIGNMENT 5

5.

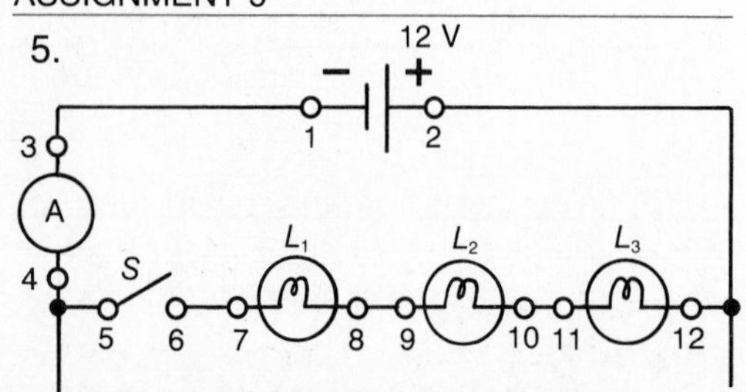

(a) Complete a wiring number sequence chart and wiring diagram for the schematic circuit in Figure 8-15.
(b) Wire the circuit. Close the switch and record the ammeter reading.
(c) With the switch closed, record the voltage drop across each lamp as measured with a DC voltmeter.
(d) In this circuit, what connection of lamps is used?

6. The teacher will wire five different circuits each with a specific type of meter connected to measure voltage, current or resistance. You will, for each circuit, record:
   (a) type of meter connected to the circuit,
   (b) how the meter is connected into the circuit,
   (c) what is being measured,
   (d) what the meter reading is.

8-8 Self Evaluation Test

1. In what way is the reading on a *digital* meter different from that on an *analog* meter?

2. Give the correct reading for each meter reading in Figure 8-16. Assume range switch to be set at range indicated.

3. The voltage and current of a lamp connected to a DC source are to be measured.
   (a) How is the voltmeter connected to measure the voltage?
   (b) How is the ammeter connected to measure the current?
   (c) Draw the schematic diagram of this circuit showing the proper connection of an ammeter and voltmeter. On this diagram indicate the polarity of the voltage source as well as the correct polarity of the meter leads.

4. When measuring voltages and currents of unknown levels, what meter range switch setting should be used?

FIGURE 8-16 SELF EVALUATION TEST 2

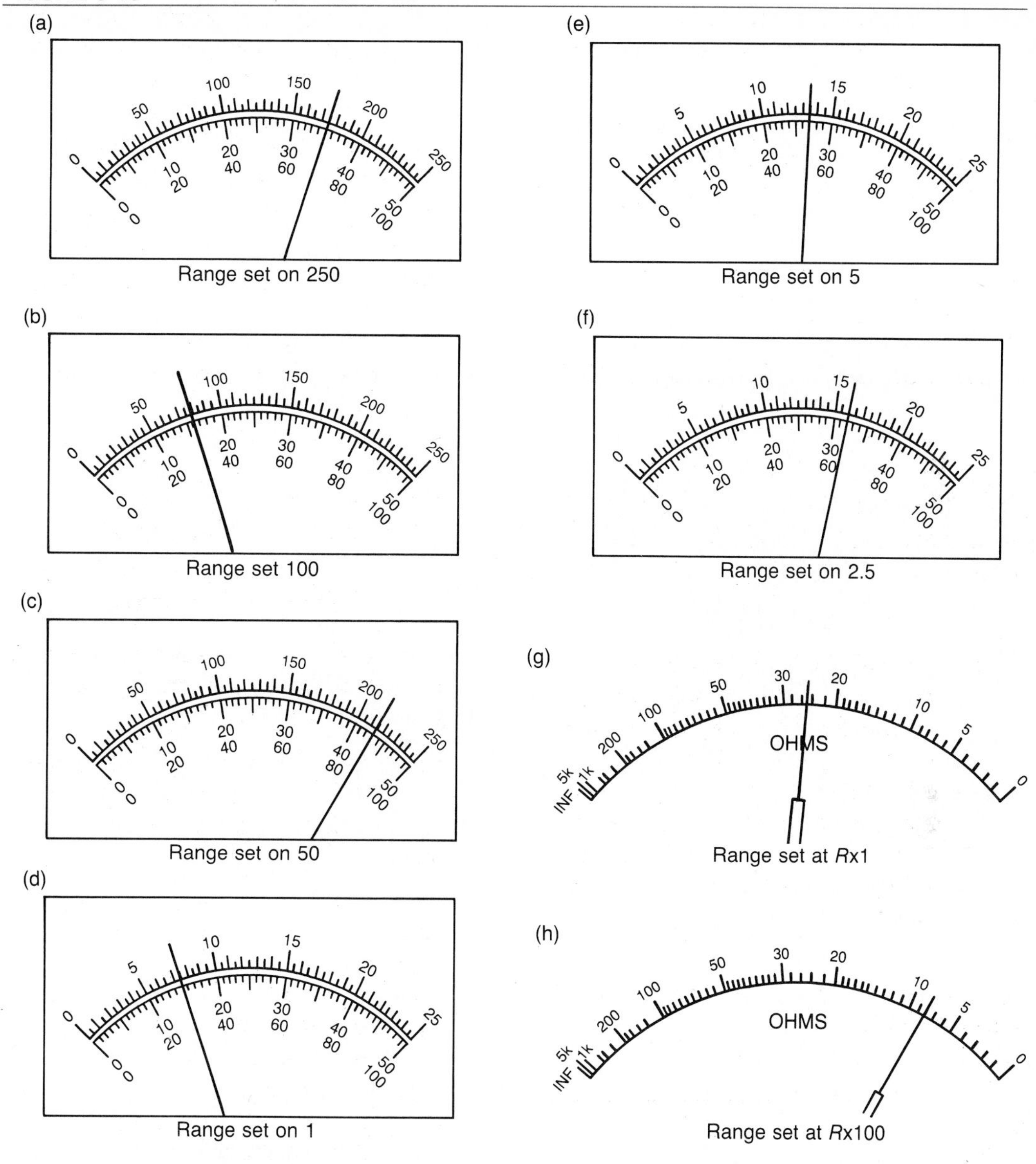

5. State one advantage and one limitation of a voltage tester.
6. A voltmeter and ammeter are connected incorrectly into a circuit. The ammeter is connected where the voltmeter should be and vice versa. Which meter will be automatically overloaded? Why?
7. Why do electricians like to use the clamp-on type of ammeter when checking circuit currents?
8. An ohmmeter is to be used to check the resistance of a resistor wired into a circuit. Before taking a reading with the ohmmeter, what two things should you check about the circuit?

9

CIRCUIT CONDUCTORS AND WIRE SIZES

OBJECTIVES

Upon completion of this unit you will be able to:

- Identify uses for different conductor forms.
- List the factors to be considered in the selection of wire insulating materials.
- State the AWG size of conductors used for common house circuits.
- Use a wire gage to determine the AWG number of a conductor.
- Define the term *conductor ampacity* rating and list the factors used in determining it.
- Define the term *line voltage drop* and explain the circuit condition which produces it.
- Measure the line voltage drop of a circuit.
- Explain the factors that determine the resistance value of a conductor.

9-1 Conductor Forms

The best electrical conductors are metals. The most popular metal used as a conductor is copper. In addition to its low resistance, copper is easy to work with and makes excellent electrical connections to terminal screws. In electronic circuits it can be easily soldered, ensuring a secure electrical connection.

Copper conductors used for wiring circuits can be made in the form of wire, cable, cord or printed circuit boards. A solid wire is a single conductor covered by some form of insulation. A stranded wire is a single conductor made up of many small diameter wires running alongside each other. The purpose of stranding conductors is to provide increased flexibility.

Both solid and stranded hook-up wire is used to complete circuits within electrical and electronic equipment (Figure 9-1).

FIGURE 9-1 SOLID AND STRANDED HOOK-UP WIRE

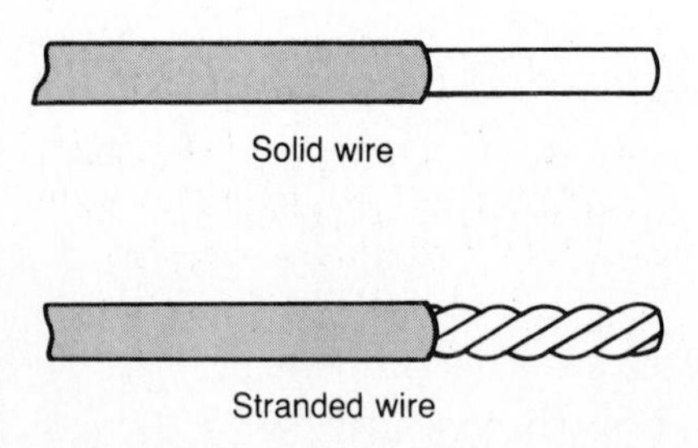

The term *cable* can refer to a larger stranded insulated wire or two or more separately insulated wires assembled within a common covering (Figure 9-2). Large single conductor cables are much more flexible than the equivalent size solid conductor and are used in circuits requiring large amounts of current. Multi-conductor nonmetallic sheathed cable is used for permanent wiring circuits in a house.

FIGURE 9-2 CABLE FORMS

A. Large stranded cable

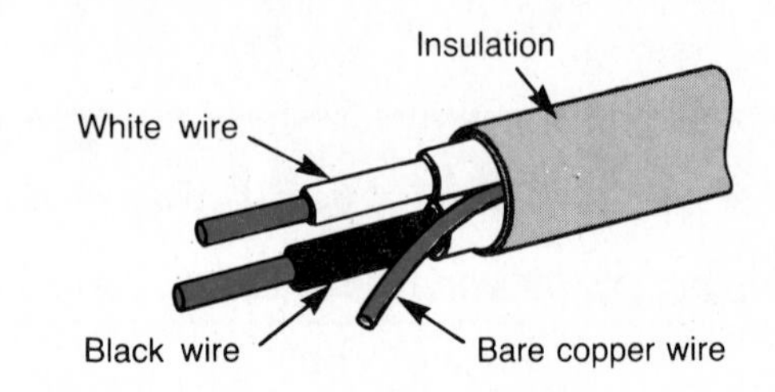

B. Multi-conductor cable

Cord is the name given to very flexible cables used to supply current to appliances and portable tools. Their construction is similar to that of a cable except that cord conductors are made of strands of very fine wire twisted together (Figure 9-3).

FIGURE 9-3 LAMP CORD

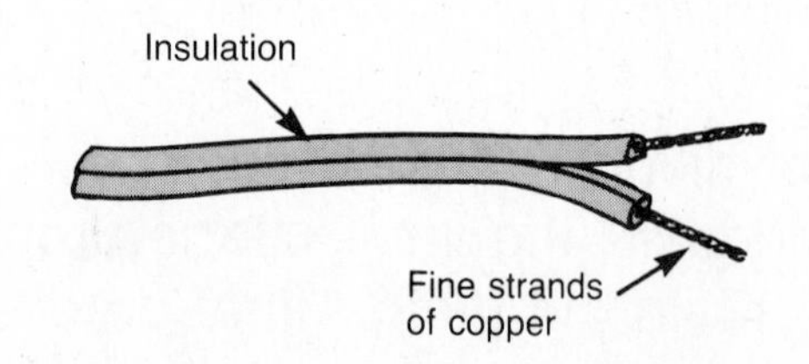

Printed circuits are used extensively in low current electronic equipment for connecting components together. A *printed circuit board* consists of conducting paths of thin copper strips etched or printed on a flat insulated plate (Figure 9-4). They also provide a convenient base for mounting small electronic components along with the means for interconnecting them.

FIGURE 9-4 PRINTED CIRCUIT BOARD

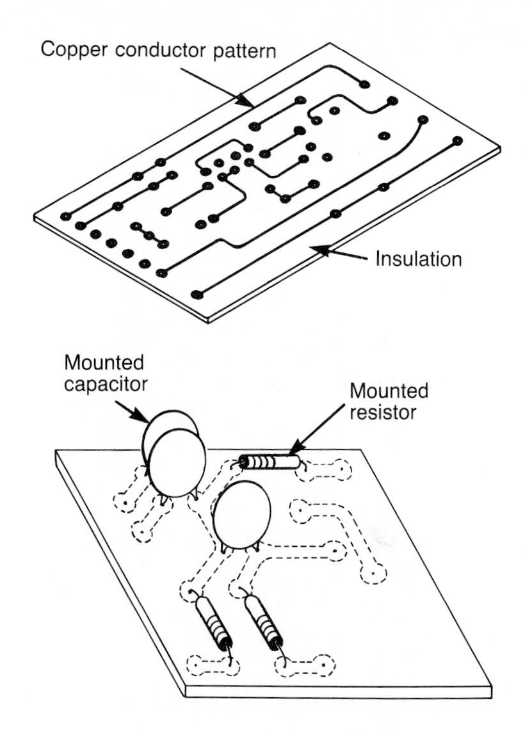

9-2 Conductor Insulation

Conductor insulators are made for a variety of applications. Some of the factors that must be considered when selecting a wire insulation are: circuit voltages, surrounding temperature conditions, moisture and conductor flexibility. Thermoplastic is one of the most commonly used insulators. Regular thermoplastic is an excellent insulator, but is sensitive to extremes of temperature. Type TW thermoplastic is a weatherproof type while type THW is weatherproof and heat-resistant.

Neoprene is a special rubber type of insulation used for power line cords on heat-producing appliances such as kettles and frypans. Another type of heat-proof insulation is asbestos (Figure 9-5). Asbestos-coated wire is used inside electric stoves for making connections to heating elements. Asbestos also has been used inside cotton-braid jackets as part of the line cord for clothes irons.

Heat-resistant plastic and rubber-based cords have generally replaced asbestos cords in household appliances.

Baked-on varnish insulation is found on the copper wire used in motor windings and relay coils. This wire is often referred to as magnet wire.

9-3 Wire Sizes

The size of a solid wire is determined by its diameter. For convenience, wire sizes are usually referred to by an equivalent gage number rather than the actual diameter.

The American Wire Gage (AWG) table is the standard used and consists of forty wire sizes ranging from AWG 40 (the smallest) all the way up to AWG 4/0. Note that

FIGURE 9-5 HEAT-PROOF IRON CORD INSULATION

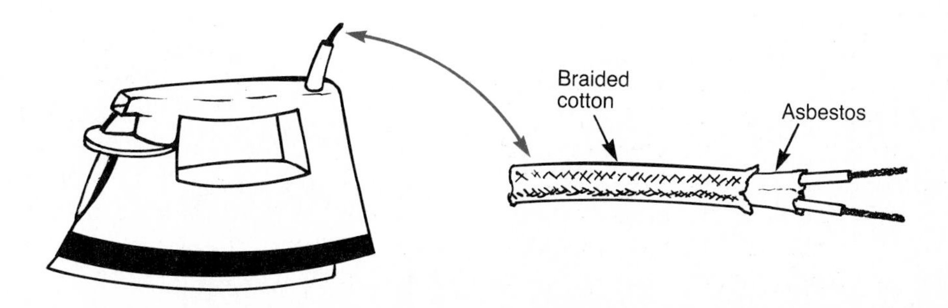

FIGURE 9-6 AWG WIRE SIZES

AWG Number	18	16	14	12	10	8	6	4	2	0	00
Approximate Area											

the larger the gage number the smaller the actual diameter of the conductor (Figure 9-6).

The AWG number for an unknown solid conductor can be determined directly by measuring it with a wire gage (Figure 9-7). A typical wire gage is in the figure. To gage the wire the insulation is carefully removed from one end. Then the bare end is inserted into the smallest slot into which it will fit without using force. The number stamped below the slot is the AWG of the wire.

9-4 Conductor Ampacity

The *ampacity* of a conductor refers to the maximum amount of current it can safely carry. This current rating or ampacity is determined by its material, gage size, type of insulation and conditions under which it is installed. Copper is a better conductor than aluminum and therefore can carry more current for a given gage. Similarly, the smaller the gage number, the larger the conductor and the more current it can carry. A conductor with a heat-resistant insulation will have a higher ampacity rating than one of equivalent size with a lower insulator temperature rating. In addition, conductors that are run singly in free air will have a higher ampacity rating than a similar conductor enclosed along with other conductors in a cable or conduit.

The National Electrical Code contains tables which list the ampacity for the ap-

FIGURE 9-7 AWG WIRE GAGE

A. Wire gage

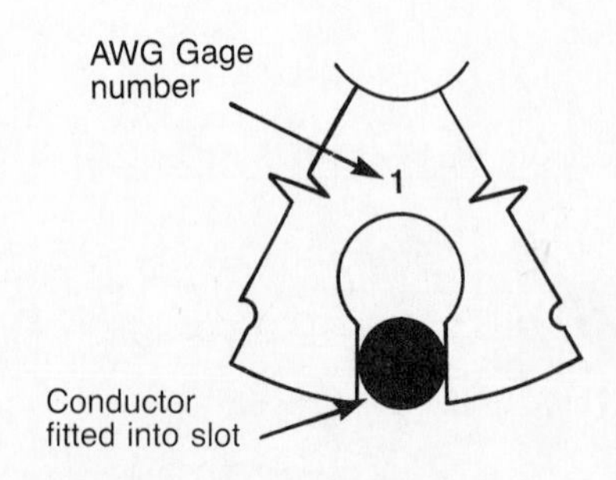

B. Using the gage

TABLE 9-1

AWG	AMPACITY	EXAMPLE APPLICATION
#14 copper	15 A	Lighting and receptacle branch circuits.
#12 copper	20 A	Electric heating circuits and 240 V water heaters rated up to 4 500 W.
#10 copper	30 A	Electric dryers rated up to 7 000 W.
#8 copper	40 A	Electric ranges rated up to 12 000 W.

proved types of conductor size, insulation and operating conditions. These tables should be referred to for specific circuit installations. Table 9-1 lists some of the common copper conductor sizes used for branch circuits in a home.

9-5 Conductor Resistance

The *resistance* of the wire conductors of the circuit is low when compared to that of the load. In most circuits the conductors are treated as being perfect conductors of electricity. As a result they are said to have zero resistance (Figure 9-8). In this case the voltage value of the source is the same as that across the load. In other words, no voltage is lost in the line.

FIGURE 9-8 ZERO LINE VOLTAGE DROP

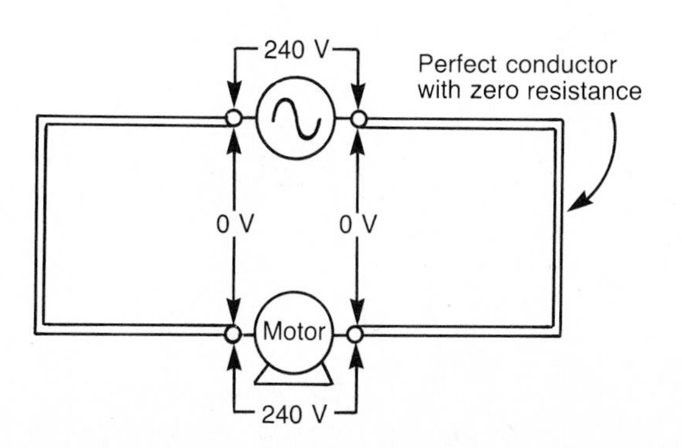

FIGURE 9-9 10 V LINE VOLTAGE DROP

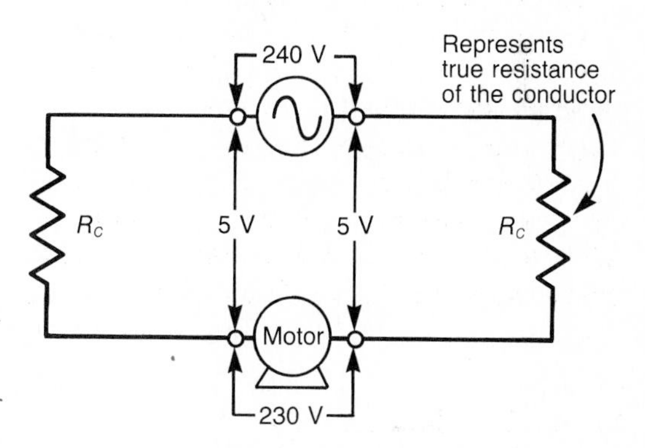

In some circuits the resistance of the conductors is important and must be taken into account. This is often the case where the load is located some distance from the voltage source. In this type of circuit the voltage at the load can be much less than what appears at the energy source. The difference in voltage between the voltage source and the load is called the **line voltage drop** (Figure 9-9). It is always desirable to keep the line voltage drop as low as possible. High line voltage drops can rob the load of energy required for its proper operation.

Line voltage drops are kept low by keeping the resistance of the line wires low. The resistance of a length of wire is determined by: the type of metal used, the operating temperature, the length of the wire, and the AWG size or cross-sectional area of the wire.

Different materials have different atomic structures which affect their ability to conduct electrons. Copper, for example, is a better conductor than aluminum but not as good as silver. However, the option of using a different conductor material to lower conductor resistance is limited.

The resistance of a conductor varies with **temperature**. For copper, the higher the temperature the higher the resistance. The two factors that determine the operating temperature of a conductor are the temperature of the surrounding air space (ambient temperature) and the amount of current flow through the conductor. When properly installed and operated at rated current, conductor temperature has a minimal effect on conductor resistance.

Conductor length is perhaps the most important factor to consider when talking about line voltage drops. The resistance of a conductor varies directly with its length. The greater the distance between the voltage source and the load, the greater the amount of line voltage drop. For very long runs the expected voltage drops are calculated before the installation and steps taken to keep the drop within acceptable limits.

The **AWG size** of the conductor is the most important factor in correcting for excessive line voltage drops. The greater the cross-sectional area of the conductor, the lower is its resistance. Wire diameter sizes are often increased above the rated current capacity required for the circuit to reduce the line voltage losses.

9-6 Practical Assignments

1. (a) Using a wire gage, measure and record the AWG number of 5 different sizes of copper wire.
   (b) How does the AWG number compare with the diameter of the wire?
2. Correctly mount and solder 15 electronic components to a printed circuit board. This will help you to develop your fine soldering skills. Use inexpensive, surplus components and printed circuit boards for this project.
3. Have the teacher demonstrate the heating effect of operating a conductor above its ampacity rating. Make note of your observations.
4. (a) Have the teacher demonstrate the line voltage drop on a full roll of AWG #14 copper wire cable. Make note of the source voltage and load voltage.
   (b) Repeat the test using the same circuit and a half roll of AWG #14 copper wire cable.
   (c) Repeat the test using the same circuit and a full roll of AWG #12 copper wire cable.
   (d) Account for the difference in line voltage drop between test (a) and (b).
   (e) Account for the difference in line voltage drop between test (a) and (c).

FIGURE 9-10 CIRCUIT FOR PRACTICAL ASSIGNMENT 5

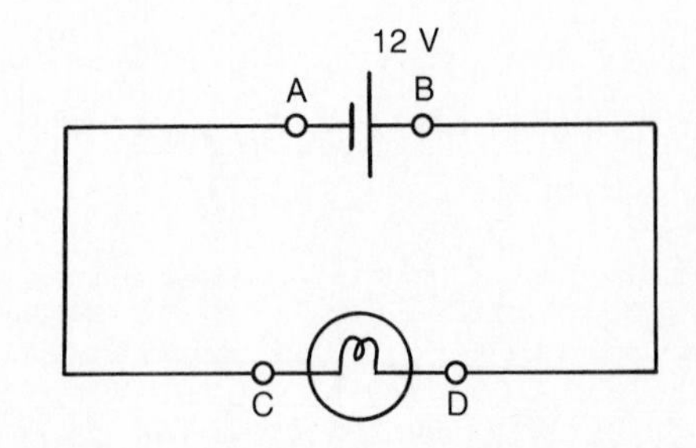

5. Wire the circuit in Figure 9-10. Using a digital voltmeter, record the line voltage drop from *A* to *C* and *B* to *D*.

9-7 Self Evaluation Test

1. Name four common conductor forms and state one practical application for each.

2. What are three important factors to be considered when selecting the type of wire insulation to be used for a particular job?

3. What is the relationship between the AWG number and the diameter of the wire?

4. State the AWG size of copper wire most commonly used for each of the following house circuits:
   (a) dryer circuit
   (b) range circuit
   (c) electrical outlets and lighting circuits
   (d) electric water heater circuit.

5. (a) Define the term conductor ampacity.
   (b) List four factors that determine the ampacity of a conductor.

6. State the effect of each of the following on the resistance value of a circuit conductor:
   (a) increasing the length of the conductor,
   (b) decreasing the diameter of the conductor,
   (c) increasing the operating temperature of the conductor,
   (d) using an aluminum conductor in place of copper.

7. (a) What is meant by the term *line voltage drop*?
   (b) Under what condition is the line voltage drop considered to be zero?

10

RESISTORS

OBJECTIVES

Upon completion of this unit you will be able to:

- Identify uses for resistance wire.
- Explain the function of a resistor.
- State the ways in which resistors are rated.
- Identify and compare different types of resistors and resistor circuits.
- Correctly apply the color code to determine the resistance and tolerance value of resistors.
- Calculate the total resistance for resistors connected in series or parallel.
- Correctly connect and measure the total resistance of resistors connected in series and parallel.

10-1 Resistance Wire

A resistance wire is used to produce heat for heating with electricity (Figure 10-1). The most popular type of resistance wire is made of a high-resistance alloy of nickel and chromium and is referred to as nichrome wire. This wire is used for the heating elements in stoves, dryers, toasters and other heating appliances.

FIGURE 10-1 RESISTANCE WIRE HEATING ELEMENT

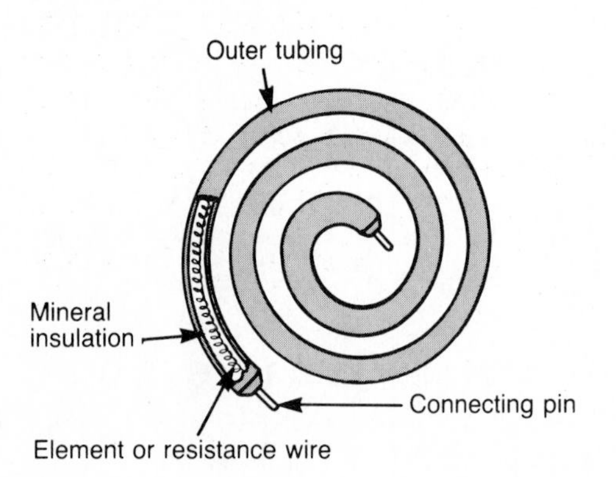

When a voltage is applied to the heating element, the high resistance of the wire converts most of the electrical energy into heat energy. In tubular elements the wire carrying current is enclosed in a tubing with a powdered mineral insulation. The powder insulation insulates the wire from the tubing as well as seals the wire off from contact with the air. This seal prevents oxidation and prolongs the life of the element.

10-2 Resistors

A resistor is an electrical component added to a circuit to lower the current flow or reduce the voltage. You can make a simple resistor by drawing a line with a lead pencil on a sheet of paper (Figure 10-2). The resistance of the line or points along it can be measured using an ohmmeter. Set the ohmmeter to its highest resistance scale to measure this resistance. You will find that the resistance varies directly with the length of the line.

FIGURE 10-2 DO-IT-YOURSELF RESISTOR

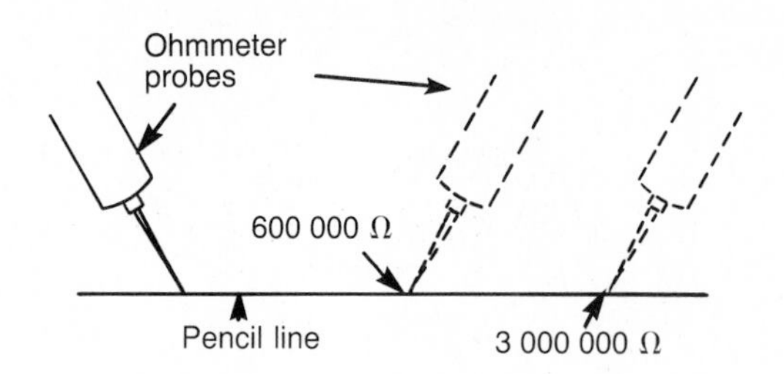

Resistors are rated in three ways. The first rating is *resistance*, measured in ohms (Ω). It is hard to manufacture a resistor with an exact number of ohms of resistance; therefore, most resistors also carry a *percentage tolerance* or *accuracy rating*. Finally, the electric current carried by a resistor produces heat, so resistors are rated according to the *amount of power* the resistor can safely handle in watts (W) (Figure 10-3). The larger the physical size of the resistor, the more heat it can safely dissipate and hence the greater the power rating.

FIGURE 10-3 RESISTOR RATINGS

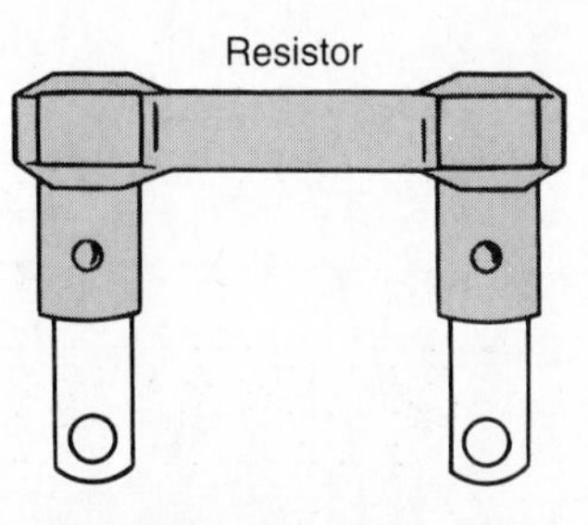

An example of resistor rating

1. 500 Ω resistance
2. ± 5% tolerance
3. 10 W

FIGURE 10-4 WIREWOUND RESISTOR

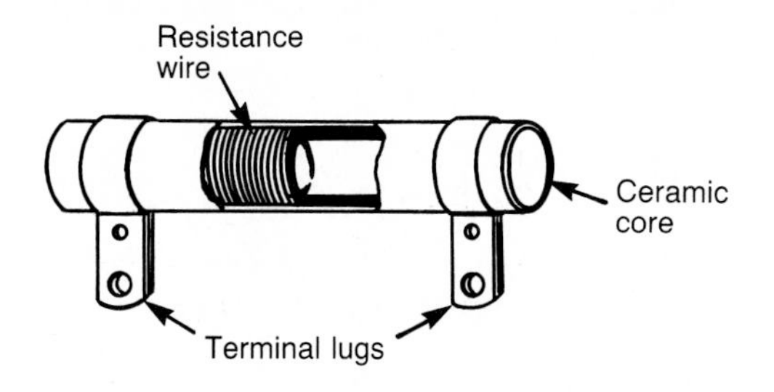

10-3 Types of Resistors

Resistors can be classified according to their construction. Wirewound resistors are made by wrapping high-resistance wire around an insulated cylinder (Figure 10-4). The rated resistance of the resistor is directly proportional to the length of the wire used. This type of resistor is expensive to manufacture. They are generally used in circuits which carry high currents or in circuits where accurate resistance values are required.

Composition or carbon resistors are made from a paste consisting of carbon and a filler material (Figure 10-5). The resistance of a carbon resistor is determined by the amount of carbon used in making the resistor. Carbon resistors are the cheapest type to manufacture and as a result are the ones most widely used. They can not handle large currents and their actual value of resistance can vary as much as 20% from their rated value.

FIGURE 10-5 CARBON RESISTOR

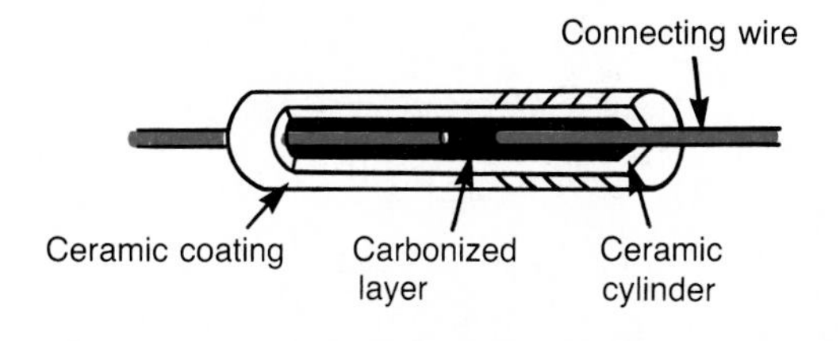

A second way in which resistors are classified is in terms of how they function. The *fixed* resistor (Figure 10-6A) is the simplest of designs and functions at one non-variable ohmic value. An *adjustable* resistor (Figure 10-6B) is one designed to provide for a range of different resistance values. This resistor contains a sliding contact that can be positioned and secured to provide different resistance values up to the maximum value of the resistor. It is, however, not designed to be continuously variable. The variable resistor (Figure 10-6C) is one designed to provide for continuous adjustment of resistance. This resistor contains a shaft connected to a moving contact. As the shaft is rotated, the moving contact provides a variable resistance which ranges from zero up to the maximum value of the resistor. The resistance material may be either resistance wire or a carbon mixture.

10-4 Rheostats and Potentiometers

A **rheostat** is a variable resistor connected using only two of its terminals to vary the resistance in a circuit (Figure 10-7). The simple auto dashboard light dimmer circuit is a good example of the rheostat connection.

A **potentiometer** (pot) is a variable resistor that makes use of all three of its terminals. Unlike the rheostat, its main purpose is not to vary resistance but to vary the voltage in a circuit. The two fixed maximum resistance leads are connected across the voltage source and the variable arm lead provides a voltage which varies from zero to maximum. Two good examples of a typical potentiometer connection are the variable voltage and speaker volume control circuits drawn in Figure 10-8.

FIGURE 10-6 RESISTORS CLASSIFIED ACCORDING TO FUNCTION

A. Fixed resistor

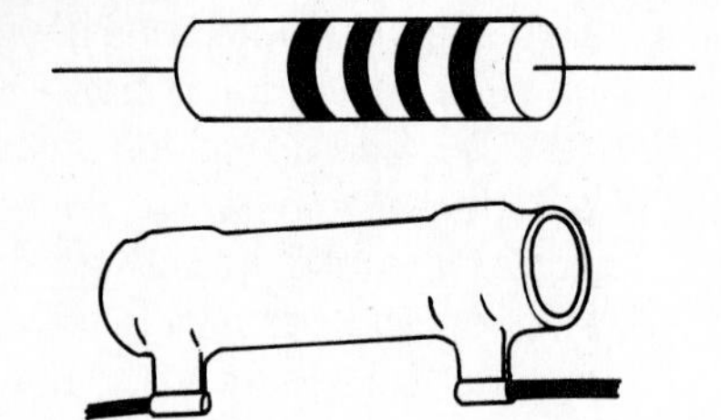

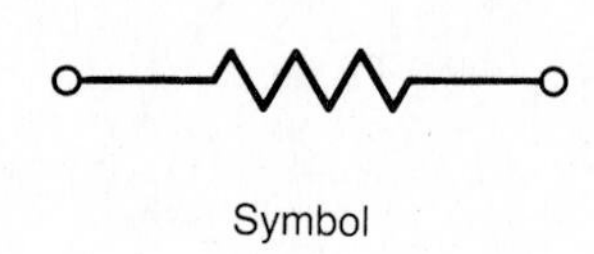

Symbol

B. Adjustable resistor

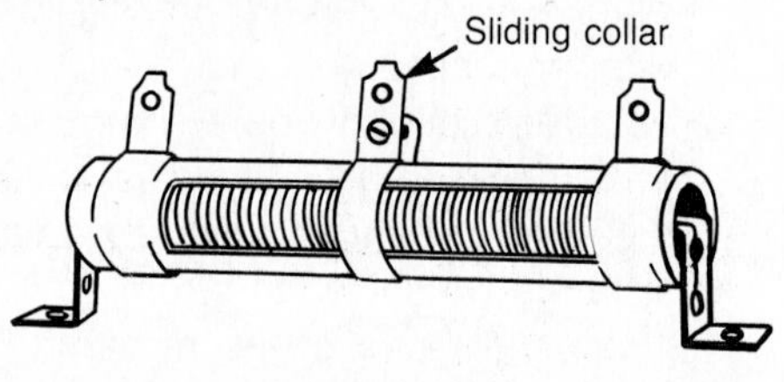

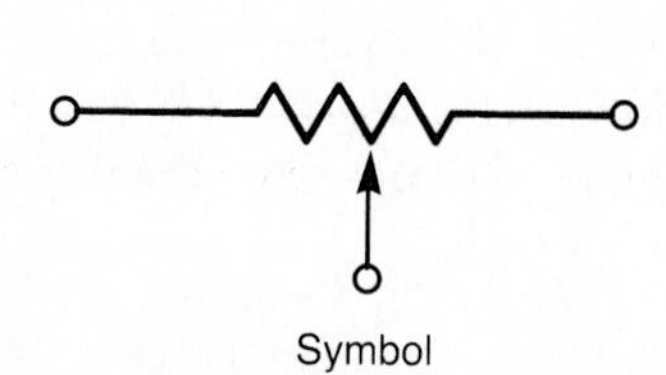

Symbol

C. Variable resistor

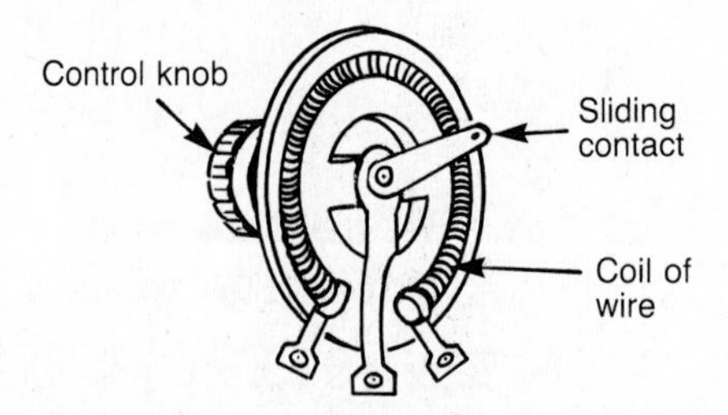

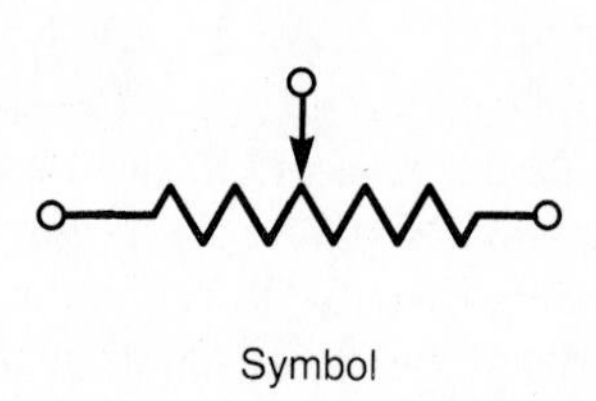

Symbol

FIGURE 10-7 VARIABLE RESISTOR CONNECTED AS A RHEOSTAT

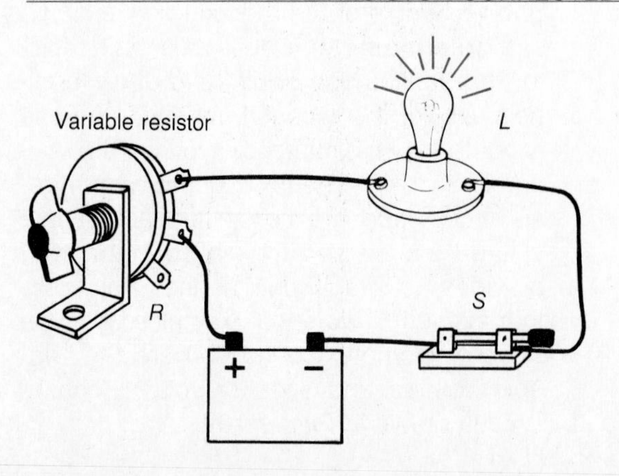

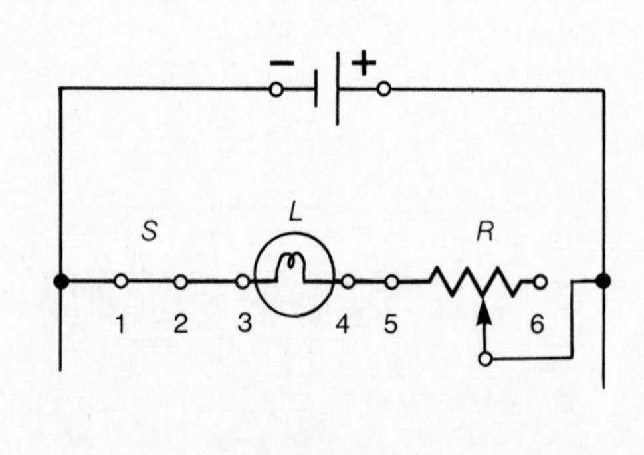

Circuit Schematic

FIGURE 10-8 VARIABLE RESISTOR CONNECTED AS A POTENTIOMETER

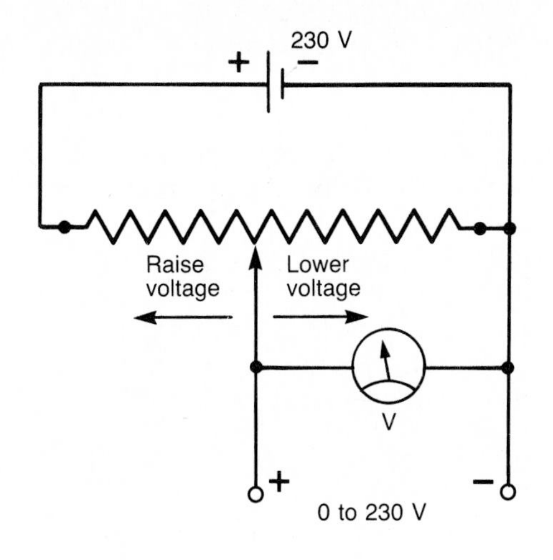

A. Variable voltage control circuit

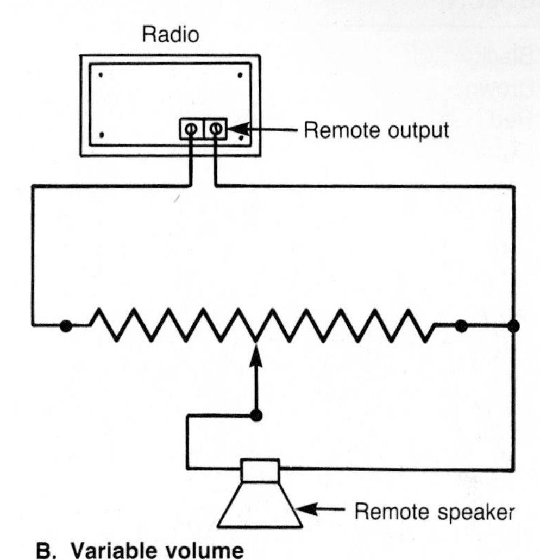

B. Variable volume control circuit

10-5 Resistor Color Code

Wirewound resistors normally have their resistance value, tolerance and power rating stamped on them. Carbon resistors, being quite small in size, use a color code system to designate resistance value and tolerance. The use of a color band code also makes it possible to read the value of the resistor regardless of the position in which the resistor is mounted in the circuit. The power rating of the resistor is indicated by the physical size of the resistor and ranges in size from 2 W to ⅛ (0.125) W (Figure 10-9).

The standard color code for carbon resistors is given in Table 10-1. Each color represents a numerical value or a multiplier.

FIGURE 10-9 POWER RATING OF CARBON RESISTORS

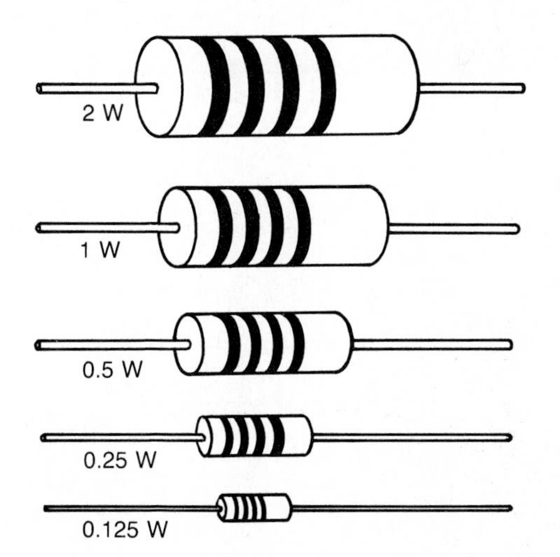

TABLE 10-1 COLOR VALUES

COLOR	NUMERIC VALUE
Black	0
Brown	1
Red	2
Orange	3
Yellow	4
Green	5
Blue	6
Violet	7
Gray	8
White	9

Color code bands on the resistor are read from left to right (Table 10-2). The color band closest to one end of the resistor is considered to be the first color band. The first two color bands indicate the first two digits in the resistance value. The third band is the multiplier band. The first two numbers are multiplied by the multiplier to obtain the actual resistance. When a fourth band is present it indicates the percent tolerance of the resistor. The tolerance band will be either a gold or silver color. A silver band indicates a tolerance of ± 10% (Figure

TABLE 10-2 READING RESISTOR COLOR BANDS

COLOR	1st NUMBER	2nd NUMBER	MULTIPLIER	TOLERANCE (PERCENT)
Black	0	0	1	
Brown	1	1	10	
Red	2	2	100	
Orange	3	3	1 000	
Yellow	4	4	10 000	
Green	5	5	100 000	
Blue	6	6	1 000 000	
Violet	7	7	10 000 000	
Gray	8	8	100 000 000	
White	9	9	1 000 000 000	
Gold			0.1	5
Silver			0.01	10
None				20
	BAND 1	BAND 2	BAND 3	BAND 4

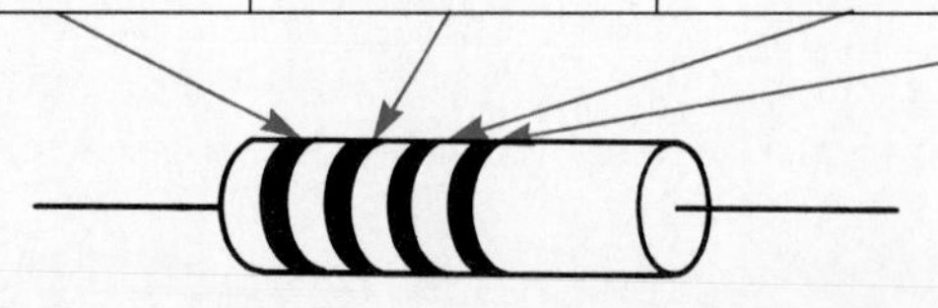

FIGURE 10-10 EXAMPLES OF RESISTOR COLOR CODES

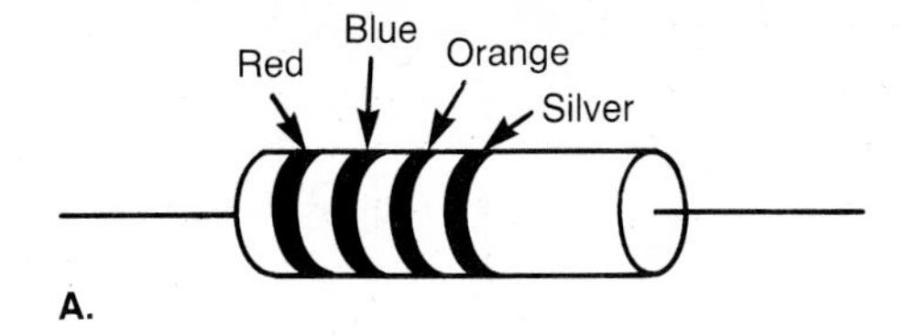

Resistance Value

Red = 2
Blue = 6
Orange = × 1 000
Silver = ± 10% tolerance
∴ Resistance Value is 26 000 Ω ± 10%
(actual resistance value may vary
2 600 Ω above or below this value)

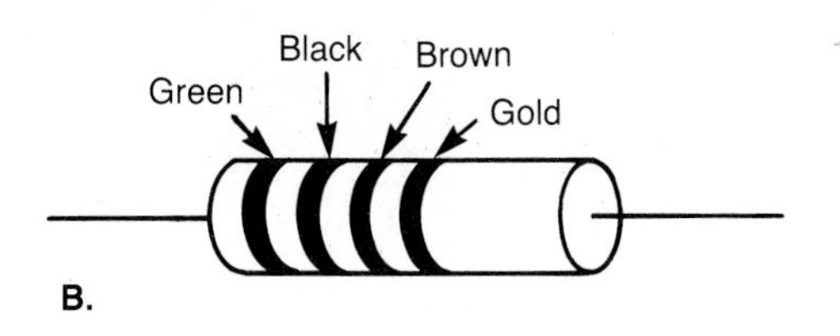

Resistance Value

Green = 5
Black = 0
Brown = × 10
Gold = ± 5% tolerance
∴ Resistance Value is 500 Ω ± 5%
(actual resistance value may vary 25 Ω above
or below this value)

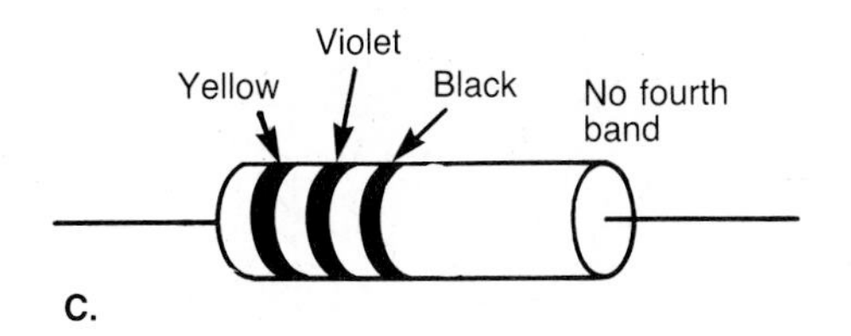

Resistance Value

Yellow = 4
Violet = 7
Black = × 1
No Fourth Band = ± 20%tolerance
∴ Resistance Value is 47 Ω ± 20%

10-10A) and a gold band a tolerance of ± 5% (Figure 10-10B). No fourth band indicates a tolerance of ± 20% (Figure 10-10C).

Resistor values of less than 10 Ω will bear the color gold or silver on the *third* band (Figure 10-11). A gold third band requires you to multiply the value indicated by the first two digits by 0.1. A silver third band indicates a multiplier of 0.01.

FIGURE 10-11 EXAMPLE OF RESISTOR VALUES LESS THAN 10 OHMS

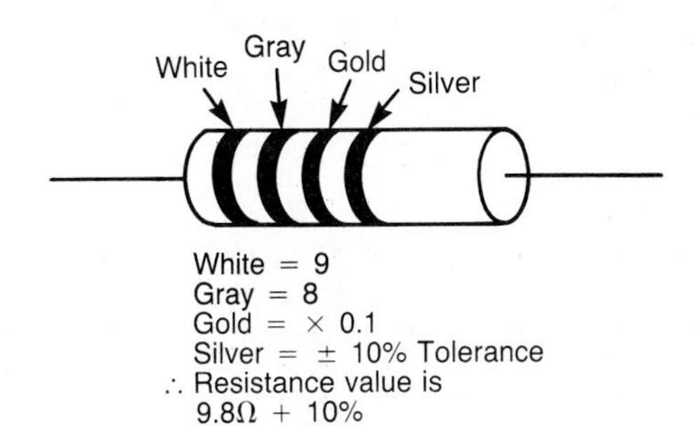

TABLE 10-3 EXAMPLES OF COLOR CODES FOR GIVEN RESISTORS

GIVEN RESISTANCE VALUE	COLOR CODE			
	1st BAND	2nd BAND	3rd BAND	4th BAND
360 Ω ± 5%	Orange	Blue	Brown	Gold
10 Ω ± 10%	Brown	Black	Black	Silver
4 700 Ω ± 20%	Yellow	Violet	Red	No Fourth Band
5 Ω ± 10%	Green	Black	Gold	Silver
8 000 Ω ± 5%	Gray	Black	Red	Gold
3 300 000 Ω ± 20%	Orange	Orange	Green	No Fourth Band

To determine the resistor color code for a given resistance and tolerance value, the same procedure that was used in reading the resistor color code is used (Table 10-3).

10-6 Series Connection of Resistors

Resistors are connected in series by connecting them end-to-end together in a line (Figure 10-12). The total resistance of the circuit formed is simply the sum of the individual resistances. If these are numbered R_1, R_2 and R_3, then the total resistance or R_T is calculated using the formula:

$$R_T = R_1 + R_2 + R_3, \text{ etc.}$$

Example

Suppose that three resistors are connected in series. R_1 is 100 Ω, R_2 is 500 Ω, and R_3 is 1 000 Ω. The total resistance is then:

$$R_T = R_1 + R_2 + R_3$$
$$R_T = 100\ \Omega + 500\ \Omega + 1\ 000\ \Omega$$
$$R_T = 1\ 600\ \Omega$$

FIGURE 10-12 RESISTORS CONNECTED IN SERIES

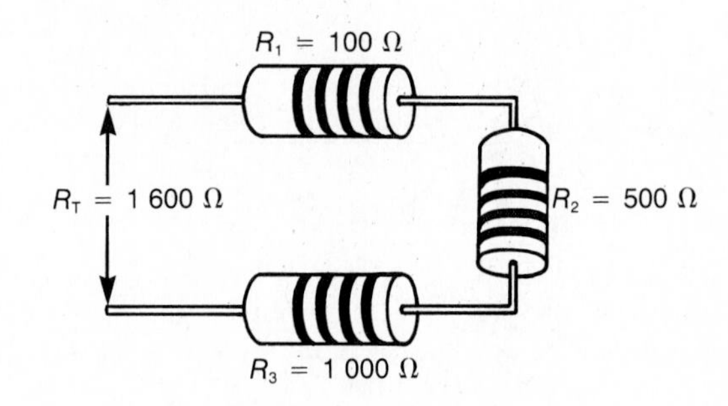

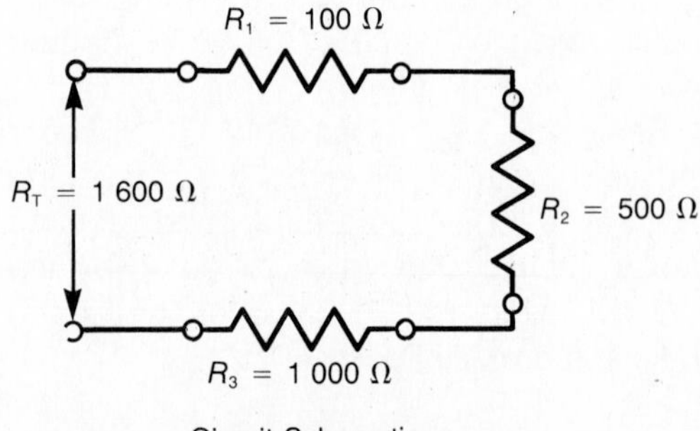

Circuit Schematic

FIGURE 10-13 RESISTORS OF THE SAME VALUE CONNECTED IN PARALLEL

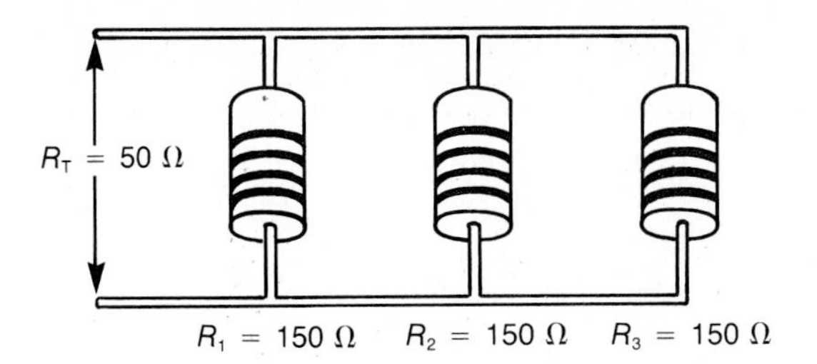

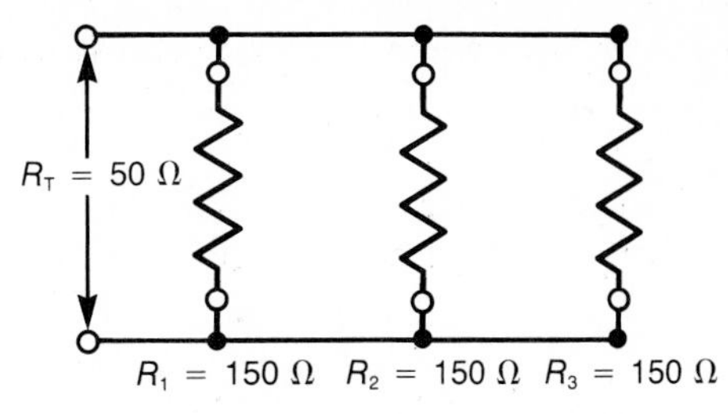

Circuit Schematic

10-7 Parallel Connection of Resistors

Resistors are connected in parallel by connecting them across a common set of line wires (Figure 10-13). The total resistance of the circuit formed is *less* than that of the lowest value of resistance present. This is because each resistor provides a separate parallel path for the current to flow. Assume all the resistors connected in parallel have the same value of resistance. The total resistance is then found by dividing the common resistance value by the total number of resistors connected.

Example
Suppose that three 150 Ω resistors are connected in parallel. The total resistance is then:

$$R_T = \frac{R\text{ (common value)}}{\text{number of resistors}}$$

$$R_T = \frac{150\ \Omega}{3}$$

$$R_T = 50\ \Omega$$

To find the total resistance of two unequal values of resistors connected in parallel (a very common use) the product over sum formula is used. This formula is:

$$R_T = \frac{R_1 \times R_2}{R_1 + R_2}$$

Example
Suppose that a 60 Ω resistor is connected in parallel with one of 40 Ω. The total resistance is then:

$$R_T = \frac{R_1 \times R_2}{R_1 + R_2}$$

$$R_T = \frac{60\ \Omega \times 40\ \Omega}{60\ \Omega + 40\ \Omega}$$

$$R_T = \frac{2\,400}{100}$$

$$R_T = 24\ \Omega$$

For more than two unequal resistors connected in parallel, the general formula for total resistance of a parallel circuit is used. This formula is:

$$R_T = \frac{1}{\frac{1}{R_1} + \frac{1}{R_2} + \frac{1}{R_3}\text{ etc.}}$$

Example
Suppose that a 120 Ω, 60 Ω, and 40 Ω resistor, are all connected in parallel. The total resistance is then:

$$R_T = \frac{1}{\frac{1}{R_1} + \frac{1}{R_2} + \frac{1}{R_3}}$$

$$R_T = \frac{1}{\frac{1}{120\ \Omega} + \frac{1}{60\ \Omega} + \frac{1}{40\ \Omega}}$$

$$R_T = \frac{1}{\frac{6}{120}}$$

$$R_T = 1 \times \frac{120}{6}$$

$$R_T = 20\ \Omega$$

10-8 Practical Assignments

1. Use an ohmmeter to measure the resistance value of each of the following heating devices:
   (a) stove element (d) toaster
   (b) clothes iron (e) electric heater
   (c) soldering iron (f) electric kettle

2. Make a simple resistor by drawing a 4 in. (10 cm) line with a lead pencil. Draw over the line several times. Measure the resistance of the line using an ohmmeter set to its highest resistance range. Record the resistance for each of the following lengths of the line:
   (a) 4 in. (10 cm)
   (b) 2 in. (5 cm)
   (c) ¾ in. (2 cm)

FIGURE 10-14 PRACTICAL ASSIGNMENT 3

3.

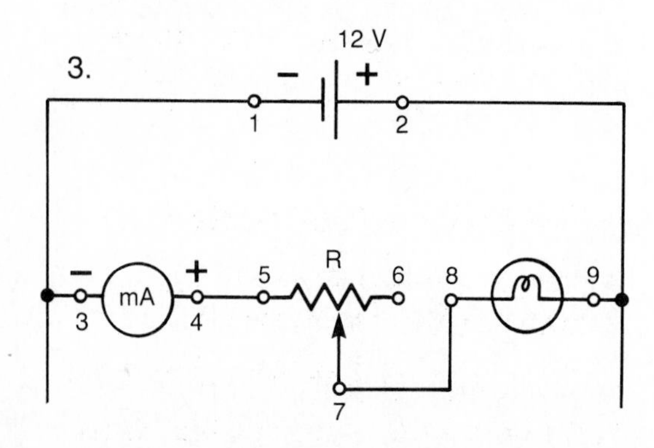

   (a) Complete a wiring number sequence chart and wiring diagram for the schematic circuit in Figure 10-14. (Use whatever value of components that are available to you.)
   (b) Wire the circuit and record the current flow for the maximum and minimum setting of the rheostat.

4. (a) Complete a wiring number sequence chart and wiring diagram for the schematic circuit in Figure

FIGURE 10-15 PRACTICAL ASSIGNMENT 4

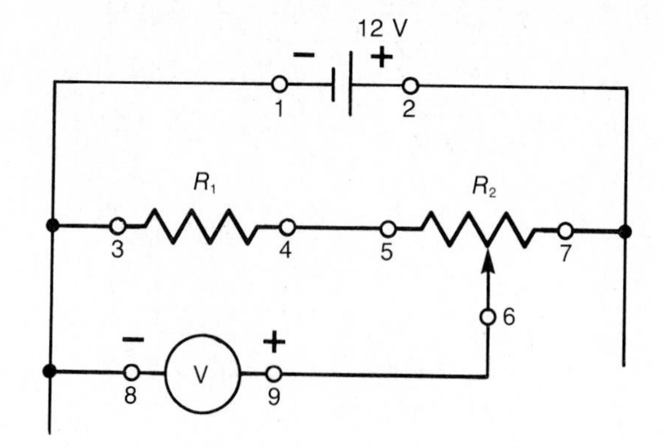

10-15. (Use whatever value of components that are available to you.)
   (b) Wire the circuit and record the voltage for the maximum and minimum setting of the potentiometer.

5. (a) Record both the color code and rated resistance value of 10 carbon resistors.
   (b) Record the resistance value of each of these resistors as measured with an ohmmeter.
   (c) Calculate the difference between the color code value and the measured value of resistance.

FIGURE 10-16 PRACTICAL ASSIGNMENT 5

6.

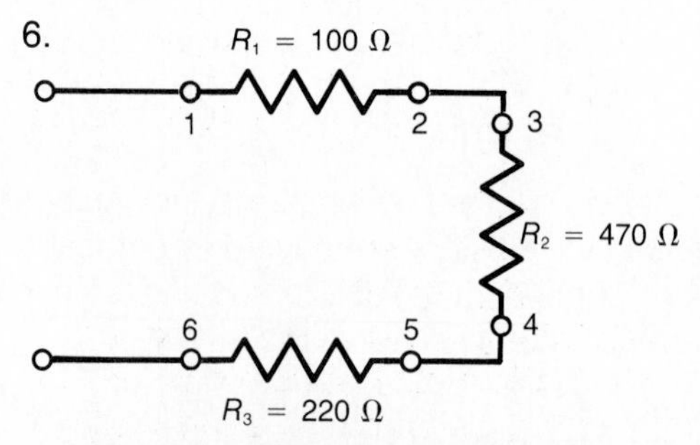

   (a) Calculate the total resistance of the schematic circuit in Figure 10-16. (You may substitute whatever resistor values are available to you.)

(b) Complete a wiring number sequence chart and wiring diagram for the circuit.
(c) Wire the circuit and record the measured value for the total resistance.

FIGURE 10-17 PRACTICAL ASSIGNMENT 6

7.

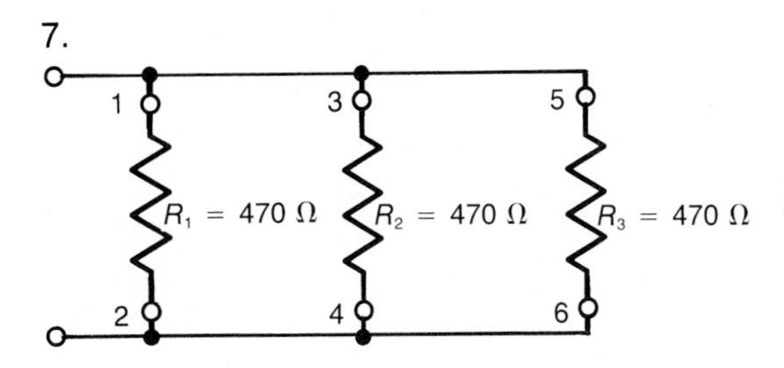

(a) Calculate the total resistance of the schematic circuit in Figure 10-17. (You may substitute whatever resistor values are available to you.)
(b) Complete a wiring number sequence chart and wiring diagram for the circuit.
(c) Wire the circuit and record the measured value for the total resistance.

FIGURE 10-18 PRACTICAL ASSIGNMENT 7

8.

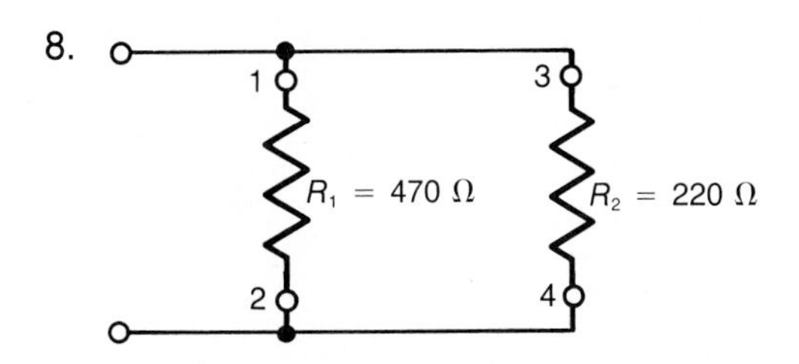

(a) Calculate the total resistance of the schematic circuit in Figure 10-18. (You may substitute whatever resistor values are available to you.)
(b) Complete a wiring number sequence chart and wiring diagram for the circuit.
(c) Wire the circuit and record the measured value for the total resistance.

9. (a) Calculate the total resistance of the schematic circuit in Figure 10-19. (You may substitute whatever resistor values are available to you.)
(b) Complete a wiring number sequence chart and wiring diagram for the circuit.

FIGURE 10-19 PRACTICAL ASSIGNMENT 8

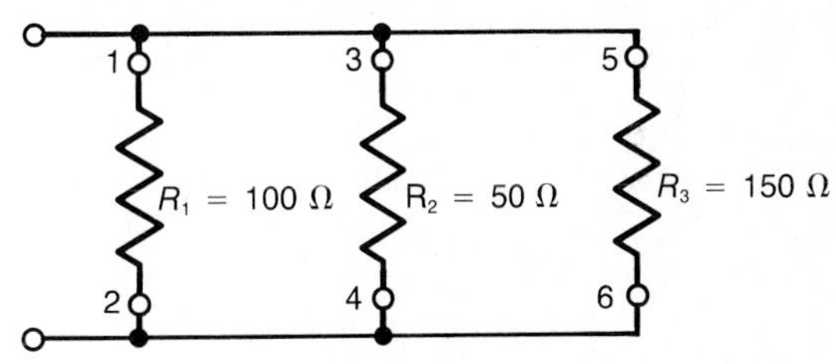

c) Wire the circuit and record the measured value for the total resistance.

10-9 Self Evaluation Test

1. What is the most common practical application for resistance wire?
2. What is the purpose of a resistor?
3. Name the three ways by which resistors are rated.
4. (a) What are the two main advantages of wirewound resistors over the carbon composition type?
   (b) What advantage do carbon composition resistors have over the wirewound type?
5. Draw the schematic for a simple rheostat lamp dimmer circuit.
6. Draw the schematic for a simple potentiometer variable voltage control circuit.

TABLE 10-4 SELF EVALUATION TEST 10

	1st BAND	2nd BAND	3rd BAND	4th BAND
(a)	Red	Green	Yellow	Silver
(b)	Orange	Blue	Brown	Gold
(c)	White	Brown	Red	none
(d)	Gray	Black	Blue	Gold
(e)	Violet	Green	Gold	Silver
(f)	Blue	Red	Black	Gold

7. Identify the color bands for each of the following resistors:
   (a) 100 Ω ± 10%
   (b) 2 200 Ω ± 5%
   (c) 47 000 Ω ± 20%
   (d) 1 000 000 Ω ± 10%
8. A 680 Ω resistor has a rated tolerance of 10%. What is the rated resistance range for this resistor?
9. Draw the schematic for the following connection of resistors:
   (a) three in series
   (b) two in parallel
10. State the resistance value and percentage tolerance for each of the color coded resistors shown in Table 10-4.
11. Calculate the total resistance for each of the following circuit connections:
    (a) series circuit: R_1 = 40 Ω, R_2 = 75 Ω
    (b) parallel circuit: R_1 = 200 Ω, R_2 = 200 Ω, R_3 = 200 Ω
    (c) series circuit: R_1 = 2 000 Ω, R_2 = 6 000 Ω, R_3 = 2 200 Ω
    (d) parallel circuit: R_1 = 14 Ω, R_2 = 32 Ω
    (e) series circuit: R_1 = 4 700 Ω, R_2 = 800 Ω, R_3 = 200 Ω
    (f) parallel circuit: R_1 = 60 Ω, R_2 = 30 Ω, R_3 = 15 Ω

11

OHM'S LAW

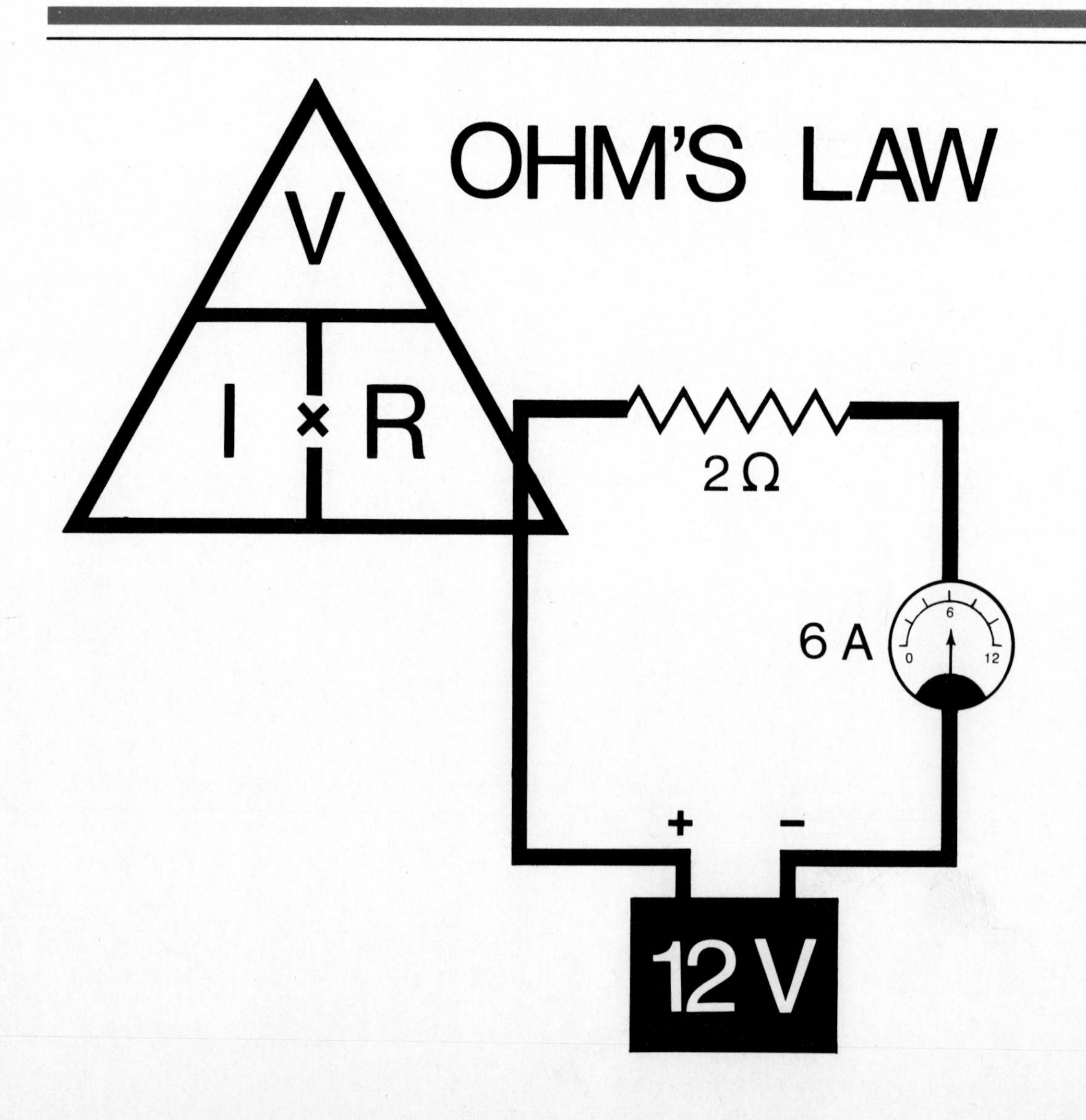

OHM'S LAW

OBJECTIVES

Upon completion of this unit you will be able to:

- Write the SI units and symbols used for common electrical quantities.
- Correctly use the metric prefixes to convert electrical quantities.
- State *Ohm's Law*.
- Carry out experiments to prove Ohm's Law.
- Solve simple circuit problems using the Ohm's Law formulas.

11-1 Electrical SI Units and Prefixes

The use of metric units is common in electricity since the ampere is a basic metric unit. Therefore, throughout this text we are using SI metric electrical units and symbols. It is important to note the correct units and symbols for this system of measurement and to use them accordingly. Table 11-1 lists the basic electrical SI metric units and symbols.

TABLE 11-1 COMMON ELECTRICAL SI METRIC UNITS AND SYMBOLS

QUANTITY	BASIC UNIT	SYMBOL
Electric potential energy difference	volt	V
Electric charge	coulomb	C
Electric current	ampere	A
Resistance	ohm	Ω
Power	watt	W
Energy	joule	J

The SI Metric system is based on the convenience of the decimal number system. Units are related by factors such as 1 000 and 1 000 000. This makes for a handy system of conversion from one unit to the other. Much of the arithmetic merely involves the shifting of the decimal point.

Metric prefixes are used to extend the basic units of electrical measurement for measuring smaller or larger quantities. For example, in a television set, the signal for the antenna may have a strength of 0.000 001 25 V while the voltage applied directly to the picture tube may be in the range of 27 000 V. Using prefixes, these values would be expressed as 1.25 μV (microvolts) and 27 kV (kilovolts). The four basic prefixes used most often in electrical and electronic measurement are: mega, kilo, milli and micro. The symbols and number values for each prefix are shown in Table 11-2.

TABLE 11-2 TABLE OF COMMON PREFIXES

PREFIX	SYMBOL	MULTIPLIER	MEANING
mega	M	1 000 000 or 10^{6}	one million
kilo	k	1 000 or 10^{3}	one thousand
milli	m	0.001 or 10^{-3}	one thousandth
micro	μ	0.000 001 or 10^{-6}	one millionth

Being able to convert from one unit to another is important when using electrical circuit formulas. Improperly mixing units results in wrong answers. To convert from one unit to another, the unit to be converted must be multiplied by the correct prefix number value. Being able to convert from one unit to another accurately comes with practice. The following examples may be of some help in mastering this skill.

1. To convert amperes (A) to milliamperes (mA), it is necessary to move the decimal point three places to the right (this is the same as multiplying the number by 1 000)

   0.012 A = ___?___ mA

   0.012 A = 0.012 mA

   0.012 A = 12 mA

2. To convert milliamperes (mA) to amperes (A), it is necessary to move the decimal point three places to the left (this is the same as multiplying by 0.001)

450 mA = _?_ A

450 mA = 450. A

450 mA = 0.45 A

3. To convert ohms (Ω) to kilohms (kΩ), it is necessary to move the decimal point three places to the left.

47 000 Ω = _?_ kΩ

47 000 Ω = 47 000. kΩ

47 000 Ω = 47 kΩ

4. To convert from megohms (MΩ) to ohms (Ω), it is necessary to move the decimal point six places to the right.

2.2 MΩ = _?_ Ω

2.2 MΩ = 2.200 000 Ω

2.2 MΩ = 2 200 000 Ω

5. To convert from microamperes (μA) to amperes (A), it is necessary to move the decimal point six places to the left.

500 μA = _?_ A

500 μA = 000 500. A

500 μA = 0.000 5 A

11-2 Ohm's Law

Ohm's Law expresses the basic relationship between current, voltage and resistance in a circuit. It is the most important Law in electricity.

Ohm's Law states that the amount of current which flows in a circuit is directly proportional to the applied voltage and inversely proportional to the resistance. In other words, when the voltage increases, the current increases; when the voltage decreases, the current decreases (Figure 11-1).

If the voltage is held constant, the current will change as the resistance changes, but in the opposite direction. The current will decrease as the resistance increases and will increase as the resistance decreases (Figure 11-2).

Ohm's Law can be expressed in the form of three formulas: one basic formula and two others derived from it. Using these three formulas, you can calculate the current, voltage or resistance value of a circuit.

FIGURE 11-1 THE EFFECT OF CHANGES IN VOLTAGE ON CURRENT

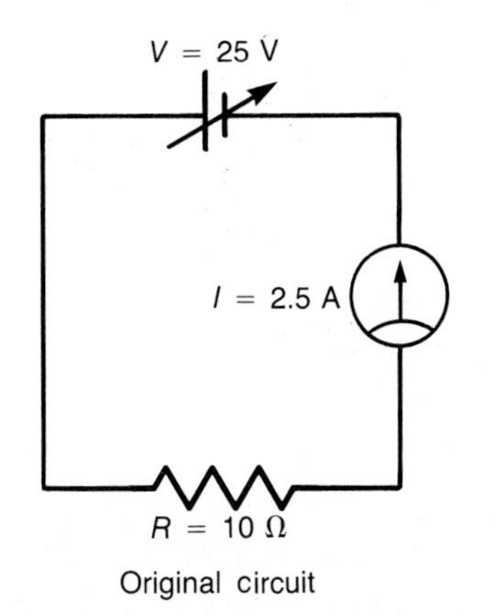

Original circuit

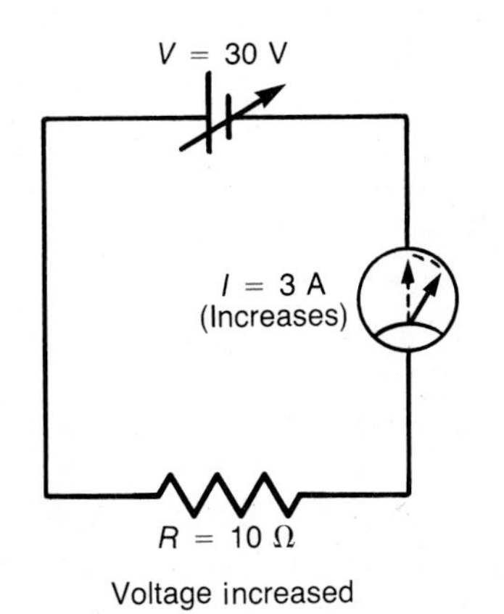

Voltage increased

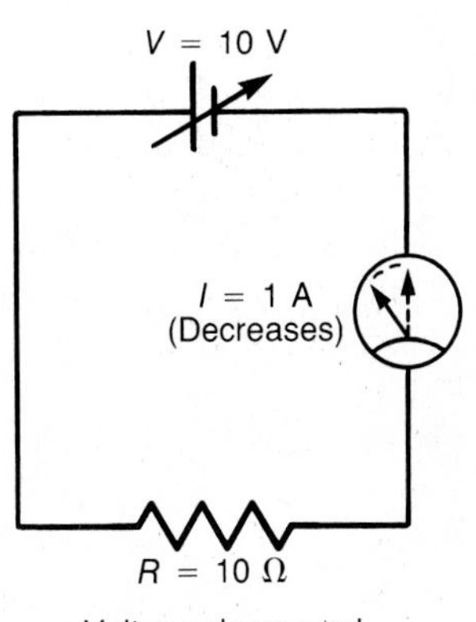

Voltage decreased

FIGURE 11-2 THE EFFECT OF CHANGES IN RESISTANCE ON CURRENT

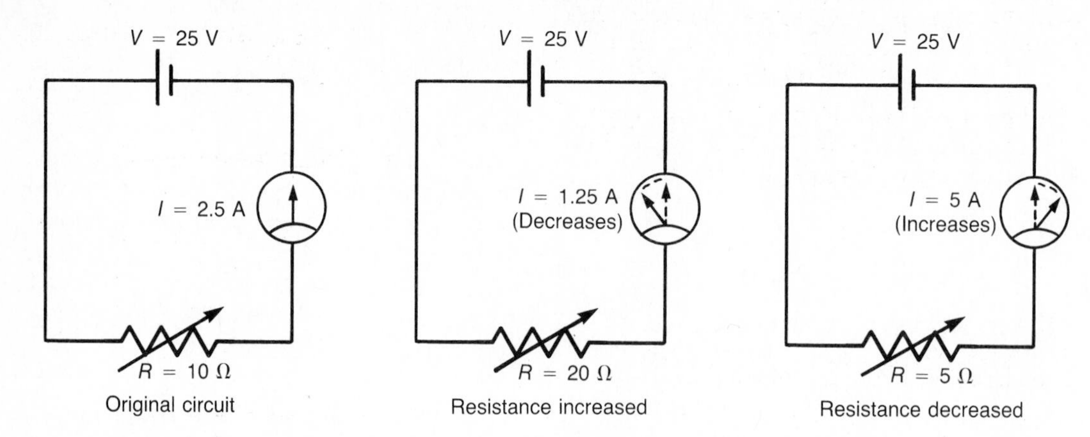

OHM'S LAW FORMULAS

1. $I = \frac{V}{R}$ 2. $V = I \times R$ 3. $R = \frac{V}{I}$

where: (common electrical circuits)

I = current in amperes (A)
V = voltage in volts (V)
R = resistance in ohms (Ω)

OR: (common electronic circuits)

I = current in milliamperes (mA)
V = voltage in volts (V)
R = resistance in kilohms (kΩ)

11-3 Applying Ohm's Law To Calculate Current

By using Ohm's Law, we can predict what is going to happen in a circuit before we wire it. When any two of the three quantities (V-I-R) are known, the third can be calculated. For example, if the voltage and resistance are known, the current can be calculated. The formula used is:

$$I = \frac{V}{R}$$

Example

Suppose that a portable electric heater with a resistance of 15 Ω is connected directly to a 120 V AC electrical outlets as shown in Figure 11-3. The current flow in this circuit is then:

$$\text{Current} = \frac{\text{Voltage}}{\text{Resistance}}$$

$$I = \frac{V}{R}$$

$$I = \frac{(120\text{ V})}{(15\ \Omega)}$$

$$I = 8\text{ A}$$

FIGURE 11-3 SOLVING FOR CURRENT

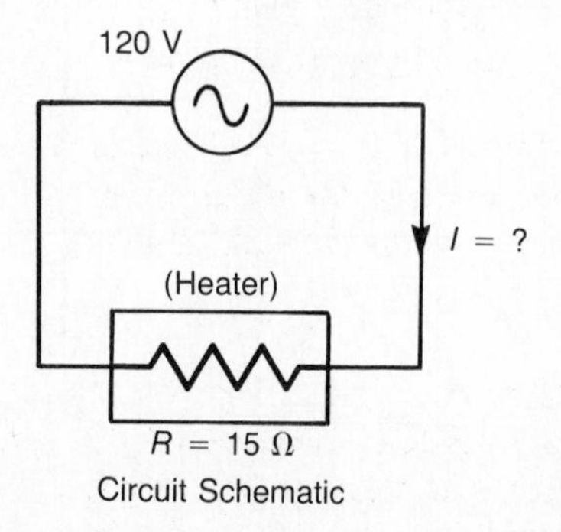

Electronic circuits operate at much lower current values than electrical circuits. This is mainly because they usually contain much higher resistance values. If the resistance of these circuits is expressed in kilohms (kΩ) the current can be calculated directly in milliamperes. The same formula,

$$I = \frac{V}{R}$$

is used.

Example
Suppose a 10 kΩ carbon resistor is connected to a 12V battery as in Figure 11-4. The current flow in this circuit is then:

$$I = \frac{V}{R}$$
$$I = \frac{(12\ \text{V})}{(10\ \text{k}\Omega)}$$
$$I = 1.2\ \text{mA}$$

FIGURE 11-4 SOLVING FOR VERY LOW CURRENT

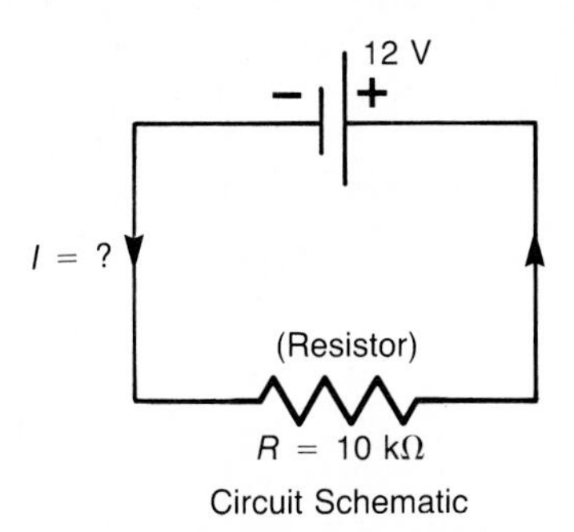

Circuit Schematic

11-4 Applying Ohm's Law To Calculate Voltage

When the current flow and resistance of a circuit are known, the voltage being applied can be calculated. The formula is

$$V = I \times R.$$

Example
Suppose a DC generator is delivering a current of 2.5 A to a lamp bank that has a combined resistance of 50 Ω as in Figure 11-5. The voltage output of the generator is then:

$$\text{Voltage} = \text{Current} \times \text{Resistance}$$
$$V = I \times R$$
$$V = (2.5\ \text{A})\ (50\ \Omega)$$
$$V = 125\ \text{V}$$

FIGURE 11-5 LAMP BANK PROBLEM

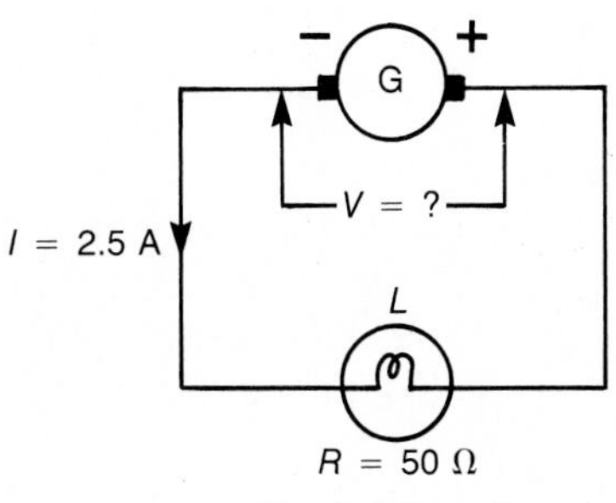

Circuit Schematic

For low current circuits the current may be expressed in milliamperes (mA) and the resistance in kilohms (kΩ). Again the same formula, namely, $V = I \times R$ is used to calculate the voltage in volts.

Example
Suppose a solar cell provides a current of 2.5 mA to a 500 Ω (0.5 kΩ) load (Figure 11-6). The voltage output of the solar cell is then:

$$V = I \times R$$
$$V = (2.5\ \text{mA})\ (0.5\ \text{k}\Omega)$$
$$V = 1.25\ \text{V}$$

FIGURE 11-6 SOLAR CELL PROBLEM

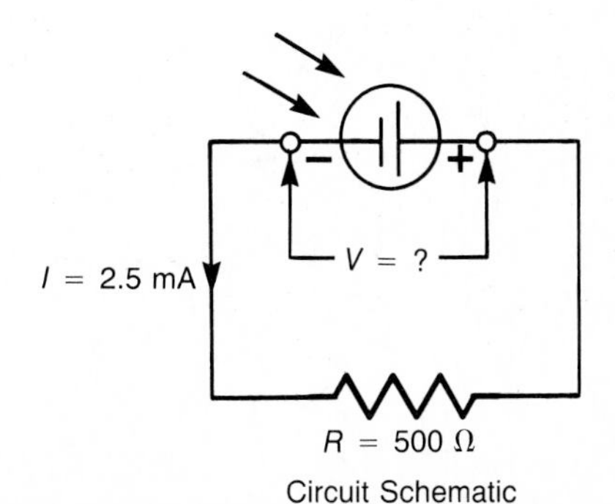

Circuit Schematic

11-5 Applying Ohm's Law To Calculate Resistance

The resistance of a load can be calculated when the applied voltage across it and the current flow through it are known. The formula used is

$$R = \frac{V}{I}.$$

Example
Suppose an electric kettle draws a current of 8 A when connected to a 120 V AC electrical outlet (Figure 11-7). The resistance of the kettle heating element is then:

FIGURE 11-7 ELECTRIC KETTLE PROBLEM

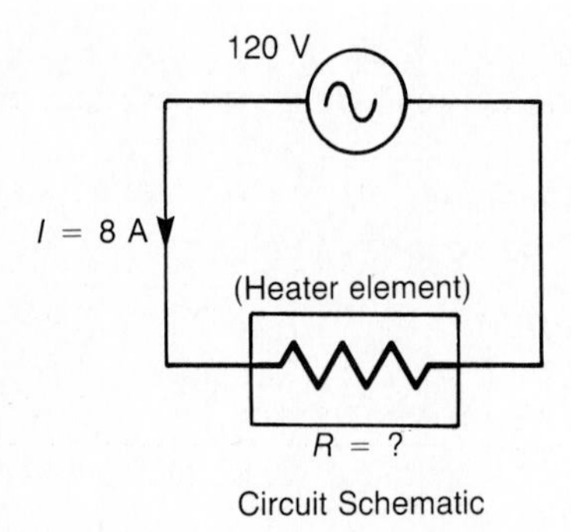

$$\text{Resistance} = \frac{\text{Voltage}}{\text{Current}}$$

$$R = \frac{V}{I}$$

$$R = \frac{(120\ \text{V})}{(8\ \text{A})}$$

$$R = 15\ \Omega$$

Example
Suppose the current flow through the resistor in Figure 11-8 is known to be 2 mA. The voltage across the resistor is measured and found to be 9 V. The resistance value of the resistor is then:

$$R = \frac{V}{I}$$

$$R = \frac{(9\ \text{V})}{(2\ \text{mA})}$$

$$R = 4.5\ \text{k}\Omega$$

$$R = 4\ 500\ \Omega$$

FIGURE 11-8 SOLVING FOR RESISTANCE

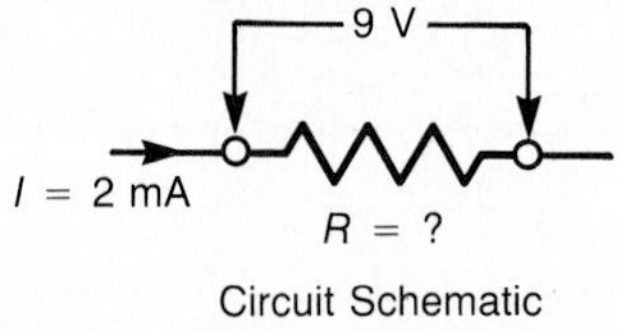

9 V

I = 2 mA

R = ?

Circuit Schematic

FIGURE 11-9 OHM'S LAW TRIANGLE

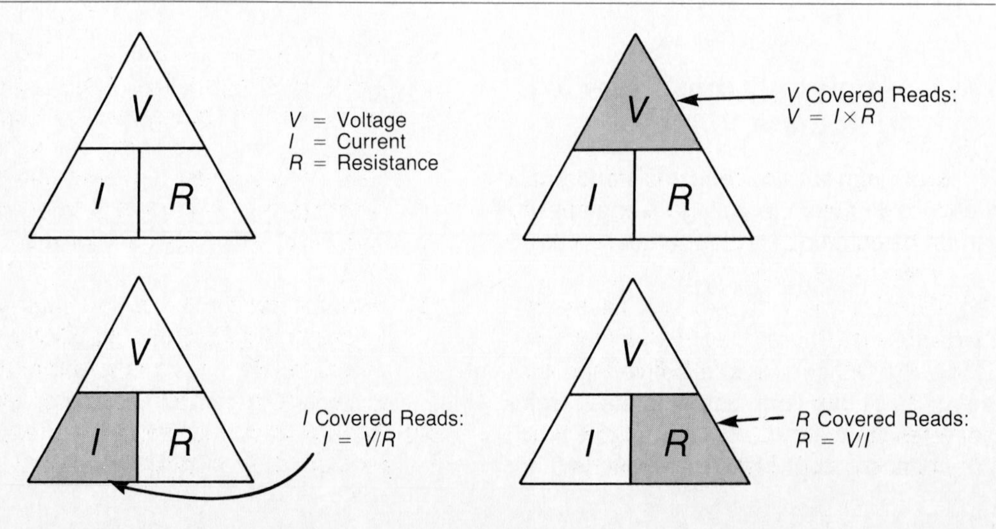

In electronic circuits, it is often more convenient to calculate the resistance values of resistors, rather than measuring them with an ohmmeter. If voltage and current values are known it is faster to calculate the resistance value than measure it.

11-6 Ohm's Law Triangle

The three Ohm's Law formulas can be easily recalled by arranging the three quantities within a triangle as shown in Figure 11-9. A finger placed over the symbol standing for the unknown quantity leaves the remaining two symbols in the correct relationship to solve for the unknown value.

11-7 Practical Assignments

FIGURE 11-10 PRACTICAL ASSIGNMENT 1

1.

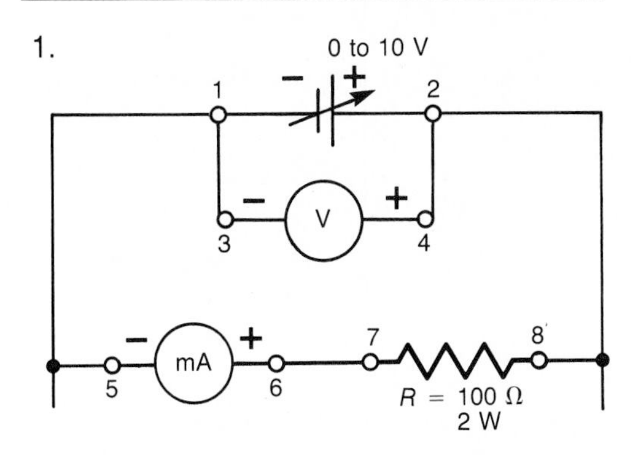

(a) The schematic circuit of Figure 11-10 is designed to show the effect on current of changes in voltage. Complete a wiring number sequence chart and wiring diagram for this circuit.
(b) Wire the circuit. Vary the voltage (in 1 V steps) from 0 to 10 V. Record (in chart form) the current flow for each voltage step.
(c) Explain, from your results, the effect on current of changes in voltage.

FIGURE 11-11 PRACTICAL ASSIGNMENT 2

2.

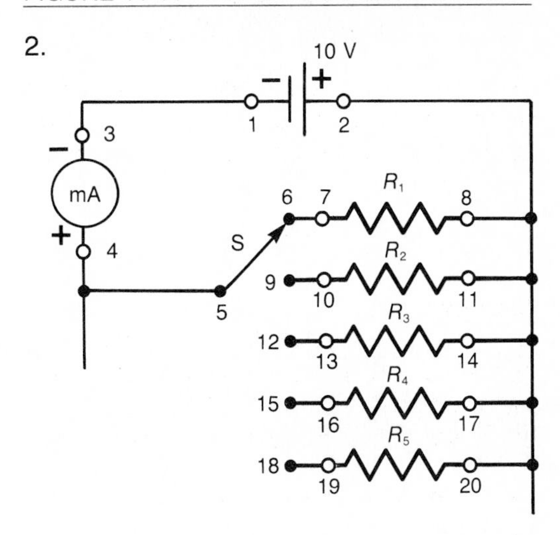

RESISTOR VALUES		
R_1	50 Ω	2 W
R_2	100 Ω	2 W
R_3	150 Ω	2 W
R_4	220 Ω	2 W
R_5	470 Ω	2 W

(a) The schematic circuit of Figure 11-11 is designed to show the effect on current of changes in resistance. (A resistance substitution box may be substituted for the switch and resistors.) Complete a wiring number sequence chart and wiring diagram for this circuit.
(b) Wire the circuit. Switch the different values of resistors into the circuit one at a time. Record (in chart form) the current flow for each resistance value.
(c) Explain, from your results, the effect on current of changes in resistance.

3.

TABLE 11-3 RESISTANCE AND VOLTAGE VALUES FOR PRACTICAL ASSIGNMENT 3

	RESISTANCE (CONNECTED)	VOLTAGE (SET TO)	CURRENT (CALCULATED)	CURRENT (MEASURED)
R_1	1 kΩ	5 V	__ mA	__ mA
R_2	1.5 kΩ	10 V	__ mA	__ mA
R_3	100 Ω	1.5 V	__ mA	__ mA
R_4	220 Ω	3 V	__ mA	__ mA
R_5	500 Ω	8 V	__ mA	__ mA

(a) The schematic circuit in Figure 11-12 is designed to prove that the current flow in a circuit can be calculated using known voltage and resistance values. The known voltage and resistance values are as given in Table 11-3. Using the formula

$$I = \frac{V}{R}$$

calculate the expected current flow in milliamperes. Record your answers in chart form.

(b) Complete a wiring number sequence chart and wiring diagram for the circuit.

(c) Wire the circuit with the resistor value and voltage setting as given in the chart. Measure and record the current for each set of resistance and voltage values.

FIGURE 11-12 PRACTICAL ASSIGNMENT 3

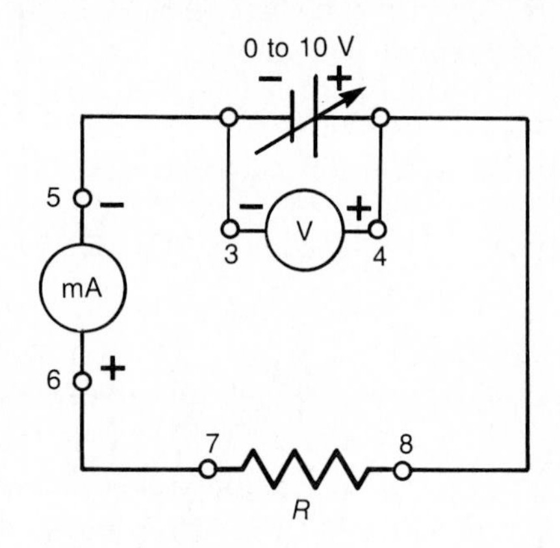

4.

TABLE 11-4 RESISTANCE AND CURRENT VALUES FOR PRACTICAL ASSIGNMENT 4

	RESISTANCE (CONNECTED)	CURRENT (SET TO)	VOLTAGE (CALCULATED)	VOLTAGE (MEASURED)
R_1	1 kΩ	5 mA	__ V	__ V
R_2	1.5 kΩ	5 mA	__ V	__ V
R_3	100 Ω	5 mA	__ V	__ V
R_4	220 Ω	5 mA	__ V	__ V
R_5	500 Ω	5 mA	__ V	__ V

(a) The schematic circuit of Figure 11-13 is designed to prove that the voltage of a circuit can be calculated using known resistance and current values. The known resistance and current values are as given in Table 11-4. Using the formula

$$V = I \times R,$$

calculate the voltage required to produce a current of 5 mA through each of the resistance values given. Record your answers in chart form.

(b) Wire the circuit with each of the resistor values indicated in the chart. For each resistor, set the variable voltage source to produce a current of 5 mA. Record, on your chart, the value of voltage required to produce this current.

FIGURE 11-13 PRACTICAL ASSIGNMENT 4

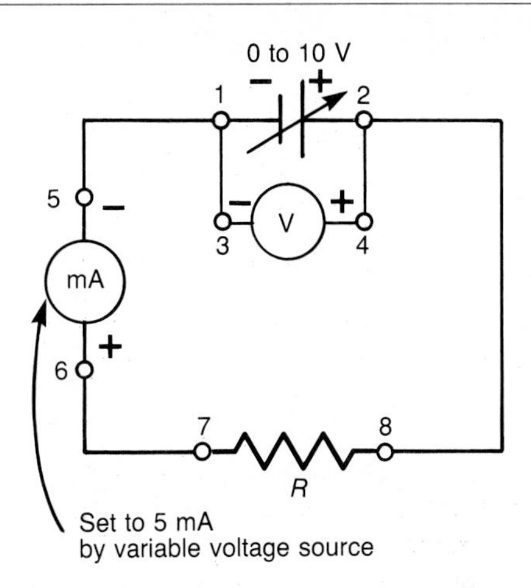

5.

TABLE 11-5 FINDING RESISTANCE VALUES IN PRACTICAL ASSIGNMENT 5

	VOLTAGE (SET TO)	CURRENT (MEASURED)	RESISTANCE (CALCULATED)	RESISTANCE (MEASURED)
R_1	10 V	—	—	—
R_2	10 V	—	—	—
R_3	10 V	—	—	—
R_4	10 V	—	—	—
R_5	10 V	—	—	—

(a) The schematic circuit of Figure 11-14 is designed to prove that the resistance of a circuit can be calculated using known voltage and current values. Wire the circuit with each of 5 unmarked resistors. With the voltage set at 10 V, record the current flow through each resistor. Record your results in Table 11-5.

FIGURE 11-14 PRACTICAL ASSIGNMENT 5

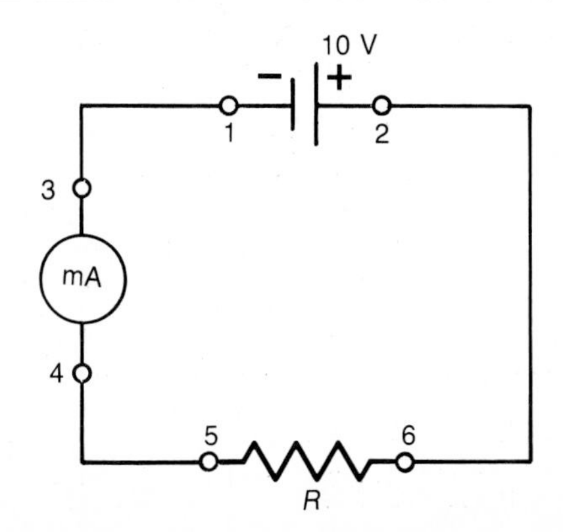

(b) Using the formula

$$R = \frac{V}{I}$$

calculate the resistance value of each unmarked resistor and record your answers in the table.

(c) Using an ohmmeter, measure the value of each unmarked resistor and record this value in the table.

11-8 Self Evaluation Test

1. Write the basic SI unit and symbol used for electric:
   (a) current
   (b) voltage
   (c) resistance
   (d) power
   (e) energy
   (f) charge
2. State the metric prefix used to represent each of the following:
   (a) $\frac{1}{1\,000}$
   (b) 1 000 000
   (c) $\frac{1}{1\,000\,000}$
   (d) 1 000
3. Convert each of the following:
   (a) 0.003 A to mA (amperes to milliamperes)
   (b) 10 kΩ to Ω (kilohms to ohms)
   (c) 0.47 MΩ to Ω (megohms to ohms)
   (d) 850 μA to A (microamperes to amperes)
   (e) 7 mA to A
   (f) 2.2 MΩ to Ω
   (g) 40 μA to A
   (h) 87 000 W to kW
   (i) 0.056 V to mV
   (j) 0.000 91 V to μV
4. State *Ohm's Law*.
5. Write the voltage, current and resistance formulas for Ohm's Law.
6. The alternator in a car delivers 12 V and has a load of 3 Ω connected across its terminals. Find the current flow in the circuit.
7. A door chime has a resistance of 120 Ω. What current will flow through it when a voltage of 16 V is applied to it?
8. A carbon resistor has a resistance of 220 Ω. The current flow through the resistor is measured and found to be 300 mA. What is the value of the voltage across the resistor?
9. A baseboard heater draws a current of 8 A when connected to a 240 V source. What is the resistance value of the heater element?
10. An electric soldering iron with a 40 Ω heating element is plugged into a 120 V outlet. How much current will be drawn by the iron?
11. The voltage across a carbon resistor is measured and found to be 1.5 V. If the resistance value of the resistor is 500 Ω, what is the value of the current flow through it in milliamperes?
12. (a) A 12 Ω resistor is connected across a 60 V DC source. Calculate the current flow through the resistor.
    (b) If the value of the applied voltage is reduced to 48 V, does the current flow increase or decrease?
    (c) Calculate the amount of change in current flow.

13. Make a copy of Table 11-6 in your notes. Calculate and record the missing values.

TABLE 11-6 SELF EVALUATION TEST 13

CURRENT	RESISTANCE	VOLTAGE
A	50 Ω	240 V
4 A	Ω	120 V
7.5 A	15 Ω	V
mA	2.5 kΩ	9 V
18 mA	kΩ	24 V
25 mA	5 kΩ	V
mA	220 Ω	18 V
0.015 A	Ω	12 V
6 mA	1 500 Ω	V

12

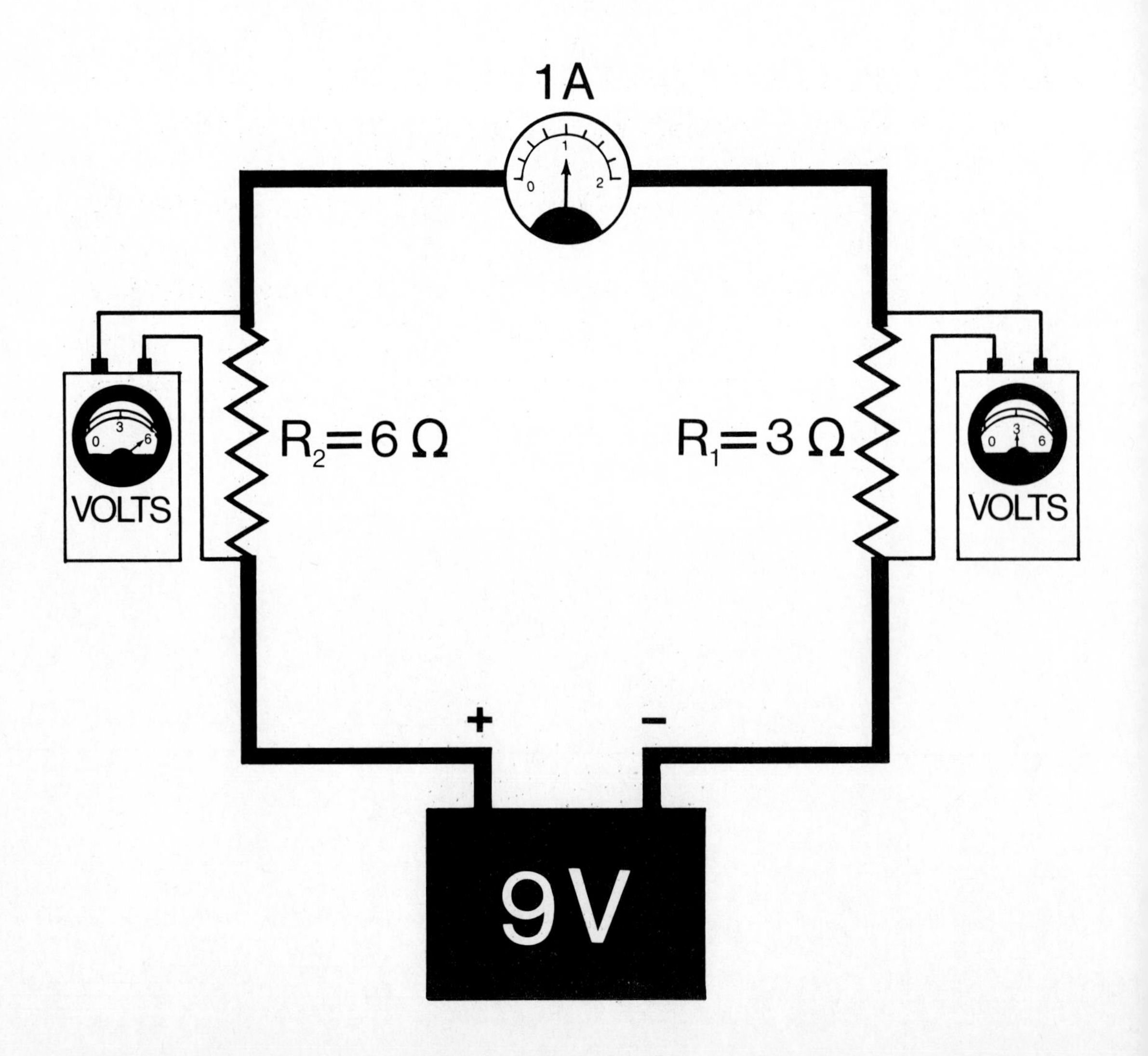

SOLVING THE SERIES CIRCUIT

OBJECTIVES

Upon completion of this unit you will be able to:

- State the voltage, current and resistance characteristics of a series circuit.
- Carry out experiments that demonstrate the characteristics of a series circuit.
- Given a series circuit with known values of voltage, current and resistance, solve for all unknown values.

FIGURE 12-1 SERIES CIRCUIT

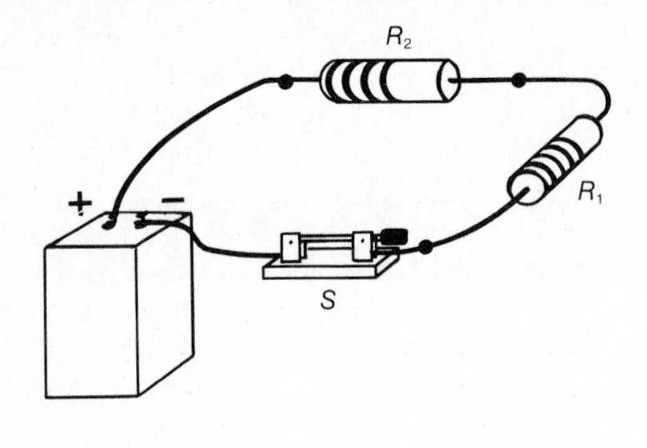

Pictorial Diagram

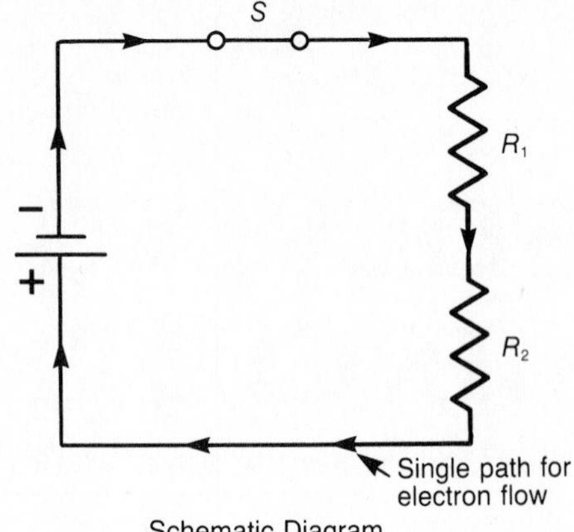

Schematic Diagram

12-1 Series Circuit Connection

A series circuit is one in which all the loads are connected in a single line end-to-end (Figure 12-1). Only one pathway can be traced from one side of the voltage source to the other for the flow of electrons. If this path is broken at any point, all current flow in the circuit stops.

12-2 Identifying Circuit Quantities

The symbols used when referring to voltage, current and resistance are V, I and R respectively. In circuits that contain more than one load resistor it is necessary to use a system of letter and number subscripts to correctly identify the different circuit quantities.

The total resistance of the circuit can be represented by the symbol R_T and the individual resistors by the symbols R_1, R_2, R_3, etc. The applied source voltage can be represented by the symbol V_T and the voltage drop across the individual resistors by the symbols V_1, V_2, V_3, etc. The total or source current can be represented by the symbol I_T and the current flow through the individual resistors by the symbols I_1, I_2, I_3, etc. (Figure 12-2).

FIGURE 12-2 IDENTIFYING QUANTITIES IN A SERIES CIRCUIT

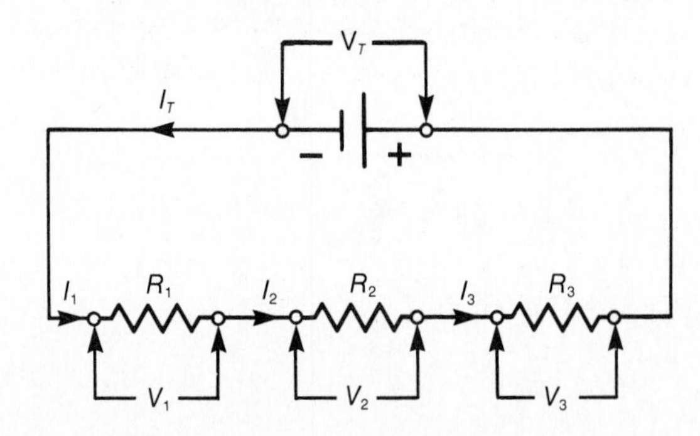

12-3 Series Circuit Current, Voltage and Resistance Characteristics

Current The current is the same value throughout a series circuit (Figure 12-3). This is due to the fact that there is only one current path. There is no loss or gain of current in a circuit. All of the electrons that flow out

FIGURE 12-3 THE CURRENT IS THE SAME AT ALL POINTS IN A SERIES CIRCUIT

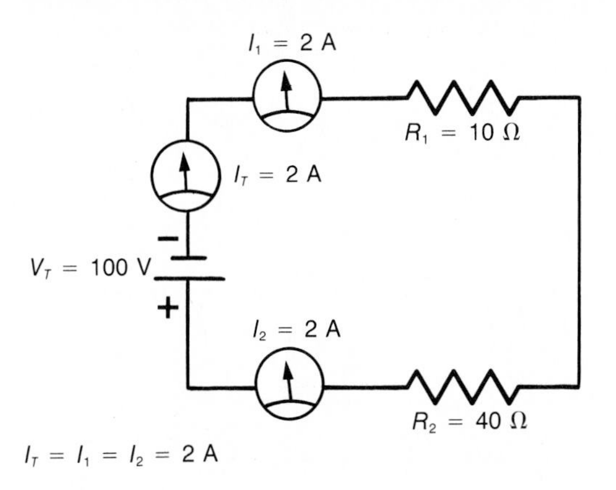

of the negative side of the voltage source must flow through each component and return to the positive side of the voltage source. Although the load resistors may be of different values, they all carry the same amount of current when connected in series.

Voltage The voltage applied to a series circuit is divided among each of the loads (Figure 12-4). The amount of voltage each load receives is directly proportional to the resistance value of the load. Total source voltage is then the sum of the voltage drops across each of the individual loads.

Resistance The total resistance of a series circuit is equal to the sum of all the individual load resistances (Figure 12-5). Since there is only one path for electrons to follow, they must travel through each of the load resistors in their journey from one side of the voltage source to the other. Therefore, all individual load resistance values must be added together to find the total circuit resistance.

FIGURE 12-4 THE SUM OF THE VOLTAGES ACROSS THE INDIVIDUAL COMPONENTS IN A SERIES CIRCUIT IS EQUAL TO THE SOURCE VOLTAGE

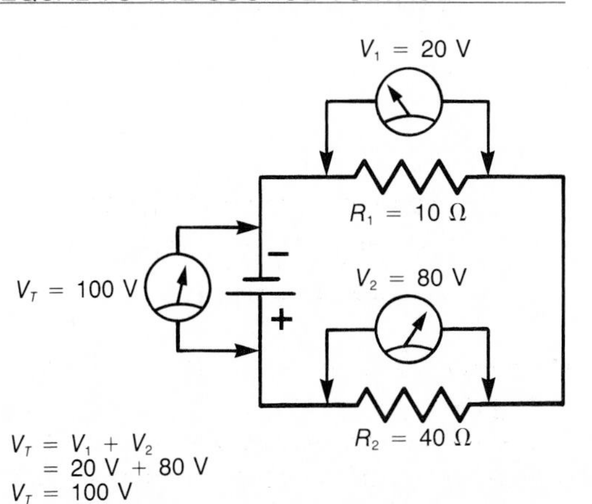

FIGURE 12-5 THE TOTAL RESISTANCE IN A SERIES CIRCUIT IS EQUAL TO THE SUM OF THE INDIVIDUAL RESISTANCES

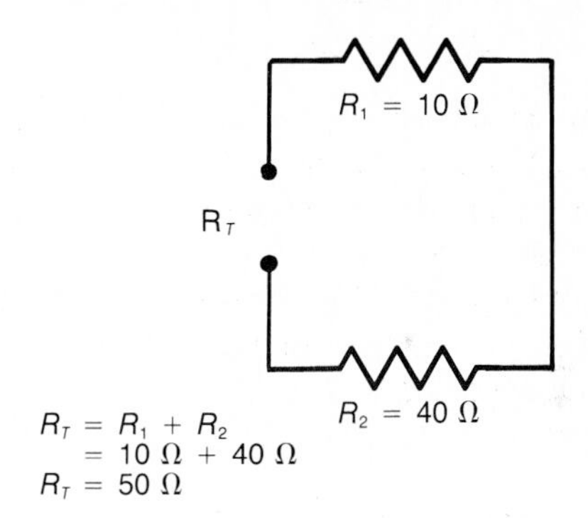

12-4 Solving Series Circuits

Given certain series circuit values of voltage, current and resistance, it is possible to calculate any unknown values of voltage, current and resistance. The ability to make these calculations is important if you are to truly understand the operation of a series circuit.

In solving a series circuit problem you must know the voltage, current and resistance characteristics of this circuit as previously presented. They are most easily remembered by the use of the following equations:

$$I_T = I_1 = I_2 = I_3 \ldots\ldots$$
$$V_T = V_1 + V_2 + V_3 \ldots\ldots$$
$$R_T = R_1 + R_2 + R_3 \ldots\ldots$$

Solving the series circuit will also require the use of Ohm's Law. Ohm's Law is most easily remembered by recalling the three basic forms of the Ohm's Law equation:

$$V = I \times R$$
$$I = V / R$$
$$R = V / I$$

Ohm's Law is applied to the circuit as a whole or to the individual loads. When applying it to the circuit as a whole it becomes:

$$V_T = I_T \times R_T$$
$$I_T = V_T / R_T$$
$$R_T = V_T / I_T$$

When applying it to the individual load it becomes:

$$V_1 = I_1 \times R_1$$
$$I_1 = V_1 / R_1$$
$$R_1 = V_1 / I_1$$

One helpful method of solving series circuits is to use a chart to record all given and calculated values of voltage, current and resistance. We shall use this method in solving each of the following typical series circuit problems:

Example 1

Problem:

Find all the unknown values of *V*, *I*, and *R* for the series circuit in Figure 12-6.

Solution:

Step 1 Make a chart and record all known values.

FIGURE 12-6 CIRCUIT FOR EXAMPLE 1

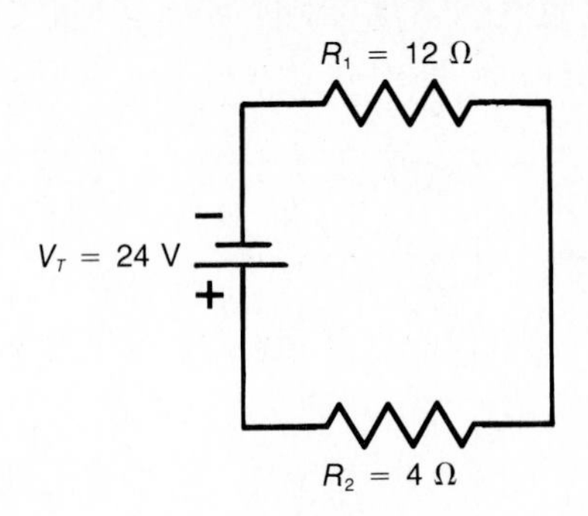

	VOLTAGE	CURRENT	RESISTANCE
R_1			12 Ω
R_2			4 Ω
TOTAL	24 V		

Step 2 Calculate R_T and enter in chart.

$$R_T = R_1 + R_2$$
$$= (12\ \Omega) + (4\ \Omega)$$
$$R_T = 16\ \Omega$$

	VOLTAGE	CURRENT	RESISTANCE
R_1			12 Ω
R_2			4 Ω
TOTAL	24 V		16 Ω

Step 3 Calculate I_T, I_1, and I_2 and enter in chart.

$$I_T = V_T / R_T$$
$$= (24\ \text{V}) \div (16\ \Omega)$$
$$I_T = 1.5\ \text{A}$$
$$I_1 = I_T$$
$$I_1 = 1.5\ \text{A}$$
$$I_2 = I_T$$
$$I_2 = 1.5\ \text{A}$$

	VOLTAGE	CURRENT	RESISTANCE
R_1		1.5 A	12 Ω
R_2		1.5 A	4 Ω
TOTAL	24 V	1.5 A	16 Ω

Step 4 Calculate V_1 and V_2 and enter in chart.

$$V_1 = I_1 \times R_1$$
$$= (1.5\ \text{A})(12\ \Omega)$$
$$V_1 = 18\ \text{V}$$
$$V_2 = I_2 \times R_2$$
$$= (1.5\ \text{A})(4\ \Omega)$$
$$V_2 = 6\ \text{V}$$

	VOLTAGE	CURRENT	RESISTANCE
R_1	18 V	1.5 A	12 Ω
R_2	6 V	1.5 A	4 Ω
TOTAL	24 V	1.5 A	16 Ω

Example 2

Problem:

Find all the unknown values of V, I, and R for the series circuit in Figure 12-7.

FIGURE 12-7 CIRCUIT FOR EXAMPLE 2

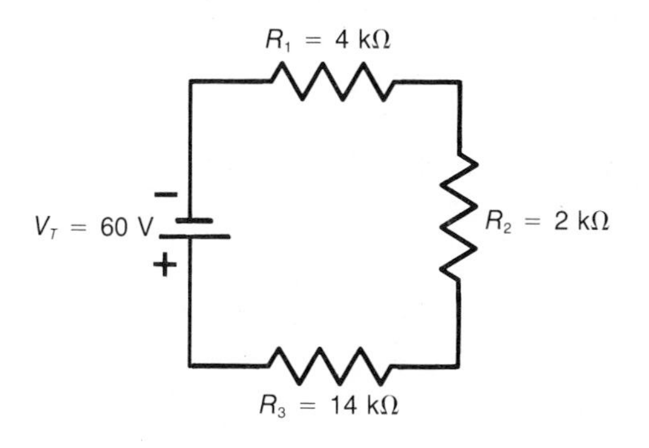

Solution:

Step 1 Make a chart and record all known values.

	VOLTAGE	CURRENT	RESISTANCE
R_1			4 kΩ
R_2			2 kΩ
R_3			14 kΩ
TOTAL	60 V		

Step 2 Calculate R_T and enter in chart.

$$R_T = R_1 + R_2 + R_3$$
$$= (4\ \text{k}\Omega) + (2\ \text{k}\Omega) + (14\ \text{k}\Omega)$$
$$= 20\ \text{k}\Omega$$

	VOLTAGE	CURRENT	RESISTANCE
R_1			4 kΩ
R_2			2 kΩ
R_3			14 kΩ
TOTAL	60 V		20 kΩ

Step 3 Calculate I_T, I_1, I_2, and I_3 and enter in chart.

$$I_T = V_T / R_T$$
$$= (60\ \text{V}) \div (20\ \text{k}\Omega)$$
$$I_T = 3\ \text{mA}$$

$$I_1 = I_T$$
$$I_1 = 3\ \text{mA}$$

$$I_2 = I_T$$
$$I_2 = 3\ \text{mA}$$

$$I_3 = I_T$$
$$I_3 = 3\ \text{mA}$$

	VOLTAGE	CURRENT	RESISTANCE
R_1		3 mA	4 kΩ
R_2		3 mA	2 kΩ
R_3		3 mA	14 kΩ
TOTAL	60 V	3 mA	20 kΩ

Step 4 Calculate V_1, V_2 and V_3 and enter in chart.

$V_1 = I_1 \times R_1$
$= (3\text{ mA})(4\text{ k}\Omega)$
$V_1 = 12\text{ V}$

$V_2 = I_2 \times R_2$
$= (3\text{ mA})(2\text{ k}\Omega)$
$V_2 = 6\text{ V}$

$V_3 = I_3 \times R_3$
$= (3\text{ mA})(14\text{ k}\Omega)$
$V_3 = 42\text{ V}$

	VOLTAGE	CURRENT	RESISTANCE
R_1	12 V	3 mA	4 kΩ
R_2	6 V	3 mA	2 kΩ
R_3	42 V	3 mA	14 kΩ
TOTAL	60 V	3 mA	20 kΩ

Example 3

Problem:

Find all the unknown values of V, I, and R for the series circuit in Figure 12-8.

Solution:

Step 1 Make a chart and record all known values.

FIGURE 12-8 CIRCUIT FOR EXAMPLE 3

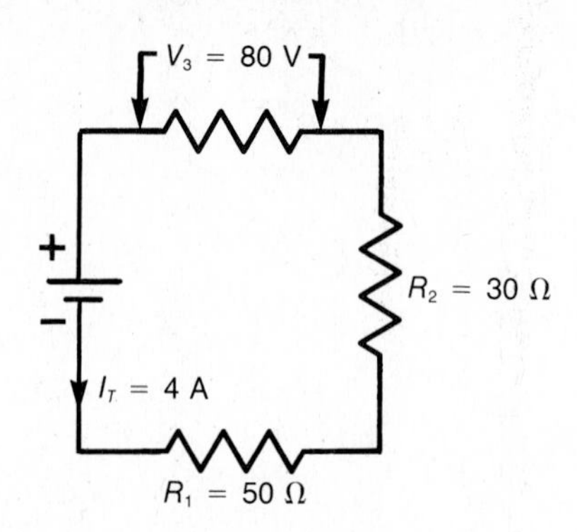

	VOLTAGE	CURRENT	RESISTANCE
R_1			50 Ω
R_2			30 Ω
R_3	80 V		
TOTAL		4 A	

Step 2 Calculate I_1, I_2 and I_3 and enter in chart.

$I_T = I_1 = I_2 = I_3 = 4\text{ A}$

	VOLTAGE	CURRENT	RESISTANCE
R_1		4 A	50 Ω
R_2		4 A	30 Ω
R_3	80 V	4 A	
TOTAL		4 A	

Step 3 Calculate V_1, V_2 and V_T and enter in chart.

$V_1 = I_1 \times R_1$
$= (4\text{ A})(50\ \Omega)$
$V_1 = 200\text{ V}$

$V_2 = I_2 \times R_2$
$= (4\text{ A})(30\ \Omega)$
$V_2 = 120\text{ V}$

$$V_T = V_1 + V_2 + V_3$$
$$= (200\text{ V}) + (120\text{ V}) + (80\text{ V})$$
$$V_T = 400\text{ V}$$

	VOLTAGE	CURRENT	RESISTANCE
R_1	200 V	4 A	50 Ω
R_2	120 V	4 A	30 Ω
R_3	80 V	4 A	
TOTAL	400 V	4 A	

Step 4 Calculate R_3 and R_T and enter in chart.

$$R_3 = V_3 / I_3$$
$$= (80\text{ V}) \div (4\text{ A})$$
$$R_3 = 20\ \Omega$$

$$R_T = V_T / I_T$$
$$= (400\text{ V}) \div (4\text{ A})$$
$$R_T = 100\ \Omega$$

	VOLTAGE	CURRENT	RESISTANCE
R_1	200 V	4 A	50 Ω
R_2	120 V	4 A	30 Ω
R_3	80 V	4 A	20 Ω
TOTAL	400 V	4 A	100 Ω

12-5 Practical Assignments

FIGURE 12-9 PRACTICAL ASSIGNMENT 1

1.

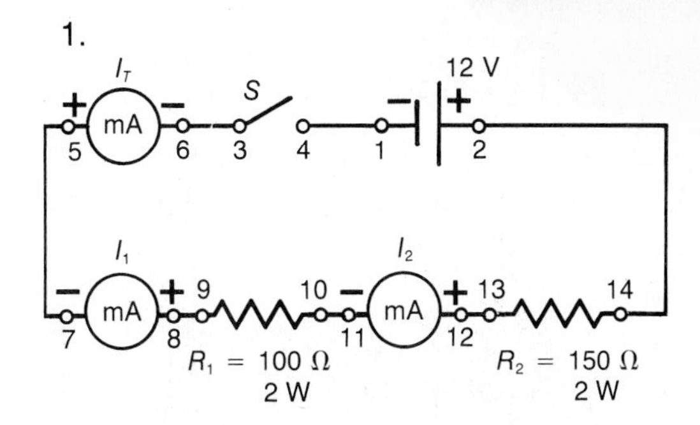

(a) The schematic circuit of Figure 12-9 is designed to show the current characteristic of a series circuit. Complete a wiring number sequence chart and wiring diagram for this circuit.

(b) Wire the circuit and record the current reading for each milliammeter.

(c) Explain, from your readings, the current characteristic of a series circuit.

FIGURE 12-10 PRACTICAL ASSIGNMENT 2

2.

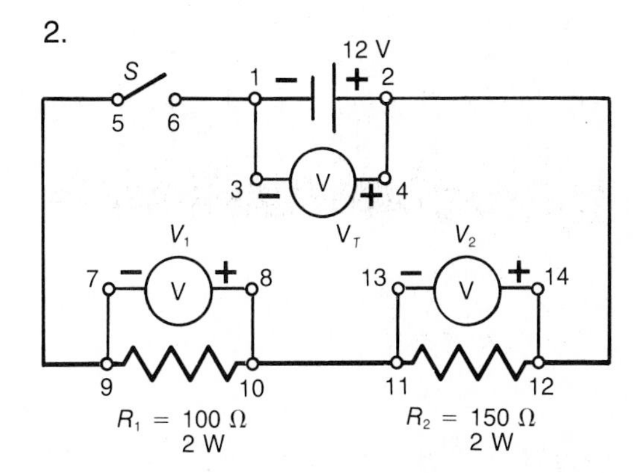

(a) The schematic circuit of Figure 12-10 is designed to show the voltage characteristic of a series circuit. Complete a wiring number sequence chart and wiring diagram for this circuit.

(b) Wire the circuit and record the voltage reading for each voltmeter.

(c) Explain, from your readings, the voltage characteristic of a series circuit.

FIGURE 12-11 PRACTICAL ASSIGNMENT 3

3.

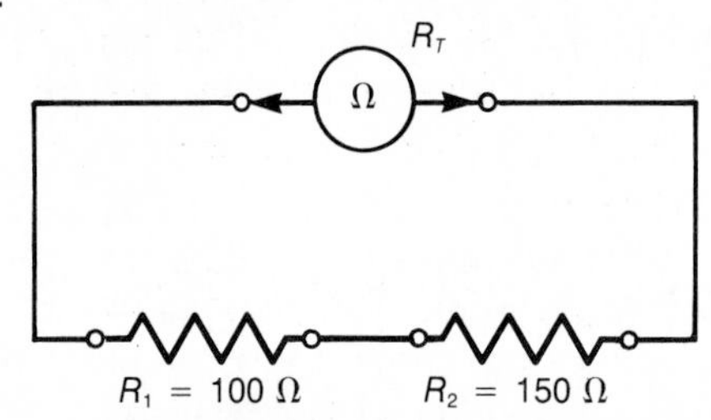

(a) The schematic circuit of Figure 12-11 is designed to show the resistance characteristic of a series circuit. Wire the circuit and record the total resistance as indicated by the ohmmeter.

(b) Explain, from your reading, the resistance characteristic of a series circuit.

FIGURE 12-12 PRACTICAL ASSIGNMENT 4

4.

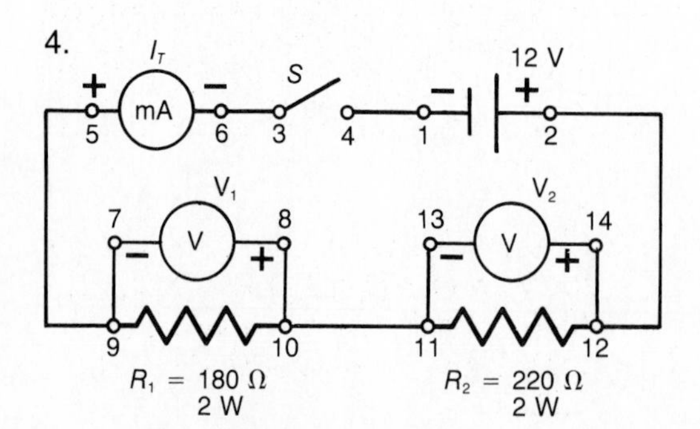

(a) Calculate all unknown values of voltage, current and resistance for the schematic circuit of Figure 12-12. Make a chart in your note-

	VOLTAGE	CURRENT	RESISTANCE
R_1	__ V	__ mA	180 Ω
R_2	__ V	__ mA	220 Ω
TOTAL	12 V	__ mA	__

book and record given and calculated circuit values.

(b) Complete a wiring number sequence chart and wiring diagram for this circuit.

(c) Wire the circuit and record the measured values of I_T, V_1, and V_2.

(d) How do the measured values compare with the calculated values?

FIGURE 12-13 PRACTICAL ASSIGNMENT 5

5.

	VOLTAGE	CURRENT	RESISTANCE
R_1	__ V	__ mA	0.5 kΩ
R_2	__ V	__ mA	1 kΩ
R_3	__ V	__ mA	1.5 kΩ
TOTAL	12 V	__ mA	__

(a) Calculate all unknown values of voltage, current and resistance for the schematic circuit of Figure 12-13. Make a chart in your note-

book and record given and calculated circuit values.
(b) Complete a wiring number sequence chart and wiring diagram for this circuit.
(c) Wire the circuit and record the measured values of I_T, V_1, V_2, and V_3.
(d) How do the measured values compare with the calculated values?

12-6 Self Evaluation Test

1. (a) State the current characteristic of a series circuit.
   (b) Express this characteristic in the form of an equation.
2. (a) State the voltage characteristic of a series circuit.
   (b) Express this characteristic in the form of an equation.
3. (a) State the resistance characteristic of a series circuit.
   (b) Express this characteristic in the form of an equation.
4. Find all unknown values of *V*, *I*, and *R* for the series circuit of Figure 12-14. Make a chart in your notebook and record given and calculated circuit values.

FIGURE 12-14 SELF EVALUATION TEST 4

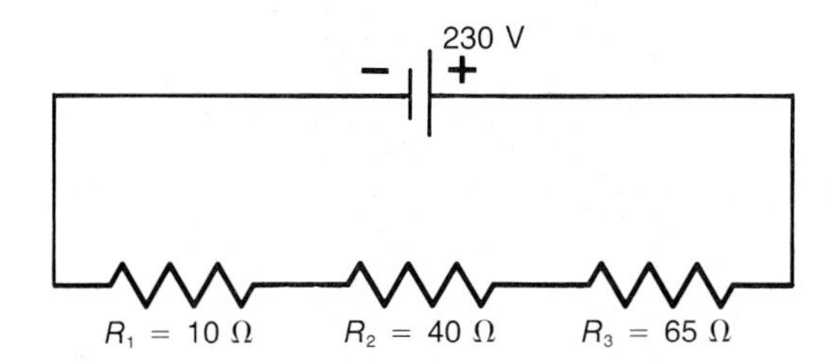

	VOLTAGE	CURRENT	RESISTANCE
R_1	__ V	__ A	__ Ω
R_2	__ V	__ A	__ Ω
R_3	__ V	__ A	__ Ω
TOTAL	__ V	__ A	__ Ω

5. Find all unknown values of *V*, *I*, and *R* for the series circuit of Figure 12-15. Make a chart in your notebook and record given and calculated circuit values.

FIGURE 12-15 SELF EVALUATION TEST 5

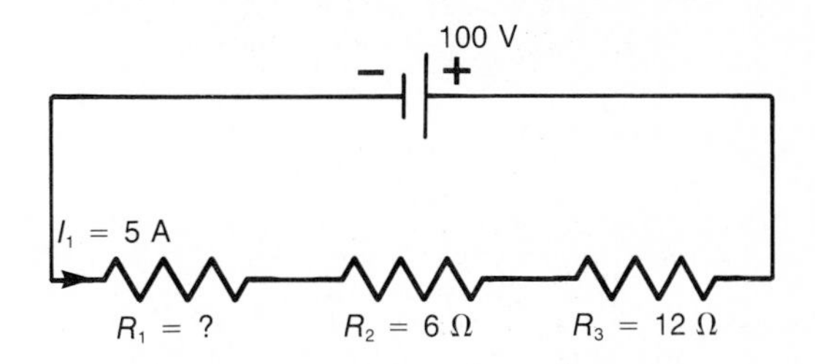

	VOLTAGE	CURRENT	RESISTANCE
R_1	__ V	__ A	__ Ω
R_2	__ V	__ A	__ Ω
R_3	__ V	__ A	__ Ω
TOTAL	__ V	__ A	__ Ω

6. Find all unknown values of *V*, *I*, and *R* for the series circuit of Figure 12-16. Make a chart in your notebook and record given and calculated circuit values.

FIGURE 12-16 SELF EVALUATION TEST 6

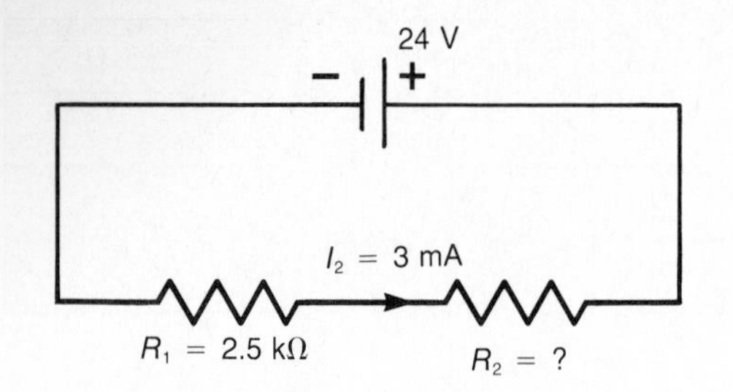

	VOLTAGE	CURRENT	RESISTANCE
R_1	___ V	___ mA	___ kΩ
R_2	___ V	___ mA	___ kΩ
TOTAL	___ V	___ mA	___ kΩ

7. Find all unknown values of *V*, *I*, and *R* for the series circuit of Figure 12-17. Make a chart in your notebook and record given and calculated circuit values.

	VOLTAGE	CURRENT	RESISTANCE
R_1	___ V	___ mA	___ Ω
R_2	___ V	___ mA	___ Ω
R_3	___ V	___ mA	___ Ω
TOTAL	___ V	___ mA	___ Ω

FIGURE 12-17 SELF EVALUATION TEST 7

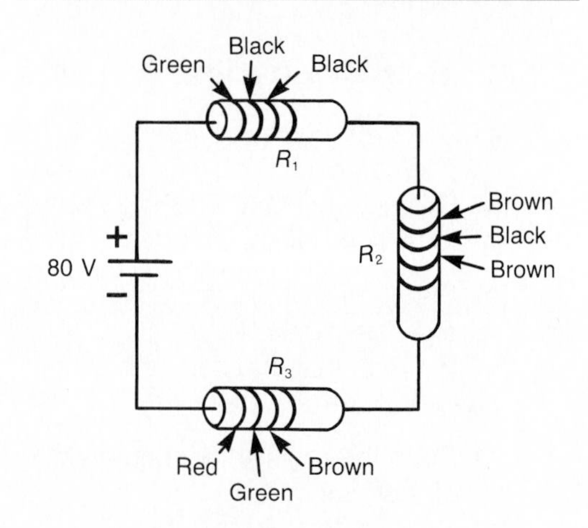

13

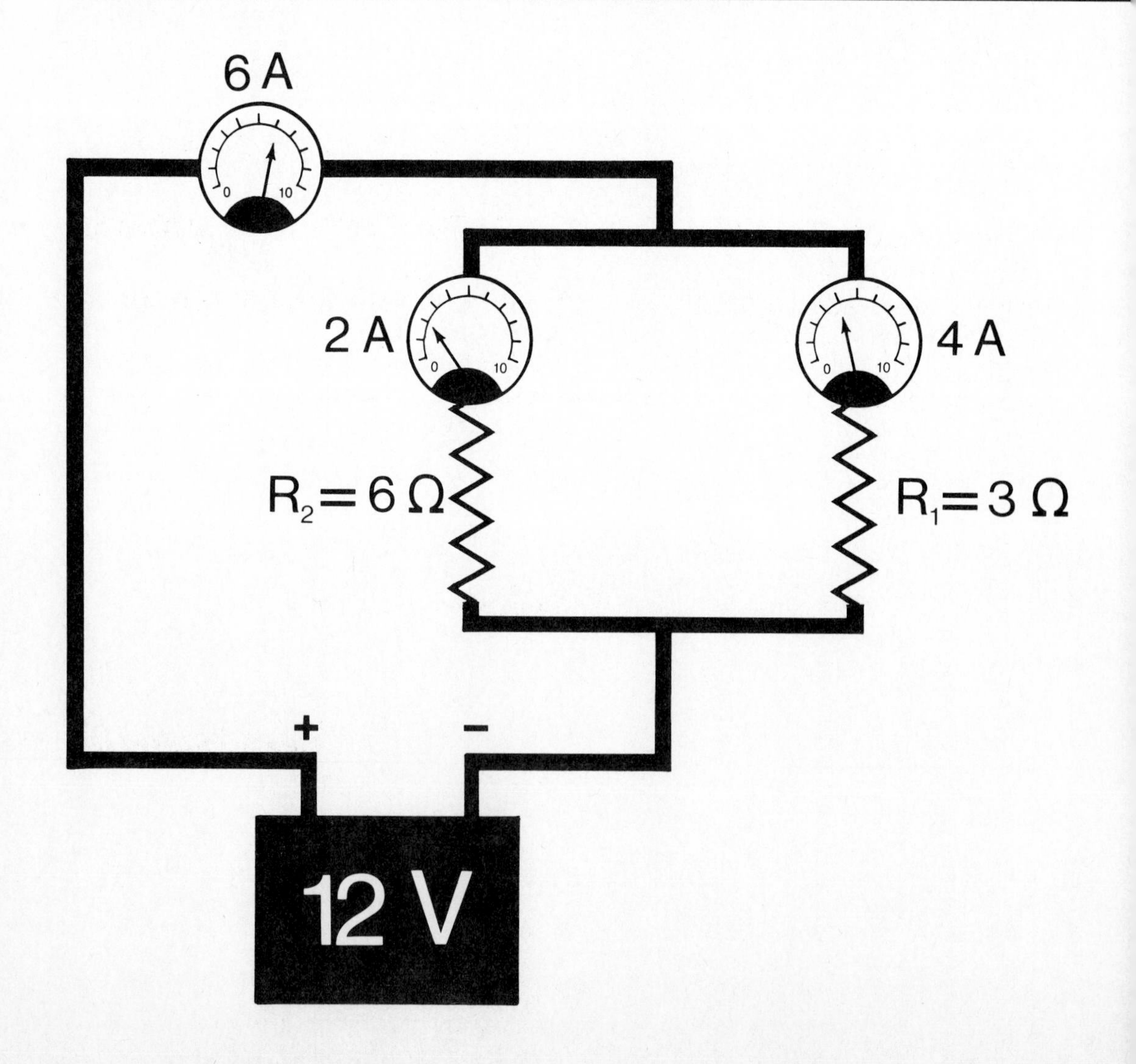

SOLVING THE PARALLEL CIRCUIT

OBJECTIVES

Upon completion of this unit you will be able to:

- State the voltage, current and resistance characteristics of a parallel circuit.
- Carry out experiments that demonstrate the characteristics of a parallel circuit.
- Given a parallel circuit with known values of voltage, current and resistance, solve for all unknown values.

FIGURE 13-1 PARALLEL CIRCUIT

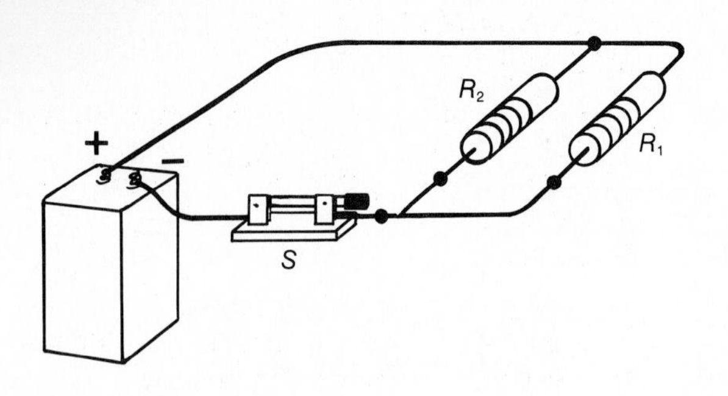

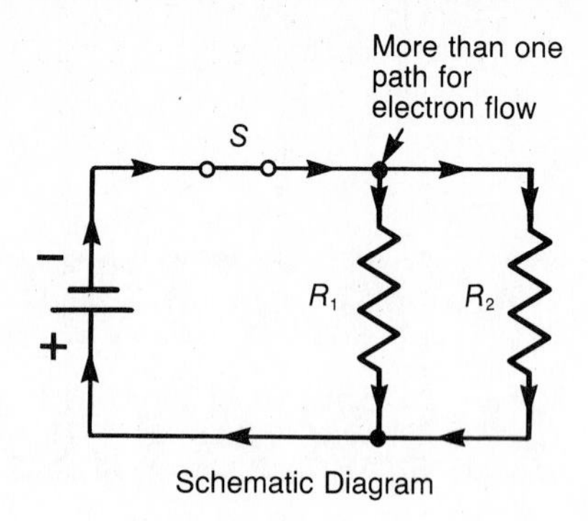

13-1 Parallel Circuit Connection

A parallel circuit is one in which all the loads are connected across the two voltage source lines (Figure 13-1). Each load acts independently of the other. This connection results in as many pathways for the current as there are load components connected in parallel.

13-2 Parallel Circuit Current, Voltage and Resistance Characteristics

Current The total current in a parallel circuit is divided among each of the loads (Figure 13-2). Each load is connected directly across the supply lines and acts independently of the other loads, as far as current is concerned. The amount of current drawn by each load is determined by the resistance value of each load. The higher the resistance of the load, the lower the amount of current taken.

Voltage The voltage across each load connected in parallel is the same and is equal in value to that of the voltage source (Figure 13-3). This is due to the fact that each load is connected directly across the two line wires.

FIGURE 13-2 THE TOTAL CURRENT IN A PARALLEL CIRCUIT IS EQUAL TO THE SUM OF THE BRANCH CURRENTS

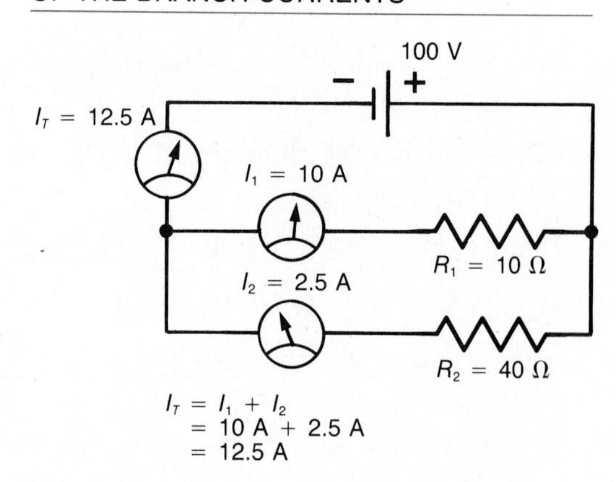

Resistance The total resistance of a parallel circuit is less than the resistance of the smallest load resistance (Figure 13-4). When you connect loads in parallel you have more paths through which the current can flow. This results in less total opposition to the total current flow and a decrease in total resistance.

13-3 Solving Parallel Circuits

The procedure to follow in solving parallel circuit values of voltage, current and resis-

FIGURE 13-3 THE VOLTAGE ACROSS EACH BRANCH OF A PARALLEL CIRCUIT IS THE SAME AS THE SOURCE VOLTAGE

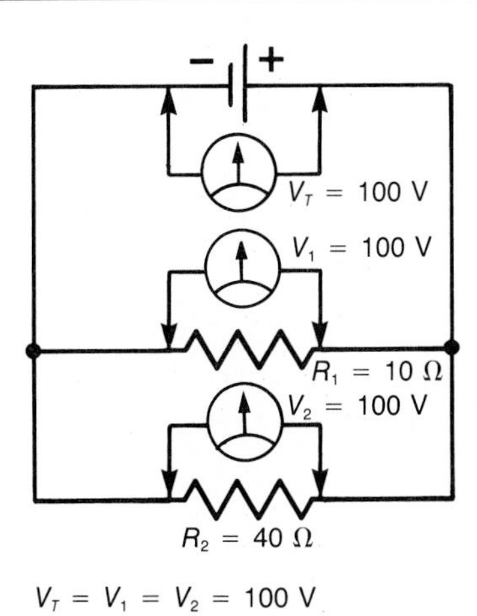

FIGURE 13-4 THE TOTAL RESISTANCE IN A PARALLEL CIRCUIT IS LESS THAN THE RESISTANCE OF ANY BRANCH

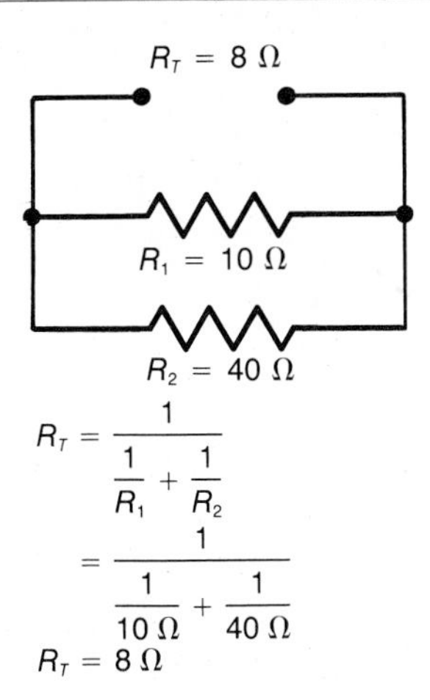

tance is similar to that used for solving series circuits. Ohm's Law as it applies to the circuit as a whole and the individual loads is used. In addition, the parallel circuit characteristics of voltage, current and resistance must also be used. The parallel circuit characteristic expressed in the form of equations are as follows:

$$I_T = I_1 + I_2 + I_3 \ldots\ldots$$

$$V_T = V_1 = V_2 = V_3 \ldots\ldots$$

$$R_T = \frac{R_1 \times R_2}{R_1 + R_2} \quad \text{(for 2 loads)}$$

$$R_T = \frac{1}{\frac{1}{R_1} + \frac{1}{R_2} + \frac{1}{R_3}} \ldots\ldots \quad \text{(for more than 2 loads)}$$

Example 1

Problem:

Find all the unknown values of V, I, and R for the parallel circuit of Figure 13-5.

FIGURE 13-5 EXAMPLE 1

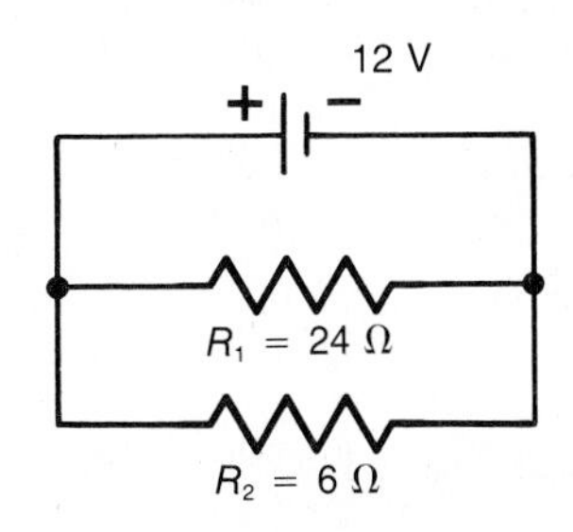

Solution:

Step 1 Make a chart and record all known values.

	VOLTAGE	CURRENT	RESISTANCE
R_1			24 Ω
R_2			6 Ω
TOTAL	12 V		

Step 2 Calculate V_1 and V_2 and enter in chart.

$V_T = V_1 = V_2 = 12$ V

	VOLTAGE	CURRENT	RESISTANCE
R_1	12 V		24 Ω
R_2	12 V		6 Ω
TOTAL	12 V		

Step 3 Calculate I_1, I_2 and I_3 and enter in chart.

$I_1 = V_1 / R_1$
$= (12\ V) \div (24\ \Omega)$
$I_1 = 0.5\ A$

$I_2 = V_2 / R_2$
$= (12\ V) \div (6\ \Omega)$
$I_2 = 2\ A$

$I_T = I_1 + I_2$
$= (0.5\ A) + (2\ A)$
$I_T = 2.5\ A$

	VOLTAGE	CURRENT	RESISTANCE
R_1	12 V	0.5 A	24 Ω
R_2	12 V	2 A	6 Ω
TOTAL	12 V	2.5 A	

Step 4 Calculate R_T and enter in chart.

$R_T = V_T / I_T$
$= (12\ V) \div (2.5\ A)$
$R_T = 4.8\ \Omega$

OR

$$R_T = \frac{R_1 \times R_2}{R_1 + R_2}$$
$$= \frac{(24\ \Omega)(6\ \Omega)}{(24\ \Omega) + (6\ \Omega)}$$
$$= \frac{144}{30}$$
$$R_T = 4.8\ \Omega$$

	VOLTAGE	CURRENT	RESISTANCE
R_1	12 V	0.5 A	24 Ω
R_2	12 V	2 A	6 Ω
TOTAL	12 V	2.5 A	4.8 Ω

Example 2

Problem:

Find all the unknown values of V, I and R for the parallel circuit of Figure 13-6.

FIGURE 13-6 EXAMPLE 2

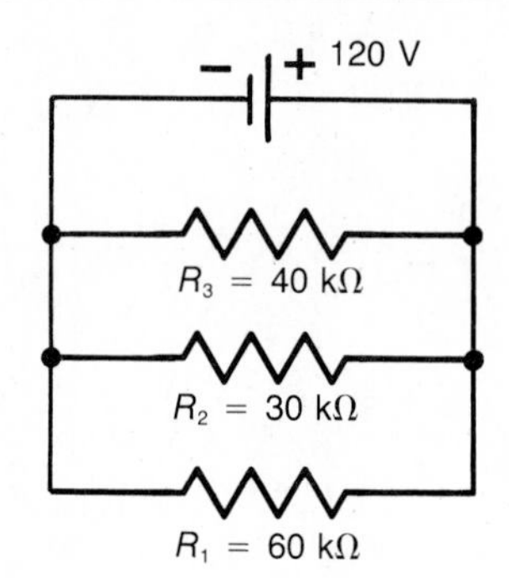

Solution

Step 1 Make a chart and record all known values.

	VOLTAGE	CURRENT	RESISTANCE
R_1	120 V		60 kΩ
R_2	120 V		30 kΩ
R_3	120 V		40 kΩ
TOTAL	120 V		

Step 2 Calculate I_1, I_2, I_3, I_T, and R_T and enter in chart.

$I_1 = V_1 / R_1$
$= (120\ V) \div (60\ k\Omega)$
$I_1 = 2\ mA$

$I_2 = V_2 / R_2$
$= (120\ V) \div (30\ k\Omega)$
$I_2 = 4\ mA$

$I_3 = V_3 / R_3$
$= (120\ V) \div (40\ k\Omega)$
$I_3 = 3\ mA$

$$I_T = I_1 + I_2 + I_3$$
$$= (2\ mA) + (4\ mA) + (3\ mA)$$
$$I_T = 9\ mA$$

$$R_T = V_T / I_T$$
$$= (120\ V) \div (9\ mA)$$
$$R_T = 13.3\ k\Omega$$

	VOLTAGE	CURRENT	RESISTANCE
R_1	120 V	2 mA	60 kΩ
R_2	120 V	4 mA	30 kΩ
R_3	120 V	3 mA	40 kΩ
TOTAL	120 V	9 mA	13.3 kΩ

Example 3

Problem:

Find all the unknown values of V, I, and R for the parallel circuit of Figure 13-7.

Solution:

Step 1 Make a chart and record all known values.

	VOLTAGE	CURRENT	RESISTANCE
R_1	24 V		
R_2	24 V		16 Ω
R_3	24 V		24 Ω
TOTAL	24 V	5.5 A	

Step 2 Calculate I_2, I_3, I_1, R_1, and R_T and enter in chart.

$$I_2 = V_2 / R_2$$
$$= (24\ V) \div (16\ \Omega)$$
$$I_2 = 1.5\ A$$

$$I_3 = V_3 / R_3$$
$$= (24\ V) \div (24\ \Omega)$$
$$I_3 = 1\ A$$

$$I_1 = I_T - (I_2 + I_3)$$
$$= (5.5\ A) - (2.5\ A)$$
$$I_1 = 3\ A$$

$$R_1 = V_1 / I_1$$
$$= (24\ V) \div (3\ A)$$
$$R_1 = 8\ \Omega$$

$$R_T = V_T / I_T$$
$$= (24\ V) \div (5.5\ A)$$
$$R_T = 4.36\ \Omega$$

	VOLTAGE	CURRENT	RESISTANCE
R_1	24 V	3 A	8 Ω
R_2	24 V	1.5 A	16 Ω
R_3	24 V	1 A	24 Ω
TOTAL	24 V	5.5 A	4.36 Ω

FIGURE 13-7 EXAMPLE 3

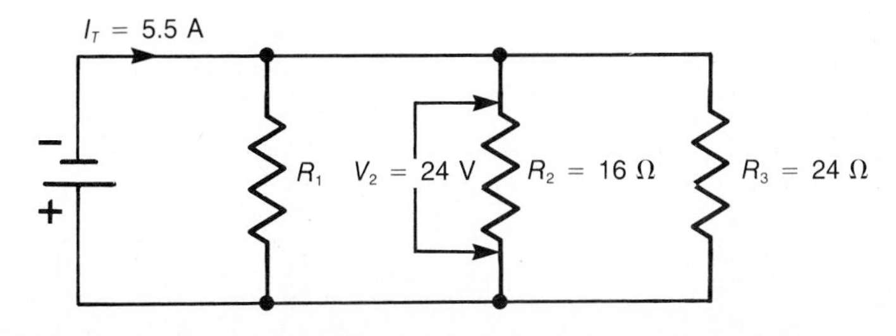

13-4 Practical Assignments

FIGURE 13-8 PRACTICAL ASSIGNMENT 1

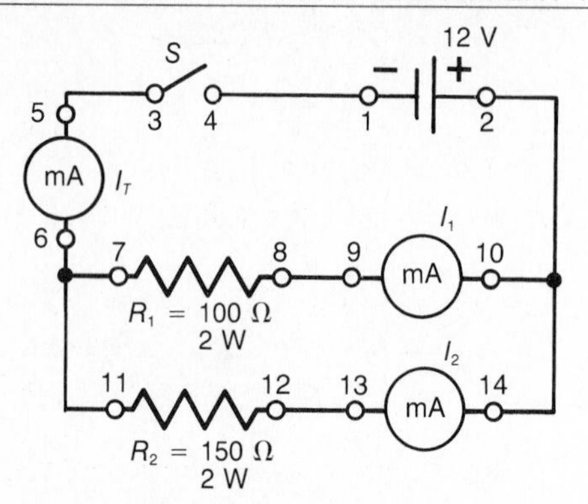

1. (a) The schematic circuit of Figure 13-8 is designed to show the current characteristic of a parallel circuit. Complete a wiring number sequence chart and wiring diagram for this circuit.
   (b) Wire the circuit and record the current reading for each milliammeter.
   (c) Explain, from your readings, the current characteristic of a parallel circuit.

FIGURE 13-9 PRACTICAL ASSIGNMENT 2

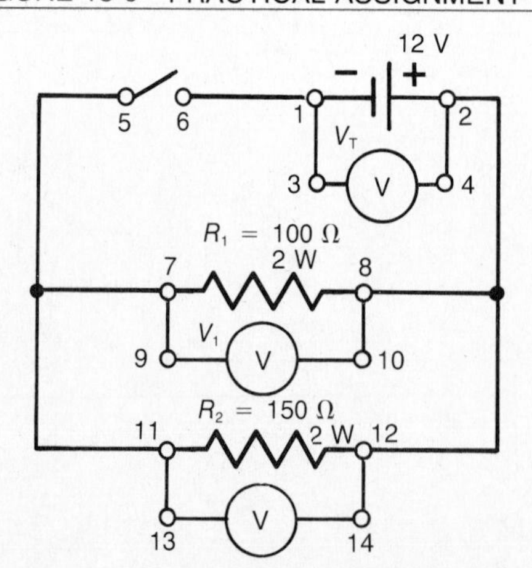

2. (a) The schematic circuit of Figure 13-9 is designed to show the voltage characteristic of a parallel circuit. Complete a wiring number sequence chart and wiring diagram for this circuit.
   (b) Wire the circuit and record the voltage reading for each voltmeter.
   (c) Explain, from your readings, the voltage characteristic of a parallel circuit.

FIGURE 13-10 PRACTICAL ASSIGNMENT 3

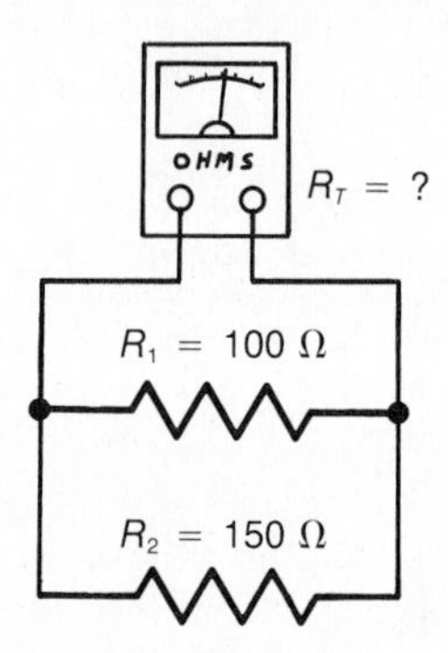

3. (a) The schematic circuit of Figure 13-10 is designed to show the resistance characteristic of a series circuit. Wire the circuit and record the total resistance as indicated by the ohmmeter.
   (b) Explain, from your reading, the resistance characteristic of a parallel circuit.

FIGURE 13-11 PRACTICAL ASSIGNMENT 4

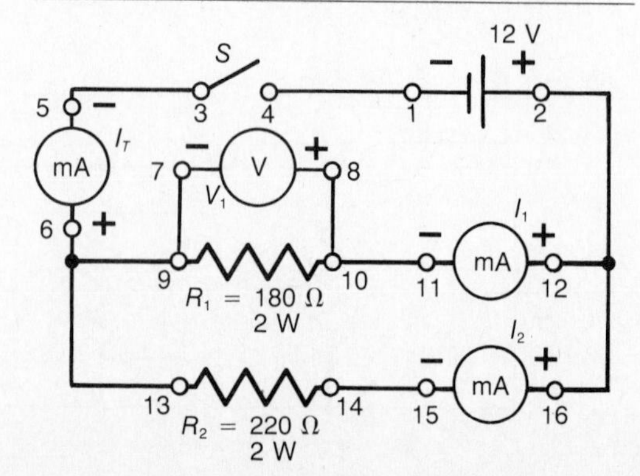

	VOLTAGE	CURRENT	RESISTANCE
R_1			180 Ω
R_2			220 Ω
TOTAL	12 V		

4. (a) Calculate all unknown values of voltage, current and resistance for the schematic circuit of Figure 13-11. Make a chart in your notebook and record given and calculated circuit values.
   (b) Complete a wiring number sequence chart and wiring diagram for this circuit.
   (c) Wire the circuit and record the measured values of I_T, I_1, I_2, and V_1.
   (d) How do the measured values compare with the calculated values?

FIGURE 13-12 PRACTICAL ASSIGNMENT 5

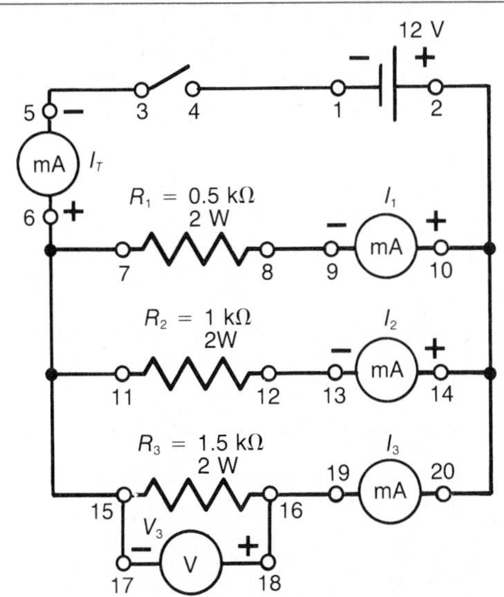

	VOLTAGE	CURRENT	RESISTANCE
R_1			0.5 kΩ
R_2			1 kΩ
R_3			1.5 kΩ
TOTAL	12 V		

5. (a) Calculate all unknown values of voltage, current and resistance for the schematic circuit of Figure 13-12. Make a chart in your notebook and record given and calculated circuit values.
   (b) Complete a wiring number sequence chart and wiring diagram for this circuit.
   (c) Wire the circuit and record the measured values of I_T, I_1, I_2, I_3, and V_3.
   (d) How do the measured values compare with the calculated values?

13-5 Self Evaluation Test

1. (a) State the current characteristic of a parallel circuit.
   (b) Express this characteristic in the form of an equation.

2. (a) State the voltage characteristic of a parallel circuit.
   (b) Express this characteristic in the form of an equation.

3. (a) State the resistance characteristic of a parallel circuit.
   (b) Express this characteristic in the form of an equation.

4. Find all unknown values of V, I, and R for the parallel circuit of Figure 13-13. Make a chart in your notebook and record given and calculated circuit values.

FIGURE 13-14 SELF EVALUATION TEST 4

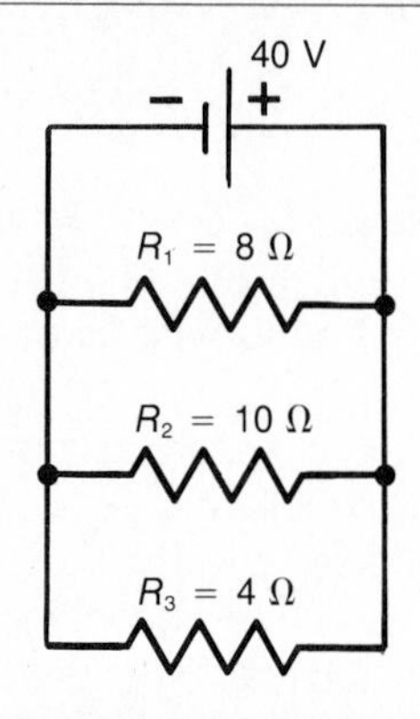

	VOLTAGE	CURRENT	RESISTANCE
R_1	__ V	__ A	__ Ω
R_2	__ V	__ A	__ Ω
R_3	__ V	__ A	__ Ω
TOTAL	__ V	__ A	__ Ω

5. Find all unknown values of V, I, and R for the parallel circuit of Figure 13-14. Make a chart in your notebook and record given and calculated circuit values.

FIGURE 13-13 SELF EVALUATION TEST 5

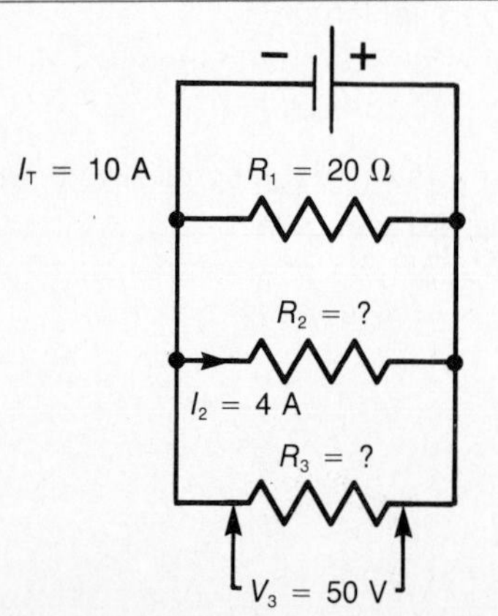

	VOLTAGE	CURRENT	RESISTANCE
R_1	__ V	__ A	__ Ω
R_2	__ V	__ A	__ Ω
R_3	__ V	__ A	__ Ω
TOTAL	__ V	__ A	__ Ω

6. Find all unknown values of V, I, and R for the parallel circuit of Figure 13-15. Make a chart in your notebook and record given and calculated circuit values.

FIGURE 13-15 SELF EVALUATION TEST 6

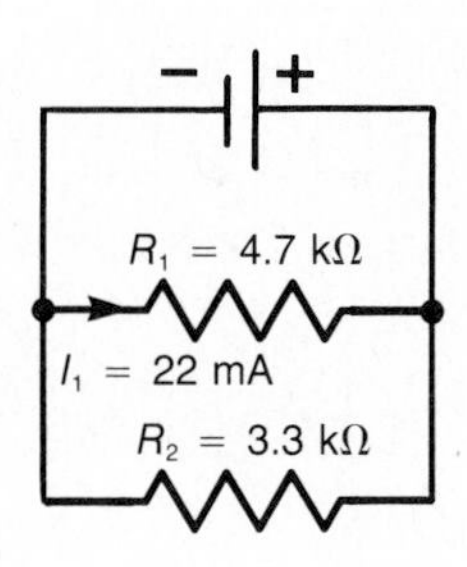

	VOLTAGE	CURRENT	RESISTANCE
R_1	__ V	__ mA	__ kΩ
R_2	__ V	__ mA	__ kΩ
TOTAL	__ V	__ mA	__ kΩ

7. Find all unknown values of V, I, and R for the parallel circuit of Figure 13-16. Make a chart in your notebook and record given and calculated circuit values.

FIGURE 13-16 SELF EVALUATION TEST 7

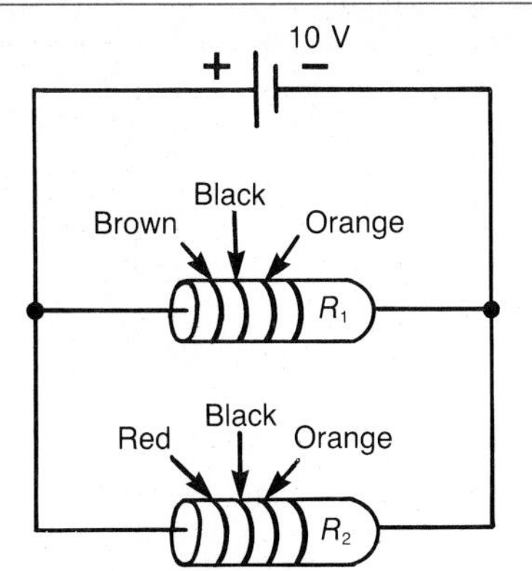

(Assume the tolerance value of each resistor is ±0%)

	VOLTAGE	CURRENT	RESISTANCE
R_1	__ V	__ mA	__ kΩ
R_2	__ V	__ mA	__ kΩ
TOTAL	__ V	__ mA	__ kΩ

14

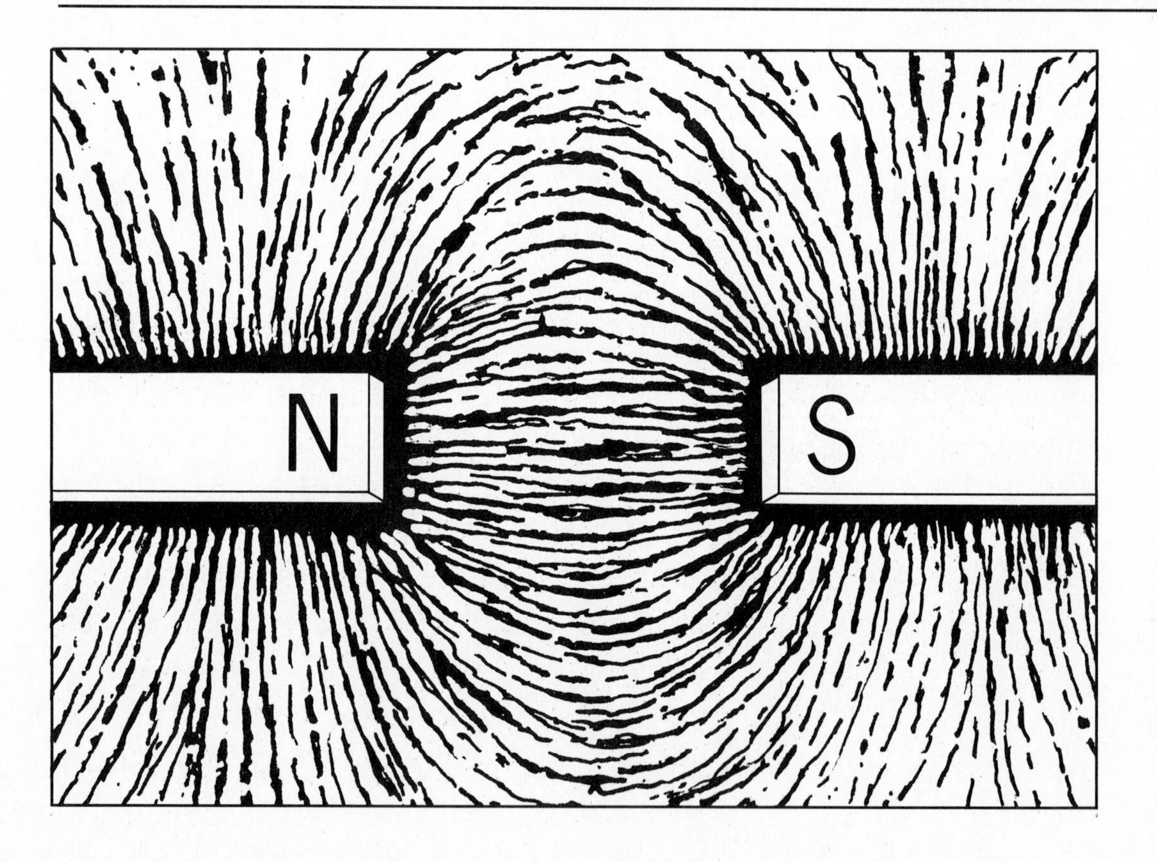

MAGNETISM

OBJECTIVES

Upon completion of this unit you will be able to:

- Define common magnetic terms.
- Identify the different types of magnets.
- Classify common materials as being magnetic or non-magnetic.
- Magnetize and demagnetize a magnetic material by electrical means.
- State the law of magnetic poles.
- Sketch common magnetic field patterns.
- Describe the characteristics of magnetic lines of force.
- Explain a theory of magnetism.
- Explain the process of magnetic shielding.

14-1 Properties of Magnets

The ability of certain materials to attract objects made of iron or iron alloys is the most familiar of all magnetic effects. This property of a material to attract pieces of iron or steel is called *magnetism* (Figure 14-1).

A magnet is a material that has the power to attract iron or its alloys. Materials that are attracted by magnets are called magnetic materials. Some common magnetic materials are iron, steel, nickel and cobalt. Nonmagnetic materials are those which are not attracted by magnets. Examples of nonmagnetic materials are wood, paper, glass, copper, aluminum and brass.

FIGURE 14-1 MAGNETIC EFFECT

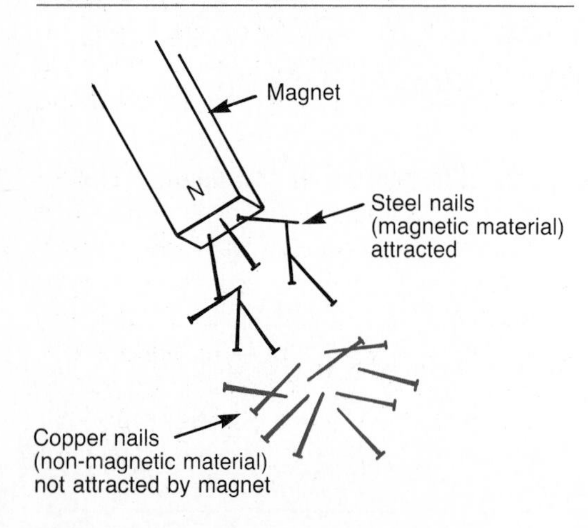

14-2 Types of Magnets

Natural and Artificial Magnets The effects of magnetism were first observed in pieces of iron ore called *lodestone* or *magnetite*. Lodestone is said to be a natural magnet because it possesses magnetic qualities when found in its natural form (Figure 14-2). Natural magnets have very little practical use because it is possible to produce much better magnets by artificial means.

FIGURE 14-2 NATURAL AND ARTIFICIAL MAGNETS

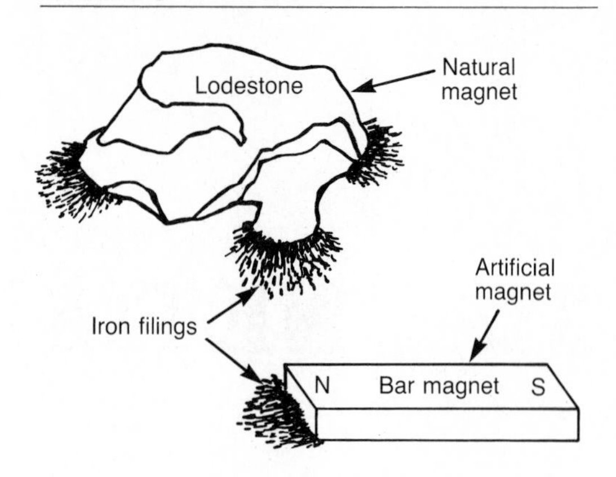

Artificial magnets are those made from ordinary unmagnetized magnetic materials. The bar magnet, horseshoe magnet and compass needle are all examples of artificial magnets (Figure 14-3).

FIGURE 14-3 COMMON ARTIFICIAL MAGNETS

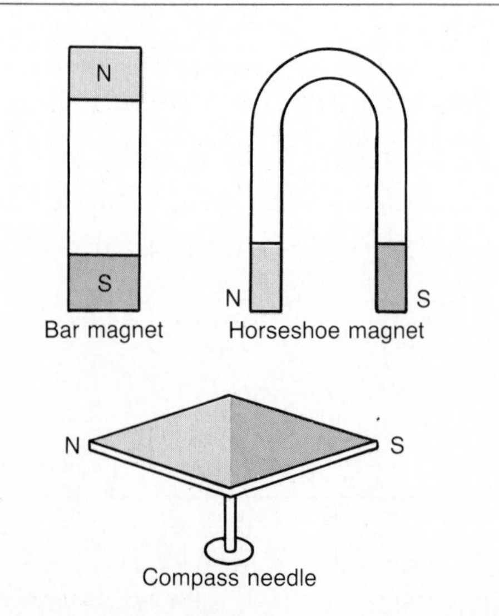

Most artificial magnets are produced electrically. The process used is a simple one. To magnetize a magnetic material electrically the material to be magnetized is first

FIGURE 14-4 THE MAGNETIZING AND DEMAGNETIZING PROCESSES

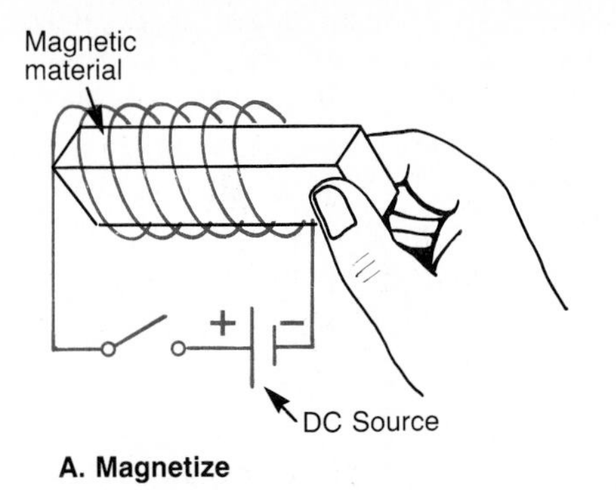

A. Magnetize

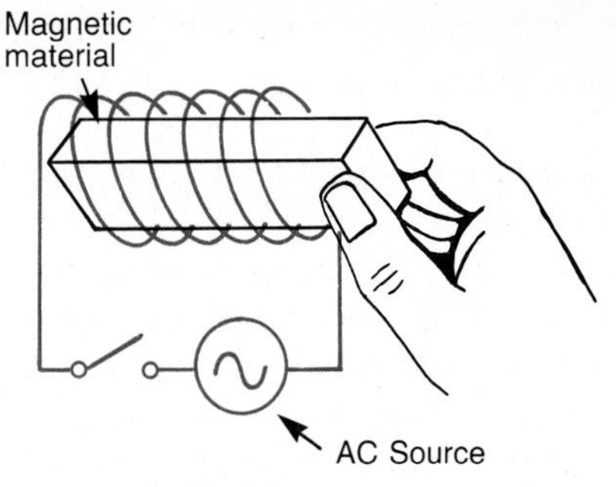

B. Demagnetize

placed in the center of a coiled piece of wire. A direct current voltage source is then applied momentarily to the coil leads (Figure 14-4A). To demagnetize an artificial magnet the same process is repeated but the voltage source used is alternating current (Figure 14-4B).

Temporary and Permanent Magnets Different magnetic materials have different abilities to retain their magnetism once magnetized. The ability of a material to retain its magnetism is determined by the *retentivity* of the material. Temporary magnets have low retentivity (Figure 14-5A). That is to say they lose most of their magnetic power once the magnetizing force is removed. Soft irons have a low retentivity factor and as such make good temporary magnets. Permanent magnets are made from hard iron and steel (Figure 14-5B). It requires more energy to magnetize them but, once magnetized, they will retain their magnetism for a long period of time.

The magnetism that remains in a magnetic material, once the magnetizing force is removed, is called *residual magnetism*. This term is usually only applied to temporary magnets. Residual magnetism is of importance in certain types of generators as it provides the initial voltage required for the generator to build up to its rated voltage.

FIGURE 14-5 RETAINING MAGNETISM

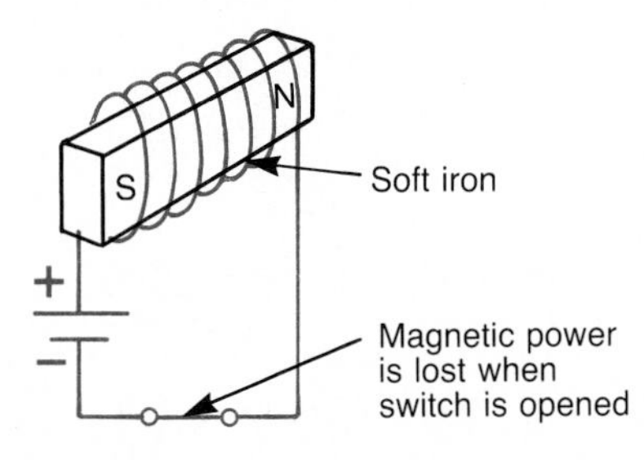

A. Temporary magnet

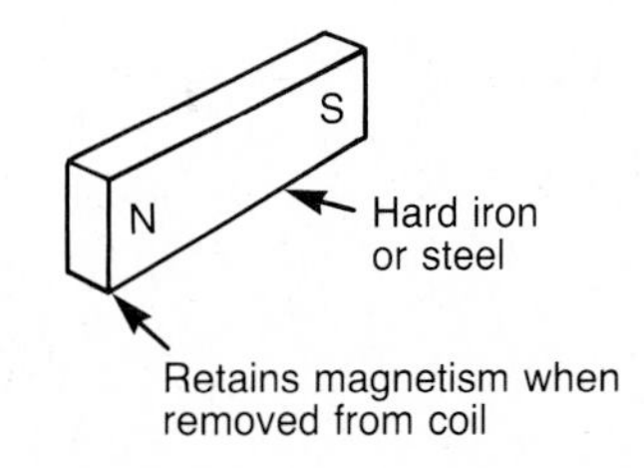

B. Permanent magnet

14-3 Law of Magnetic Poles

The effects of magnetism are strong at the ends of the magnet and weak in the middle. The ends of the magnet, where the magnetic effects are the greatest, are called the *poles* of the magnet. Every magnet has two such poles. These poles are identified as the *north* and *south* poles of the magnet (Figure 14-6).

FIGURE 14-6 MAGNETIC POLES

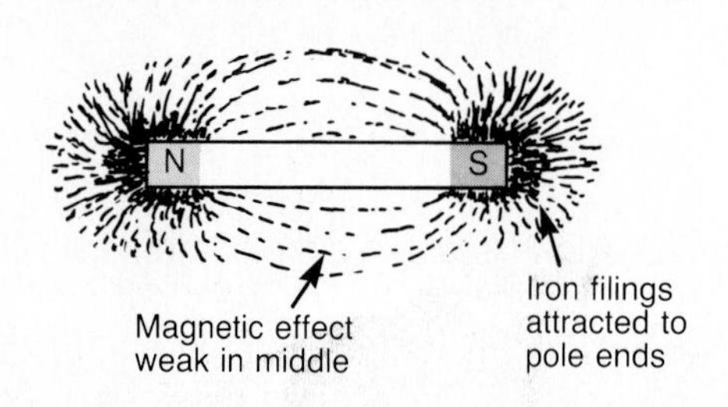

The law of magnetic poles states that: **like poles repel and unlike poles attract**. Placing the north pole of a suspended magnet near the south pole of a second magnet will cause the two pole ends to come together or be attracted to each other (Figure 14-7A). Repeating this experiment using the two north pole ends will cause the two pole ends to move apart or produce a repelling effect (Figure 14-7B).

If one bar magnet is placed on a desk and a second magnet is moved slowly towards it, you will observe that the attracting or repelling force will increase as the distance between the poles of the magnet decreases. Actually, this magnetic force varies inversely as the square of the distance between poles. For example, if the distance between two unlike poles is increased to twice the distance, the attracting force will be reduced to one-quarter of its former value (Figure 14-8).

14-4 Using a Compass to Identify Poles

The Earth itself acts like a huge fixed magnet with magnetic poles located near the north and south geographic poles. A compass is simply a permanent magnet pivoted at its midpoint so that it is free to move in a horizontal plane. Due to the magnetic attraction between poles the compass will always come to rest with the same end pointing towards the north. The end of the compass that points to the geographic north was established as being the north-seeking pole of the compass.

The compass can be used to identify the poles of a magnet (Figure 14-9). First, identify the north- and south-seeking poles of the compass. Remember the north-seeking pole of the compass points to the geographic north pole. Next, place the compass at one of the pole ends of the magnet. Apply the law of magnetic poles to identify the unmarked pole of the magnet. If the north-

FIGURE 14-7 LAW OF MAGNETIC POLES

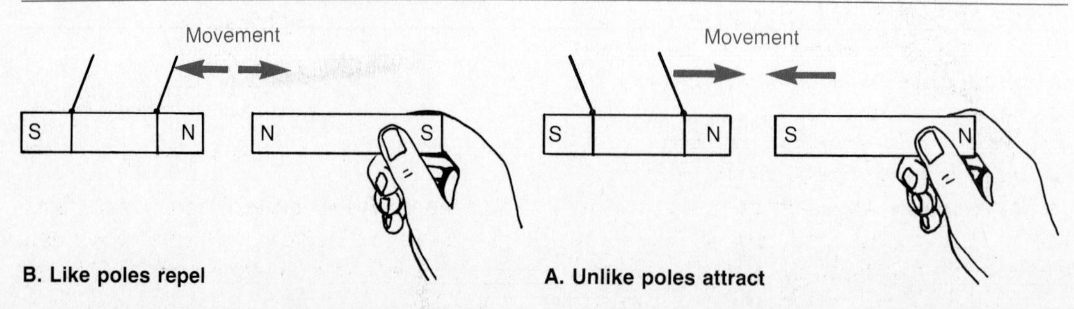

FIGURE 14-8 DISTANCE AND FORCE BETWEEN POLES

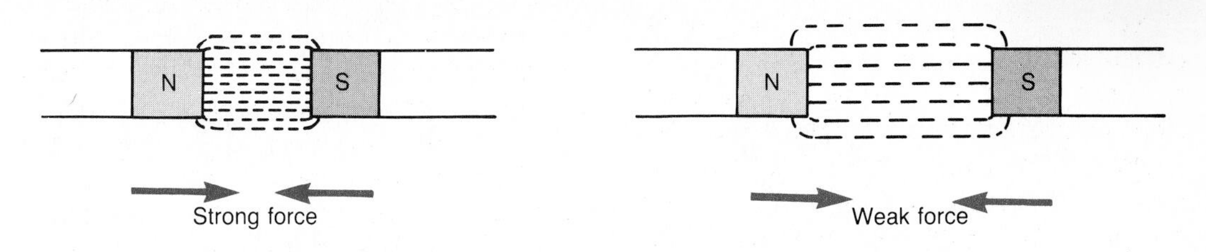

FIGURE 14-9 USING A COMPASS TO IDENTIFY POLES OF A MAGNET

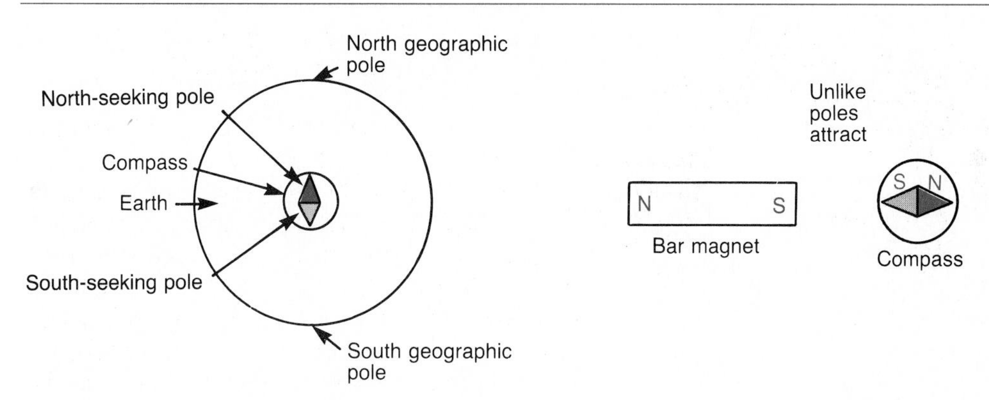

seeking pole of the compass is attracted to the pole, then it is a north magnetic pole. If the south-seeking pole of the compass is attracted to the pole then it is a south magnetic pole.

14-5 Magnetic Fields

The area surrounding a magnet, in which the invisible magnetic force is evident, is called the *magnetic field* of the magnet. A representation of this magnetic field pattern can be observed through the use of iron filings sprinkled over the area around the magnet (Figure 14-10).

At times, it is necessary to illustrate the direction and intensity of magnetic field patterns. A commonly used method of representing the forces in a magnetic field is by use of lines called *magnetic lines of force*.

Although lines of force are invisible, they are assumed to have certain characteristics. These characteristics are summarized as follows:

- lines of force form closed loops
- lines of force exit from the magnet's N pole and enter at its S pole
- lines of force travel most easily through soft iron
- lines of force repel each other
- lines of force cannot intersect or cross each other
- there is no known insulator for lines of force.

14-6 Magnetic Shielding

Certain types of electrical equipment, such as meters, are affected in their operation and accuracy by stray magnetic lines of force.

FIGURE 14-10 MAGNETIC FIELD PATTERNS

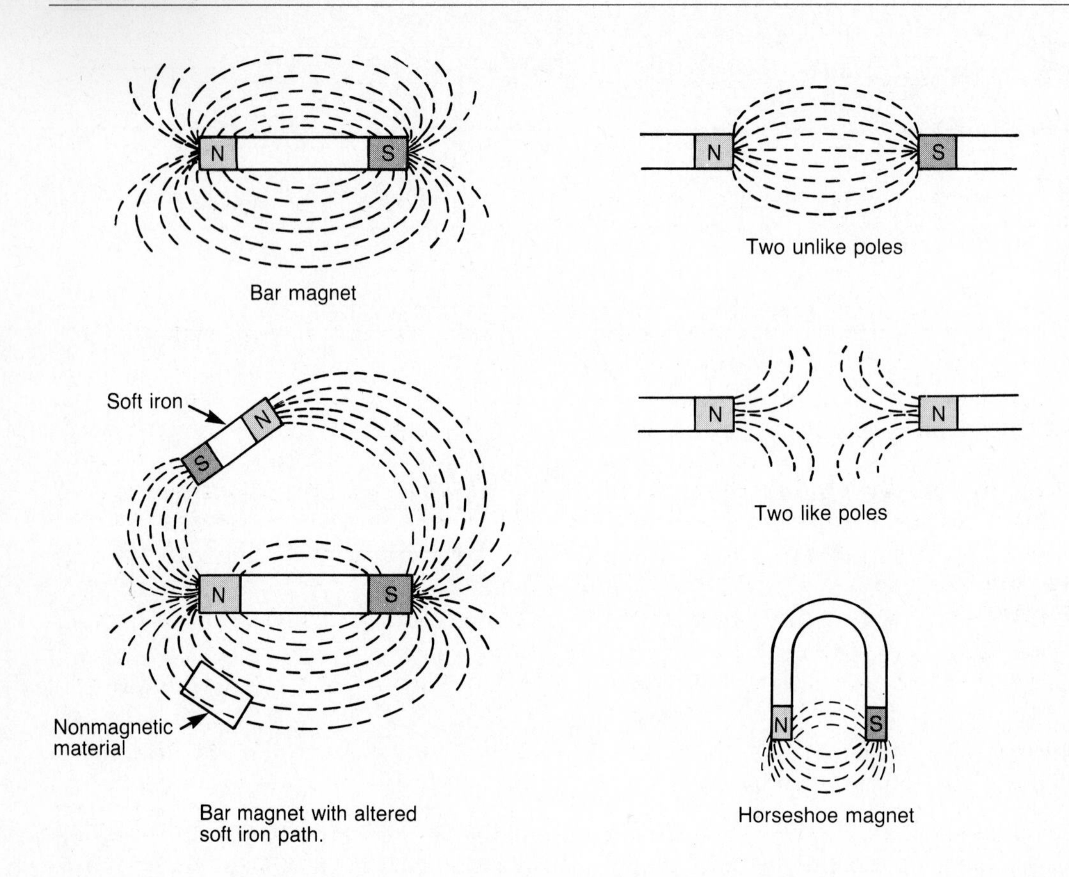

FIGURE 14-11 SHIELDING AGAINST LINES OF FORCE

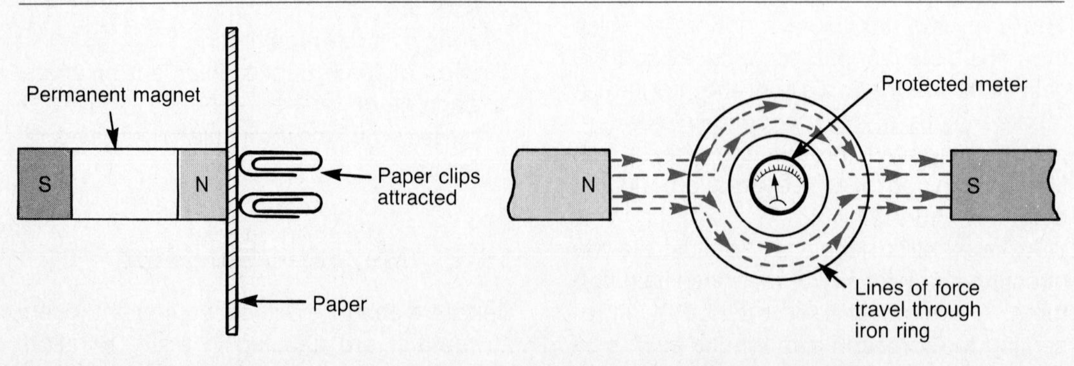

A. No insulator for lines of force

B. Protecting against lines of force

FIGURE 14-12 MOLECULAR THEORY OF MAGNETISM

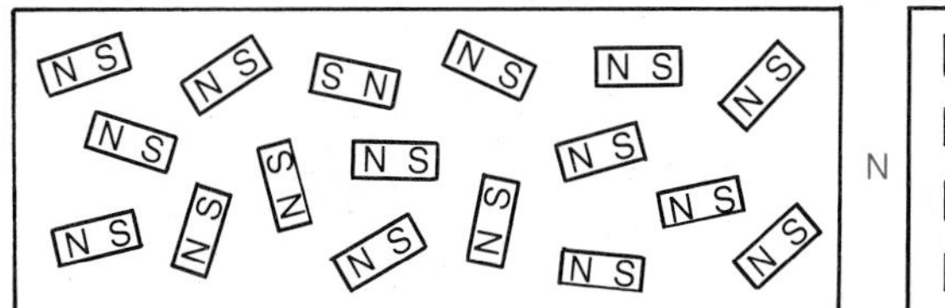

A. Unmagnetized bar

B. Magnetized bar

As mentioned, one of the characteristics of lines of force is that there is no known insulator for them and this presents a problem as far as protecting devices from stray magnetic fields (Figure 14-11A). The problem is overcome by making use of another characteristic. That characteristic is that lines of force travel most easily through soft iron. Meters to be protected are surrounded by a low resistance soft iron magnetic path so that any stray magnetic lines of force travel around rather than through the meter (Figure 14-11B). The same principle of design is applied in motors and transformers to minimize the radiation of lines of force from the magnetic fields of these devices.

14-7 Theories of Magnetism

Different theories have been developed over the years in an attempt to explain what causes magnetism. The *molecular theory* of magnetism assumes that each molecule (group of atoms) of a substance is, in fact, a small magnet. When a material is unmagnetized its molecular magnets are arranged in a random fashion (Figure 14-12A). The net result is a cancellation of the magnetic effect. In a magnetized bar the molecular magnets are arranged so that their magnetic fields are aligned in the same direction (Figure 14-12B).

The *electron theory* of magnetism assumes that each spinning electron is a small magnet. Unmagnetized materials have electrons spinning in different directions causing the cancellation of the magnetic effect (Figure 14-13A). Magnetized materials tend to have more of their electrons spinning in the same direction (Figure 14-13B).

There is a definite limit to the amount of magnetism a material can have. This limit

FIGURE 14-13 ELECTRON THEORY OF MAGNETISM

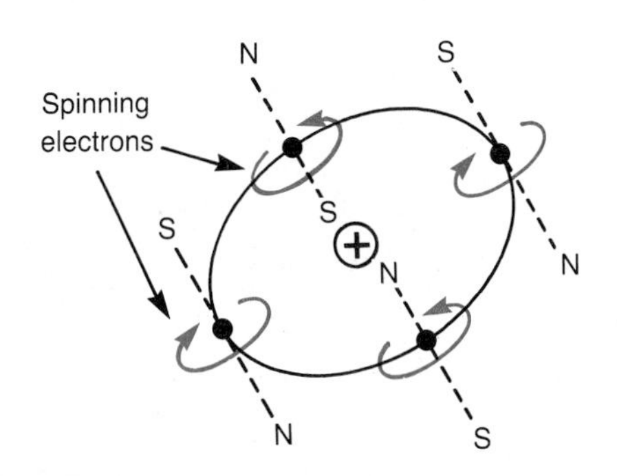

A. Unmagnetized material

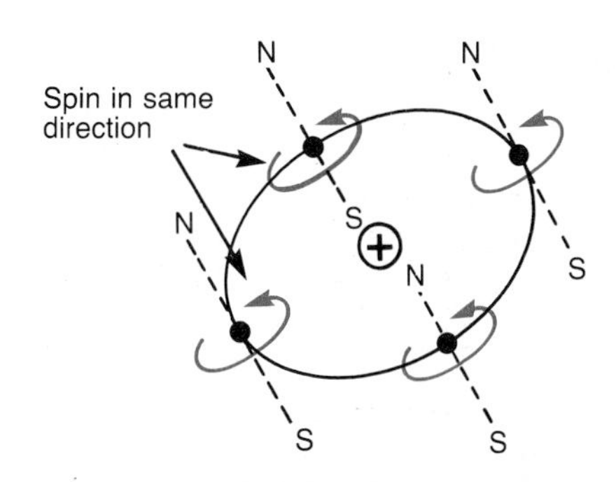

B. Magnetized material

is reached when all the molecular magnets are aligned or all the electrons are spinning in the same direction. When maximum magnetic strength is reached the material is said to be magnetically saturated.

14-8 Uses for Permanent Magnets

Permanent magnets of various shapes are used extensively in electrical and electronic equipment. Simple bar magnets are often used in lab work to study the effects of magnetism. Horseshoe shaped magnets provide a much stronger magnetic field than a bar magnet of equal material. This is due to the fact that the poles are closer together, thus concentrating the magnetic field in a smaller space. Horseshoe magnets are often used in the construction of analog type deflection meters (Figure 14-14).

Both small wheel-operated bicycle generators and hand-operated flashlight generators use permanent magnets to generate the voltage used to operate lights. With the flashlight generator, rotating the crank of the generator provides the mechanical force or motion required for generator action while the internal permanent magnet provides the required magnetism (Figure 14-15). The faster the crank is turned, the more voltage is generated and the more brightly the lamp will glow. These generators are designed for AC or DC output and make a very useful voltage source when experimenting with electricity and electromagnetism.

FIGURE 14-14 PERMANENT MAGNET METER

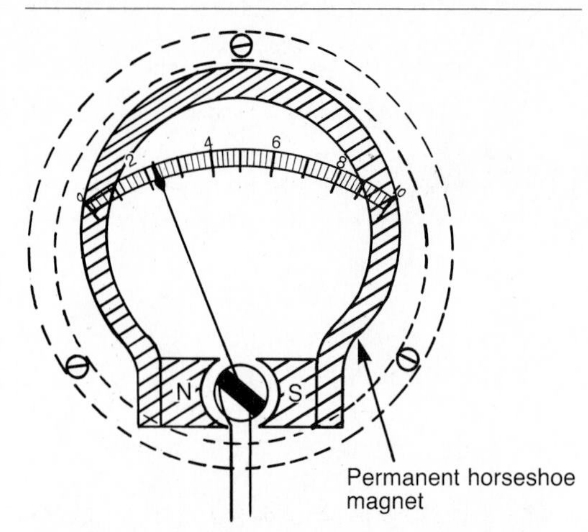

FIGURE 14-15 PERMANENT MAGNET HAND GENERATOR

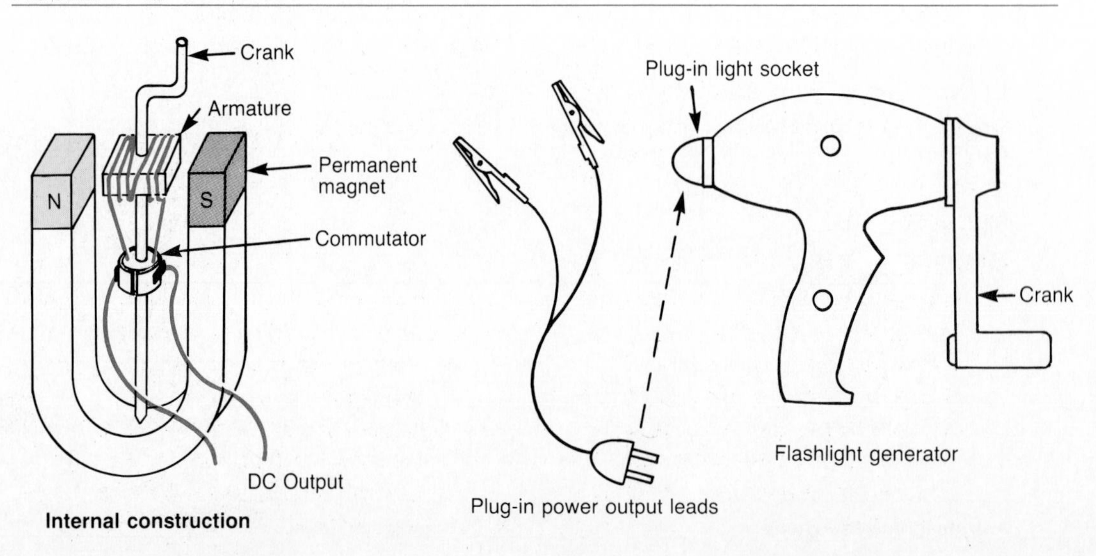

FIGURE 14-16 PERMANENT MAGNET SPEAKER

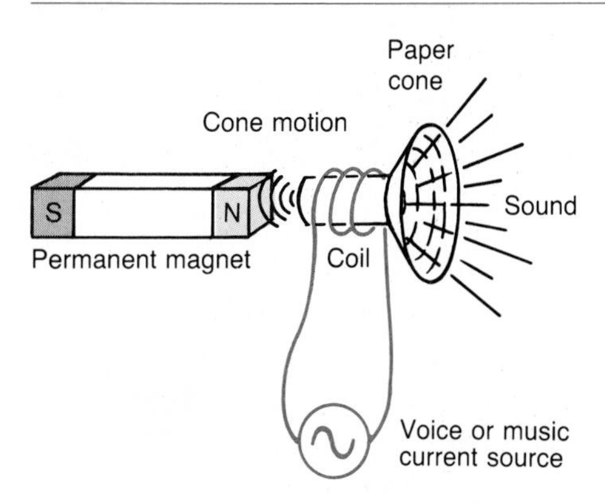

Other electrical and electronic devices that use permanent magnets include speakers, microphones, and small permanent magnet motors (Figure 14-16).

14-9 Practical Assignments

1. (a) Outline the procedure used to electrically magnetize and demagnetize a magnetic material.
   (b) Magnetize and demagnetize a small screwdriver using a coil and the proper voltage source.
   (c) Test the magnetic state of the screwdriver using the head of a small machine screw.
2. Using a permanent magnet, test and classify each of the following as being magnetic or non-magnetic materials: steel, brass, copper, iron and aluminum.
3. (a) Using one fixed and one suspended bar magnet, demonstrate that (i) like poles repel, and (ii) unlike poles attract.
   (b) Using the same apparatus, demonstrate the relationship between the distance between the two poles and the amount of magnetic attraction.
4. Given a compass and an unmarked bar magnet, correctly identify the North and South poles of the bar magnet.
5. Using paper and iron filings, observe and sketch the shape of the magnetic field patterns of:
   (i) a bar magnet
   (ii) two unlike poles
   (iii) two like poles
   (iv) horseshoe magnet

14-10 Self Evaluation Test

1. Re-write each of the following terms in your notebook. Beside each term, write the most appropriate definition from the list provided.

   TERMS
   (a) magnet
   (b) magnetic material
   (c) non-magnetic material
   (d) natural magnet
   (e) artificial magnet
   (f) permanent magnet
   (g) temporary magnet
   (h) residual magnetism
   (i) magnetic pole
   (j) magnetic lines of force
   (k) magnetic saturation
   (l) magnetic field

   DEFINITIONS
   - has magnetic qualities in original state
   - maximum magnetic strength of a magnet
   - magnets made by electrical means
   - area surrounding a magnet
   - ends where the magnetic effects are greatest
   - magnetism that remains once the magnetizing force is removed
   - retains its magnetism for a long period of time

- material that has the property to attract iron
- represents the forces in a magnetic field
- materials that are attracted by magnets
- lose most of their magnetism once the magnetizing force is removed
- materials not attracted by magnets

2. Classify each of the following in terms of magnetic or non-magnetic material:
   (a) brass
   (b) iron
   (c) steel
   (d) copper
   (e) aluminum

3. State the Law of Magnetic Poles.

4. State the relationship between the distance between two unlike poles and the amount of magnetic attraction between them.

5. Sketch the magnetic field patterns for:
   (a) a bar magnet
   (b) two unlike poles
   (c) two like poles

6. List five characteristics of magnetic lines of force.

7. Describe how the magnetic effect is explained according to the molecular theory of magnetism.

8. Describe how the magnetic effect is explained according to the electron theory of magnetism.

9. Explain how instruments are shielded against stray magnetic fields.

15

ELECTROMAGNETISM

OBJECTIVES

Upon completion of this unit you will be able to:

- Correctly apply the conductor and coil left-hand rules.
- Carry out experiments to prove the conductor and coil left-hand rules.
- Sketch common electromagnetic field patterns.
- State the factors that determine the strength of an electromagnet.
- Explain Ohm's Law for the magnetic circuit.
- Describe the basic operation of practical electromagnetic devices.

15-1 Magnetic Field around a Current-Carrying Conductor

Whenever electrons flow through a conductor, a magnetic field is created around the conductor (Figure 15-1). If DC current flows, the magnetic field will act in one direction, either clockwise or counter-clockwise, around the conductor. An AC current flow will produce a magnetic field that varies in direction with the direction of the electron flow.

FIGURE 15-1 MAGNETIC FIELD AROUND A CURRENT-CARRYING CONDUCTOR

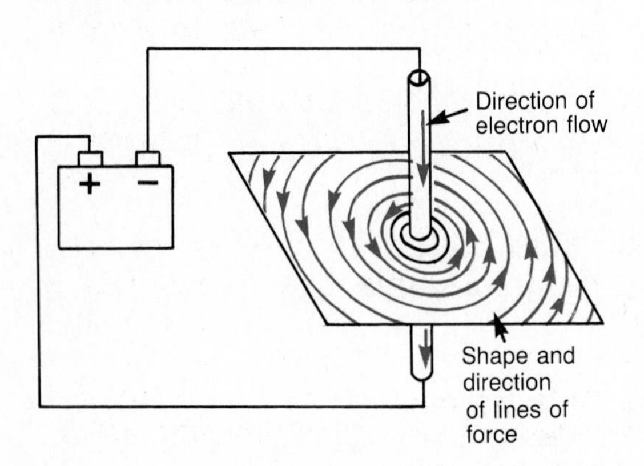

FIGURE 15-2 USING A COMPASS TO TRACE THE MAGNETIC FIELD

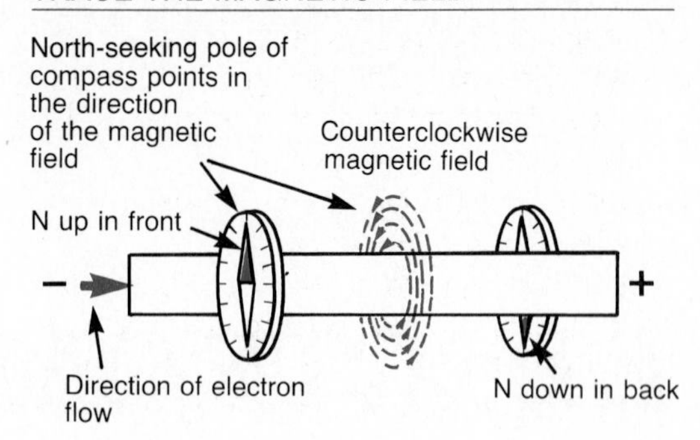

The strength of the magnetic field around a single conductor is usually weak and therefore goes undetected. A compass can be used to reveal both the presence and direction of this magnetic field (Figure 15-2). When the compass is brought close to a conductor carrying DC current, the north-seeking pole of the compass needle will point in the direction the magnetic lines of force are travelling. As the compass is rotated around the conductor a definite circular pattern will be indicated.

FIGURE 15-3 USING IRON FILINGS TO DETECT THE PRESENCE OF A MAGNETIC FIELD

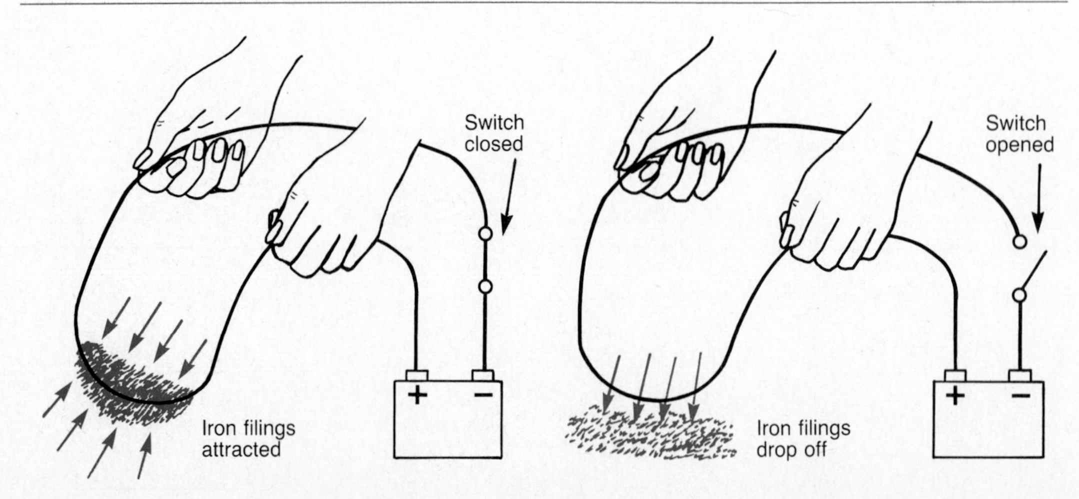

The amount of current flowing through a single conductor determines the strength of the magnetic field produced around the conductor. The greater the current flow, the stronger the magnetic field produced. Currents of two or three amperes can be produced by momentarily shorting a single piece of wire across an ordinary "D" cell. The presence of the magnetic field about the shorted conductor can be detected by dipping the wire into a pile of iron filings (Figure 15-3). The filings will be attracted to the wire and cling to it as long as a complete circuit is maintained to create electron flow.

FIGURE 15-4 LEFT-HAND CONDUCTOR RULE

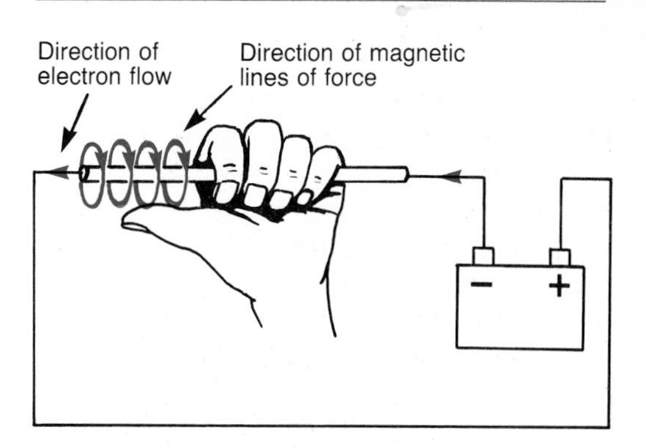

15-2 Left-Hand Conductor Rule

A definite relationship exists between the direction of current flow through a conductor and the direction of the magnetic field created. A simple rule has been established for determining the direction of the magnetic field when the direction of the electron flow is known (Figure 15-4). This rule is known as the **left-hand conductor rule** and uses electron flow from negative to positive for the current direction. The rule is stated as follows: *Grasp the conductor in the left hand, with the thumb pointing in the direction of the electron flow and the fingers will point in the direction of the magnetic lines of force.* Using this rule, if either the direction of the lines of force or the current is known, the other can be determined.

An end view of the wire is often used to simplify the drawing of a conductor that is carrying current (Figure 15-5). The end of the wire is represented by a circle. Current flow into the conductor is represented by a cross, and current flow out of the conductor is represented by a dot.

FIGURE 15-5 END VIEW OF CONDUCTOR AND MAGNETIC FIELD

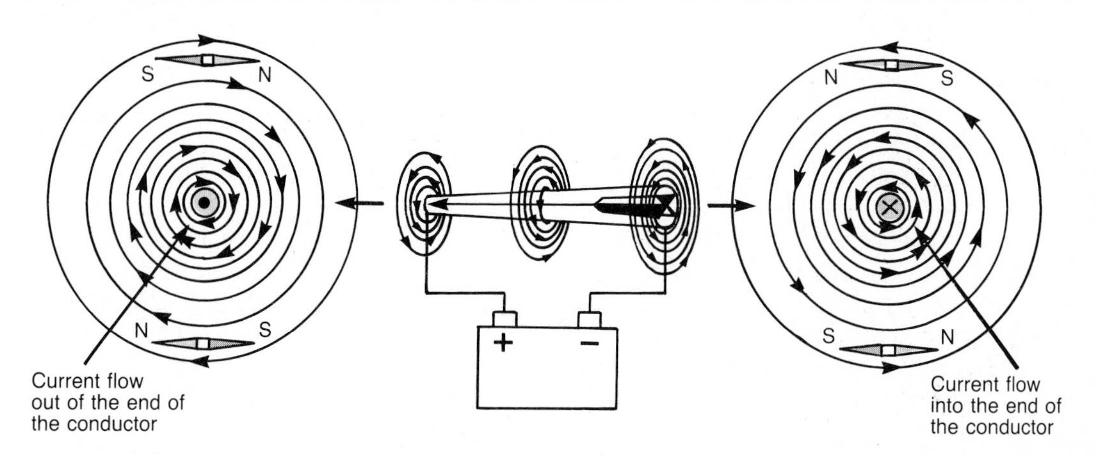

15-3 Magnetic Field about Parallel Conductors

The resulting magnetic field produced by current flow in two adjacent conductors tends to cause the attraction or repulsion of the two conductors. If the two parallel conductors are carrying current in opposite directions, the direction of the magnetic field is *clockwise* around the one conductor and *counter-clockwise* around the other (Figure 15-6). This sets up a repelling action between the two individual magnetic fields and the conductors would tend to move apart.

FIGURE 15-6 PARALLEL CONDUCTORS WITH CURRENT FLOWS IN OPPOSITE DIRECTIONS

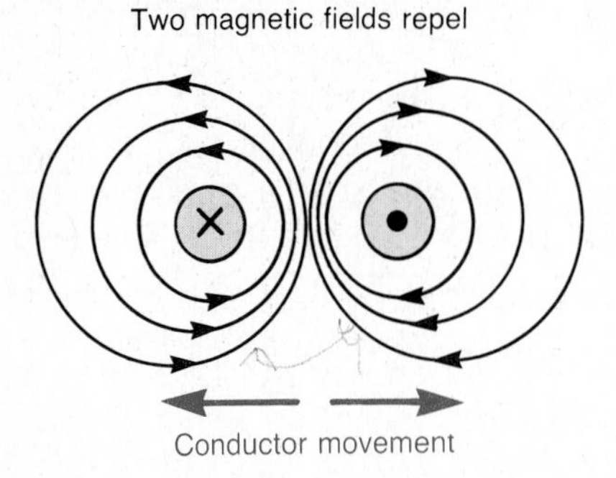

When two parallel conductors are carrying current in the same direction, the direction of the magnetic field is the same around each field (Figure 15-7). The magnetic lines of force between the conductors oppose each other leaving essentially no magnetic field in this area. At the top and bottom of the conductors, the lines of force act in the *same direction* and link together and act around *both* conductors. As a result, the two conductors will tend to move towards each other. The two conductors under this condition will create a magnetic field equivalent to one conductor carrying twice the current.

FIGURE 15-7 PARALLEL CONDUCTORS WITH CURRENT FLOWS IN THE SAME DIRECTION

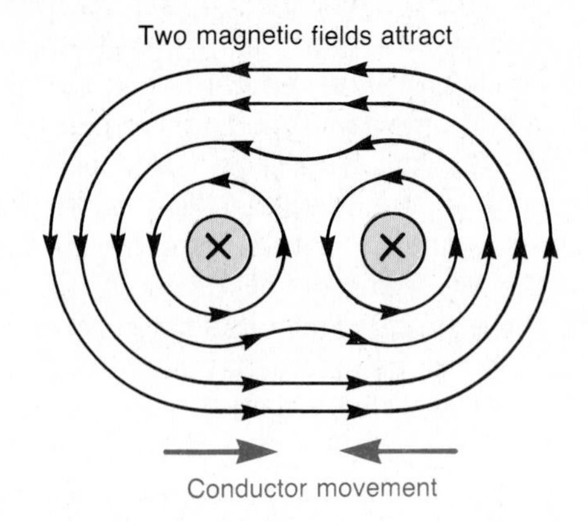

The magnetic force between parallel conductors must be taken into consideration when designing large pieces of electrical equipment that handle very high current flows. In this situation, increased stress can result in damage to the conductors if they are not properly secured.

15-4 Magnetic Field about a Coil

As mentioned previously, two conductors lying alongside each other carrying currents in the same direction create a magnetic field twice as strong as that of the single conductor. If we take a single piece of wire and wind it into a number of loops to form a coil we create the equivalent of several parallel conductors carrying current in the same direction (Figure 15-8). The total resulting magnetic field is the sum of all the single loop magnetic fields. A coil so formed will have a magnetic field pattern similar to that of a bar magnet with a definite North and South pole.

FIGURE 15-8 MAGNETIC FIELD PRODUCED BY A CURRENT-CARRYING COIL

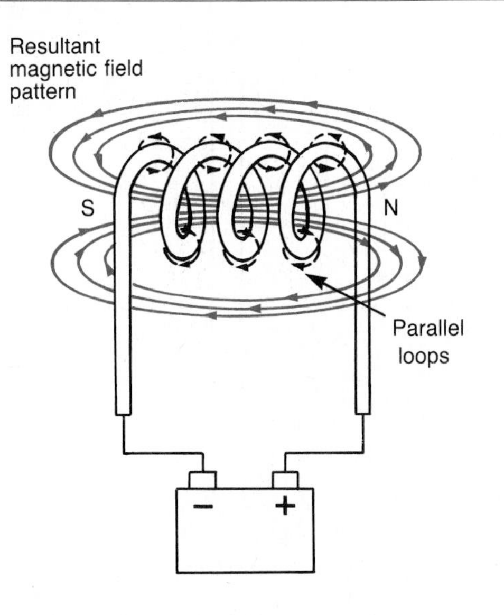

15-5 Left-Hand Coil Rule

A definite relationship exists between the direction of current flow through a coil, the direction in which the wire is wound to form the coil and the location of the north and south poles (Figure 15-9). The *Left-Hand Coil Rule* has been established to determine any one of the three factors when the other two are known. The current flow direction used is electron flow from negative to positive. The rule is stated as follows: *If the coil is grasped in the left hand with fingers pointing in the direction of the current flow, the thumb will point in the direction of the north pole of the magnet.*

15-6 The Electromagnet

When a coil is wound over a core of magnetic material such as soft iron, the device becomes a practical electromagnet (Figure 15-10). The strength of the magnetic field is greatly increased by adding the iron ore. This increase in magnetic strength is the result of magnetism *induced* into the core. When current flows through the coil the core becomes magnetized by induction. The magnetic lines of force produced by the magnetized core align themselves with those of the coil to produce a much stronger mag-

FIGURE 15-9 LEFT-HAND COIL RULE

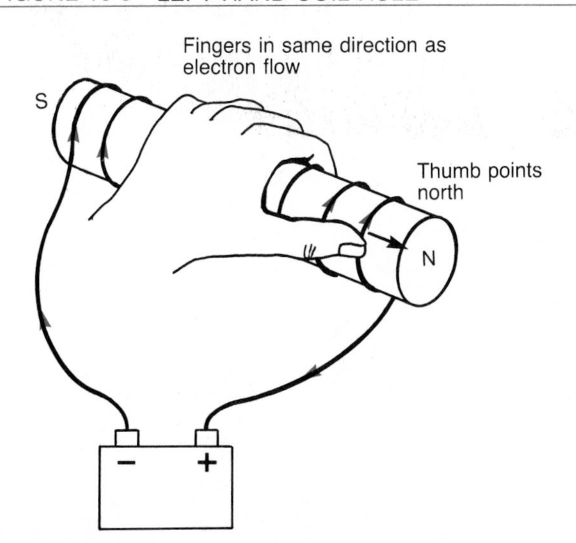

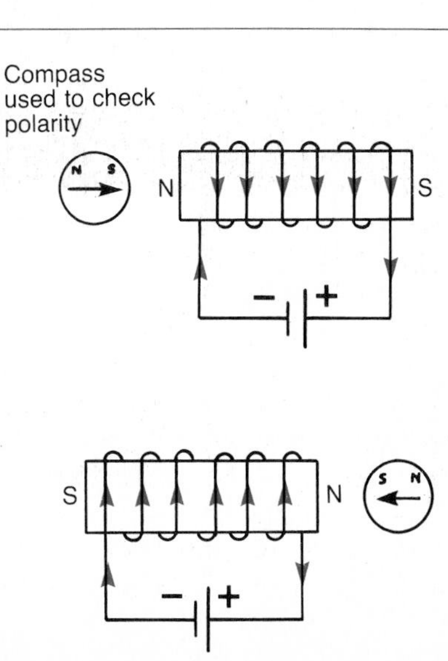

FIGURE 15-10 BASIC ELECTROMAGNET

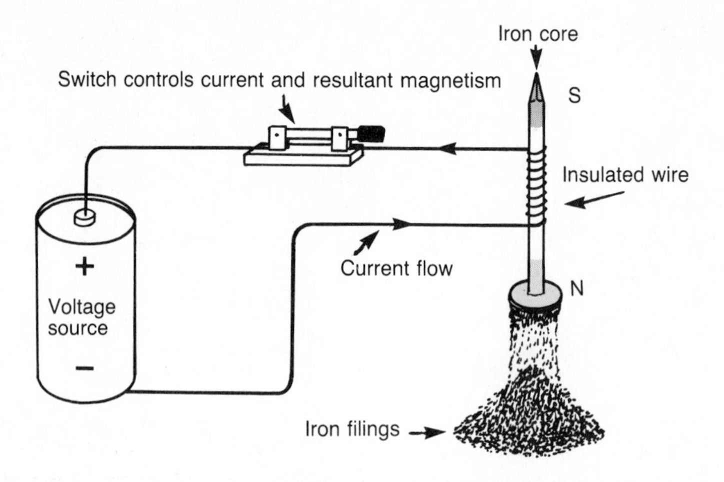

netic field. Once the current stops flowing in the coil, both the coil and the soft iron core lose their magnetism.

In addition to the type of core used, the number of turns on the coil and the amount of current flowing through each turn determines the strength of the electromagnet (Figure 15-11). Current is the most important electrical control we have over the electromagnet. By wiring a switch in series with the electromagnet and voltage source, we can switch the magnetism on and off. This

FIGURE 15-11 FACTORS WHICH DETERMINE THE STRENGTH OF AN ELECTROMAGNET

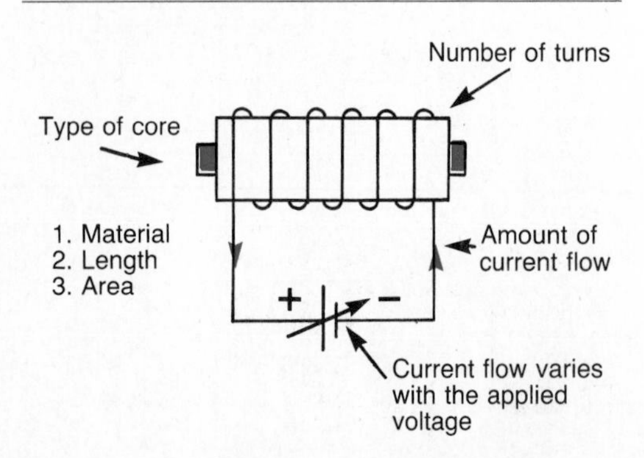

is an important feature in many magnetically operated electrical devices.

15-7 The Magnetic Circuit

The magnetic circuit is similar to the electrical circuit. Basically, a magnetic circuit is a closed-loop path for magnetic lines of force, much like an electric circuit is a closed-loop path for the flow of electrons. In the electrical circuit, electrons travel from the negative to the positive terminal of the voltage source (Figure 15-12A). In the magnetic circuit, lines of force travel from the North pole to the South pole of the electromagnet (Figure 15-12B). The rate of flow of electrons in the electrical circuit is called *current* (I) and is measured in *amperes* (A). The total number of lines of force in the magnetic circuit is called *magnetic flux* (Φ). The SI unit used to measure magnetic flux is the *weber* (Wb).

Just as an electric current is the result of an *electromotive force* (emf) or voltage acting on the electric circuit, so a magnetic flux is produced by a *magnetomotive force* (mmf)

FIGURE 15-12 ELECTRIC CURRENT AND MAGNETIC FLUX

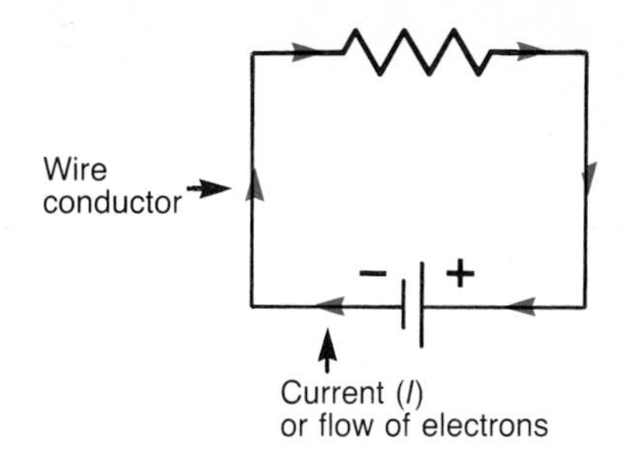

A. Electric circuit

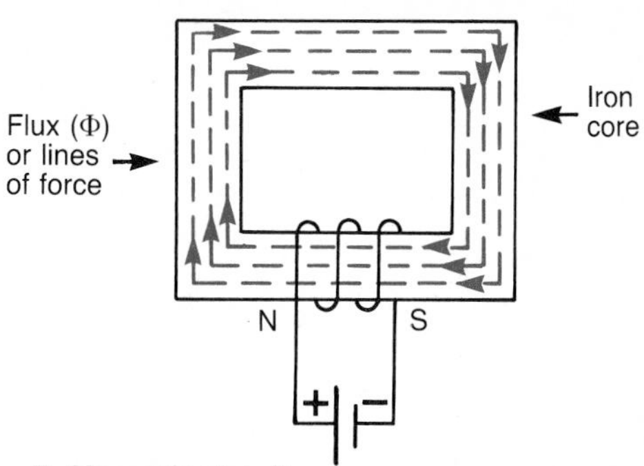

B. Magnetic circuit

FIGURE 15-13 MAGNETOMOTIVE FORCE

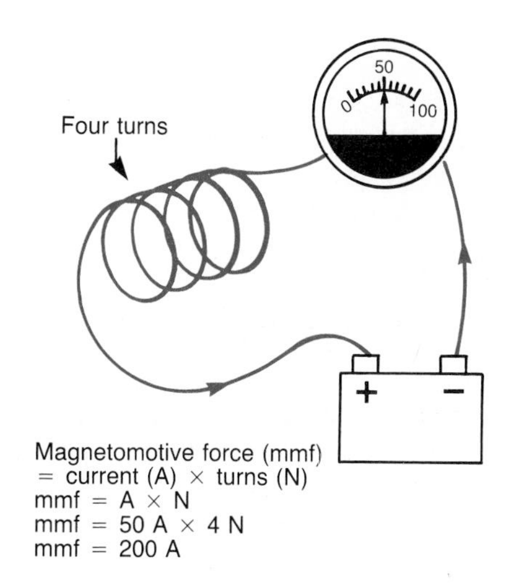

acting upon the magnetic circuit (Figure 15-13). The magnetomotive force of a current-carrying coil is the product of the current in amperes (A) and the number of turns (N) on the coil. The approved SI unit of mmf is the ampere (A); however, the more descriptive term *ampere-turns* is sometimes used.

The magnetic circuit counterpart of resistance in an electrical circuit is called *reluctance* (*R*) (Figure 15-14). Therefore, reluctance is a measure of the opposition offered by a magnetic circuit to the setting up of flux, just as resistance is the opposition to current flow in an electric circuit. The reluctance of a magnetic circuit depends upon the type of material(s) used in the circuit, its

FIGURE 15-14 RESISTANCE AND RELUCTANCE

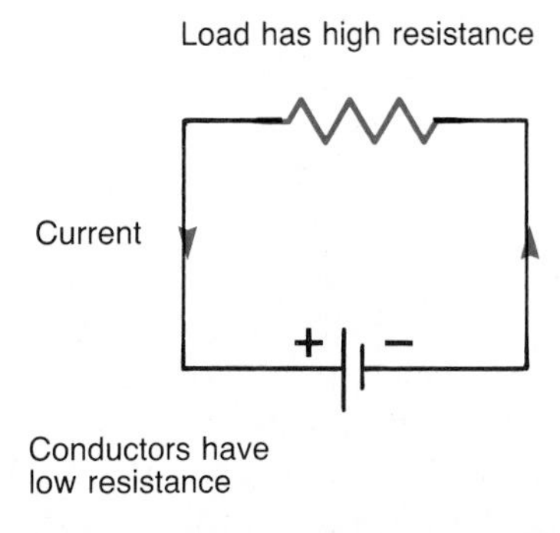

A. Electrical resistance

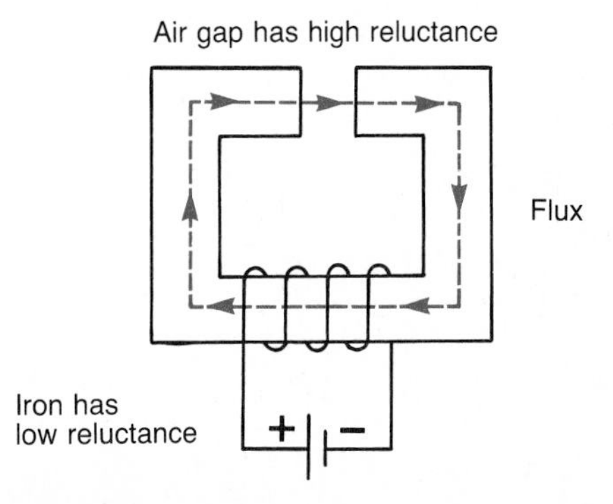

B. Magnetic reluctance

FIGURE 15-15 OHM'S LAW FOR THE ELECTRIC AND MAGNETIC CIRCUIT

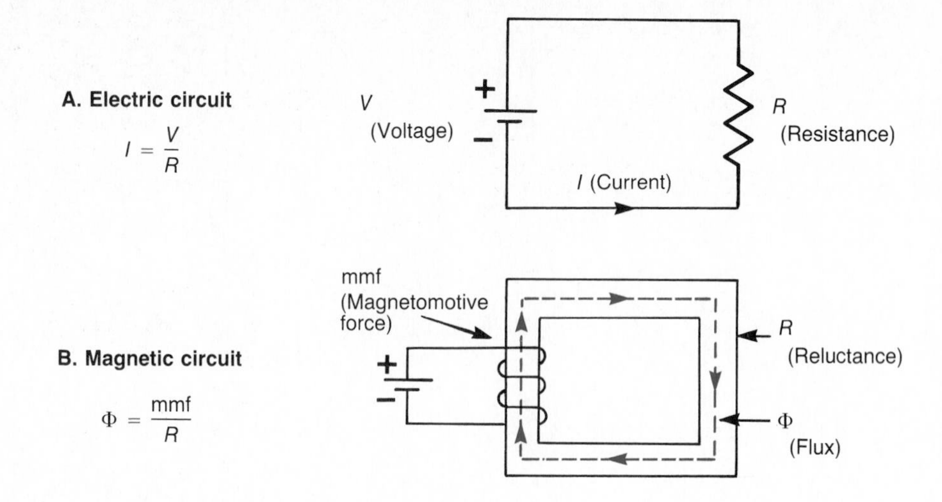

length and cross-sectional area. The SI unit used to measure reluctance is the *reciprocal henry*.

The similarity between magnetic and electric circuits extends to Ohm's Law (Figure 15-15A, B). The same relationship exists between magnetomotive force, flux and reluctance as between voltage, current and resistance. Ohm's Law for the magnetic circuit states that the flux produced by a magnetic circuit is directly proportional to the magnetomotive force and inversely proportional to the reluctance.

Calculation and measurement of voltage, current and resistance in the electrical circuit is relatively easy to do and very useful to the electrician or electronic technician. In practical situations, the same is not true for the magnetic circuit. Although meters do exist for measuring the different quantities of the magnetic circuit, they are most often used in the lab, not on the job.

15-8 Uses for Electromagnets

Electromagnets can be made much more powerful than permanent magnets. In addition, the strength of the electromagnet can be easily controlled from zero to maximum by controlling the current flowing through the coil. For these reasons electromagnets have many more practical applications than do permanent magnets.

Large industrial electromagnets are used with cranes for moving scrap iron. This type of electromagnet has the capability of lifting heavy loads of magnetic scrap metal (Figure 15-16). Lift and drop control is easily accomplished by the connection and disconnection of voltage to the electromagnet.

Most motors and generators make use of electromagnets. In these machines the strength of the electromagnet is varied to change the generated voltage or the speed of the motor. In a typical generator circuit,

FIGURE 15-16 LIFTING MAGNET CIRCUIT

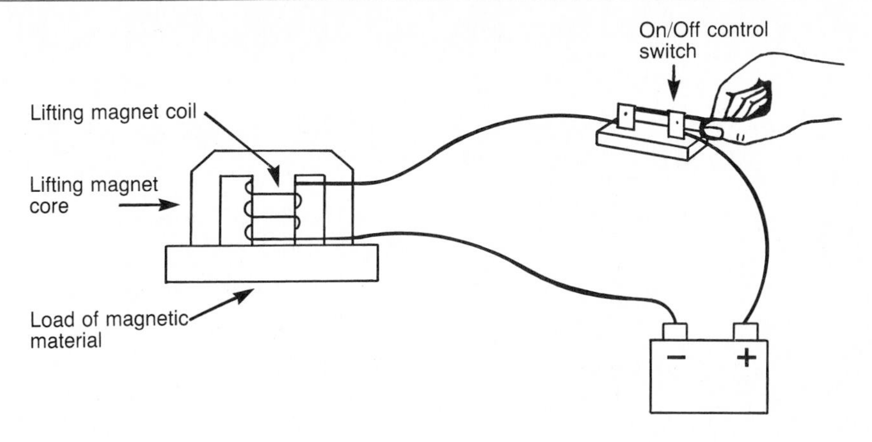

the current flow through the field coils is adjusted by means of a variable resistor or *rheostat* connected in series with the coils and DC voltage source (Figure 15-17). Varying the current in turn varies the strength of the magnetic field.

A *solenoid* is an electromagnet with a moveable iron core or plunger (Figure 15-18). When power is applied to the electromagnet the plunger is pulled into the coil. These electromagnets can be used to open and lock doors electrically. They are also used to control water flow in appliances such as automatic dishwashers and washing machines.

FIGURE 15-17 GENERATOR MAGNETIC FIELD CIRCUIT

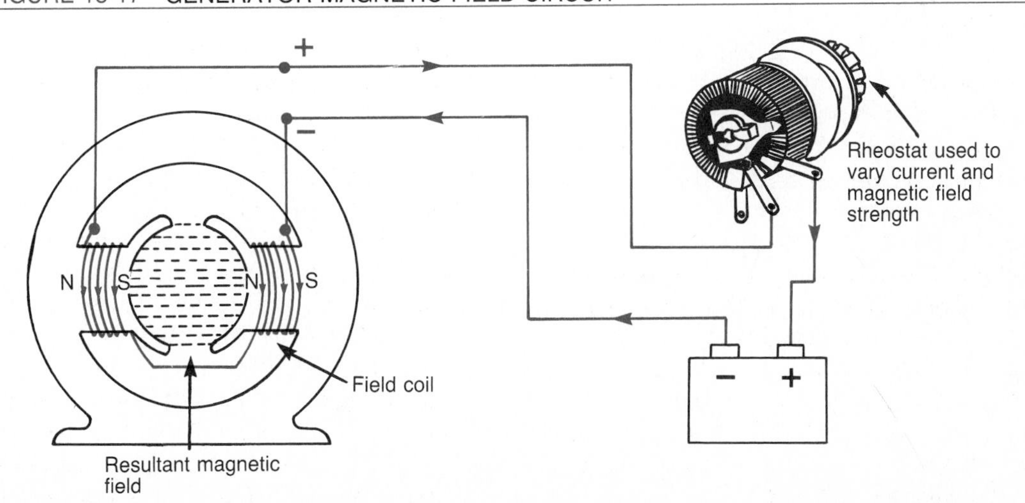

FIGURE 15-18 SOLENOID CONTROL CIRCUIT

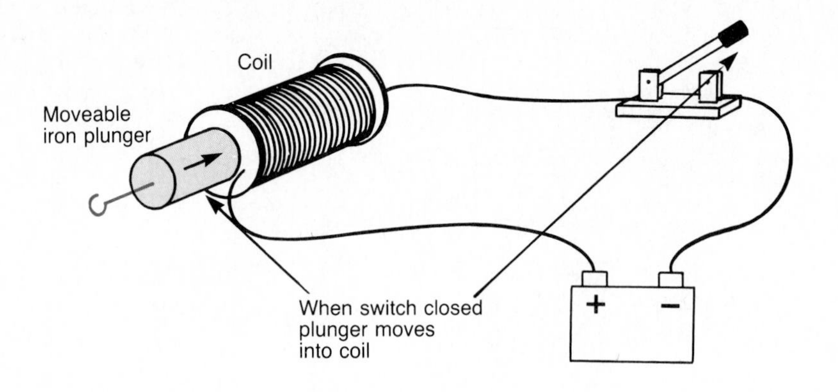

FIGURE 15-19 TRANSFORMER CIRCUIT

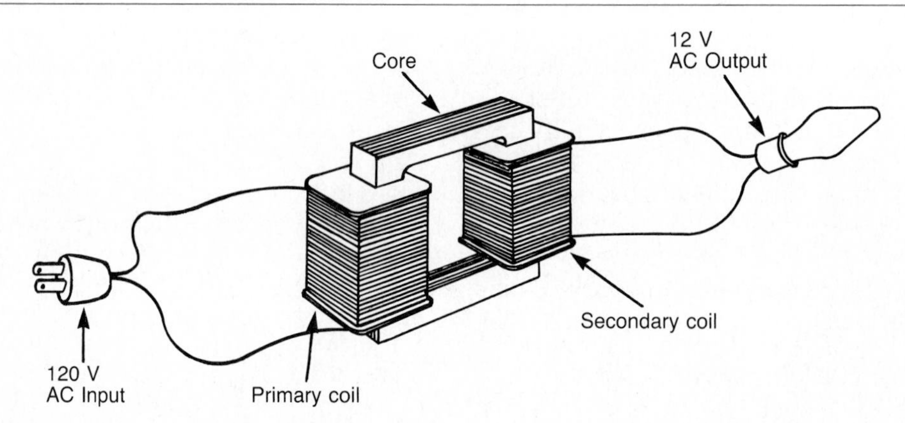

Transformers are electrical devices that are used to raise or lower AC voltages (Figure 15-19). This device uses two electromagnetic coils to transform or change AC voltage levels.

The simple electric bell uses two small coils wound on a core to form a horseshoe magnet (Figure 15-20). Current passing through the coils magnetizes the core, thus attracting the moveable armature to it causing the hammer to strike the bell.

15-9 Practical Assignments

1. (a) Apply the *left-hand single conductor rule* to determine the direction of the magnetic lines of force around a current carrying conductor. Use a DC hand-generator as the voltage source and a compass to trace the direction of the lines of force.
   (b) Illustrate your results by means of a simple diagram.

FIGURE 15-20 VIBRATING BELL CIRCUIT

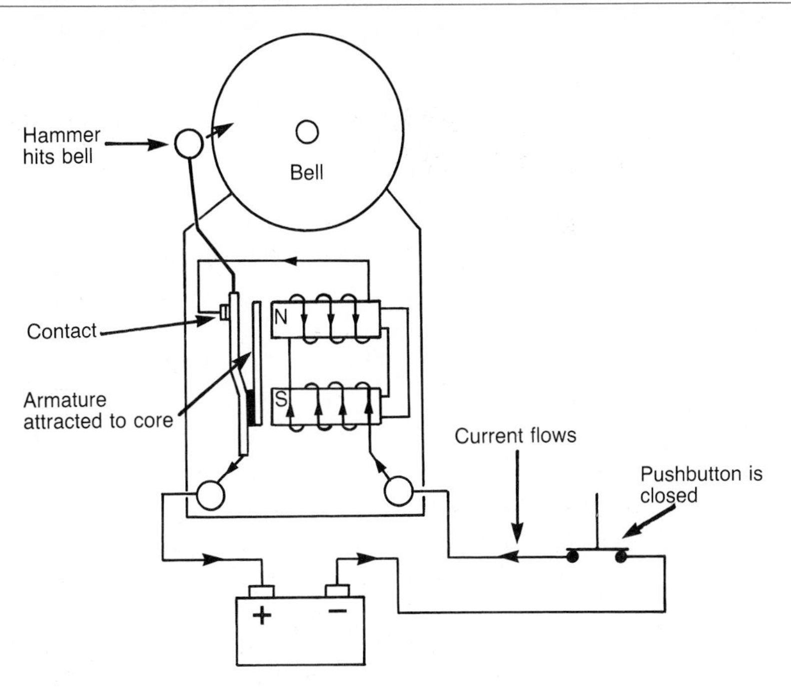

2. (a) Make a simple electromagnet by winding a number of turns of insulated copper wire on a soft iron core.
   (b) Using the DC hand-generator as the voltage source, demonstrate its magnetic capabilities by testing its ability to attract paper clips.
   (c) Demonstrate how the amount of current flow can be used to control the strength of the electromagnet by varying the speed at which you crank the generator.
   (d) Apply the left-hand coil rule to determine the North and South poles of the electromagnet. Verify with a compass. Illustrate your results by means of a simple diagram.
3. (a) Demonstrate the electromechanical action of a low voltage DC solenoid.
   (b) Vary the voltage and current to this solenoid from zero to its rated value and make note of how this affects the amount of magnetic pull on the plunger.
4. Examine the construction of the electromagnets in the following devices:
   (a) motor
   (b) generator
   (c) transformer
   (d) speaker
   (e) meter movement
   (f) electric bell

15-10 Self Evaluation Test

1. What is the relationship between electricity and magnetism?
2. Apply the left-hand conductor and coil rule to each of the following problems:

(a) Redraw, indicating the direction and shape of the lines of force.

(b) Redraw, indicating the direction and shape of the lines of force

(c) Redraw, indicating the direction of the electron flow, the North and South poles, and the shape and direction of the lines of force.

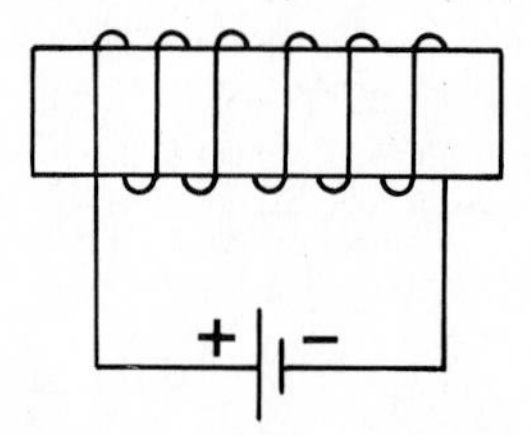

(d) Redraw, indicating the direction of electron flow and the North and South poles.

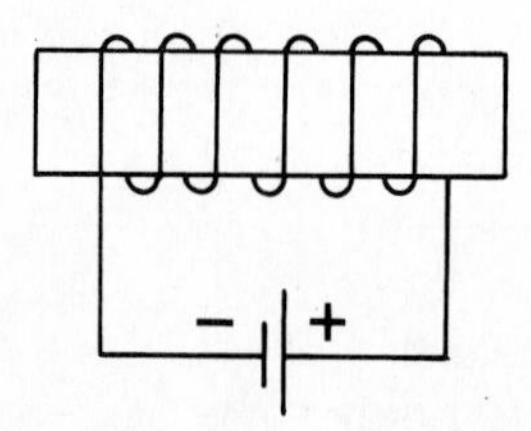

(e) Redraw, indicating the direction of electron flow and the North and South poles.

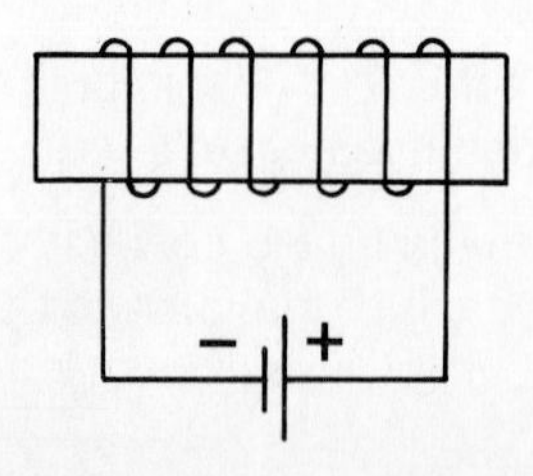

(f) Redraw, indicating the direction and shape of the resultant lines of force, and the direction in which the conductors would tend to move.

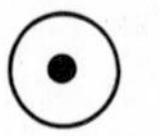 

(g) Redraw, indicating the direction and shape of lines of force, and the direction in which the conductors would tend to move.

3. (a) Redraw the electrical generator field circuit in Figure 15-21 with all coils properly connected in series to produce the poles indicated.
   (b) On your diagram indicate the pattern of the magnetic lines of force.

FIGURE 15-21 SELF EVALUATION TEST 3

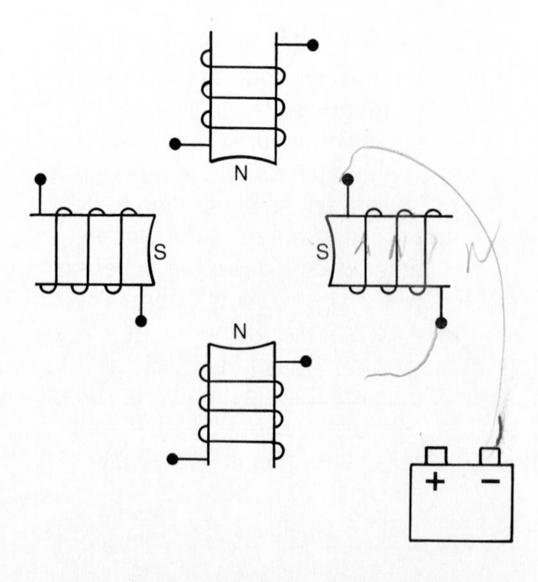

4. What two factors determine the North and South poles of an electromagnet?

5. What are the three main factors that determine the strength of an electromagnet?

6. Explain each of the following terms as they apply to the magnetic circuit:
   (a) Magnetomotive Force
   (b) Magnetic Flux
   (c) Reluctance

7. State Ohm's Law for the magnetic circuit.

8. What two methods can be used to show the presence of a magnetic field around a conductor carrying a DC current?

9. Name two advantages that electromagnets have over permanent magnets.

10. (a) Describe the construction of a solenoid.
    (b) State one practical application for this device.

11. (a) How is *lift* and *drop* control of a crane-operated lifting magnet accomplished?
    (b) Why is this type of crane application useless for moving scrap copper material?

12. How is the strength of the magnetic field in a typical generator controlled?

13. Redraw, indicating the correct polarity of the DC voltage source.

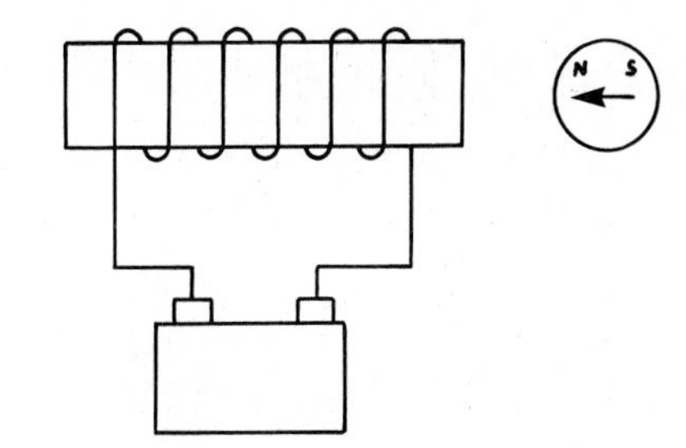

16

DURACELL
PILE ALCALINE
DURACELL
ALKALINE BATTERY
DURACELL
ALKALINE BATTERY
DURACELL
PILE ALCALINE
DURACELL
PILE ALCALINE

BATTERIES

OBJECTIVES

Upon completion of this unit you will be able to:

- Explain the ways in which cells and batteries are rated.
- Compare the advantages and disadvantages of carbon-zinc, alkaline and mercury cells.
- Design a suitable connection of cells to meet a given battery requirement.
- Explain how cells and batteries are tested.
- Compare the operation of a primary and secondary cell.
- Test primary dry cells.
- Properly charge a rechargeable battery.

16-1 The Voltaic Cell

A voltaic cell is the basic device for converting chemical energy into electrical energy (Figure 16-1). It contsists of two different metal plates immersed in a solution. The metal plates are called *electrodes* and the solution is called the *electrolyte*.

One type of voltaic cell uses copper and zinc as the two electrodes and sulfuric acid as the electrolyte. When placed together, a chemical reaction occurs between the electrodes and the sulfuric acid. This reaction produces a negative charge on the zinc (surplus of electrons) and a positive charge on the copper (deficiency of electrons). If an external circuit is connected across the two electrodes, electrons will flow from the negative zinc electrode to the positive copper electrode. The electric current will flow as long as the chemical action continues. In this type of cell the zinc electrode is eventually consumed as part of the chemical reaction.

Frequently, a single cell is incorrectly called a battery. By strict definition, a battery consists of two or more cells connected together. These cells are usually enclosed in one case.

FIGURE 16-1 VOLTAIC CELL

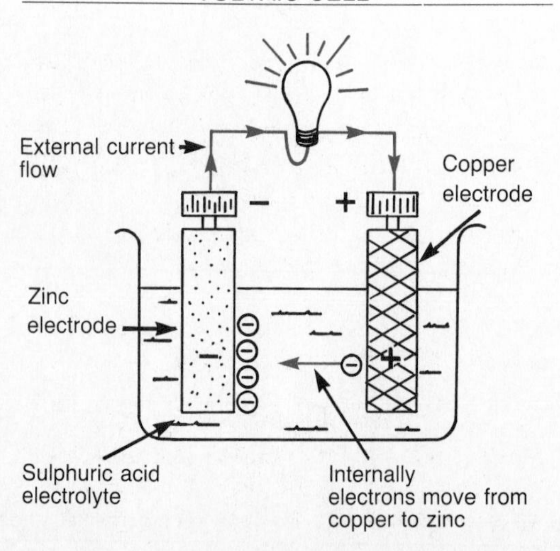

Primary cell is the name given to any cell in which an electrode is consumed gradually during normal use. It cannot be restored to its original useful state by electrical recharging. A typical primary cell is the kind used in flashlights.

The **secondary cell** has the ability to be used over a longer period of time. Its chemical energy is replenished by recharging it at specific intervals. A typical example of the use of a secondary cell is the battery used in automobiles.

16-2 Cell and Battery Ratings

Voltage All cells and batteries are rated for their normal output voltage. The voltage output from a single cell (Figure 16-2A) is between 1 and 2 V and depends on the material used for the electrodes and the type of electrolyte. Batteries (Figure 16-2B) with higher output voltages contain cells connected in series.

FIGURE 16-2 SINGLE CELL AND BATTERY

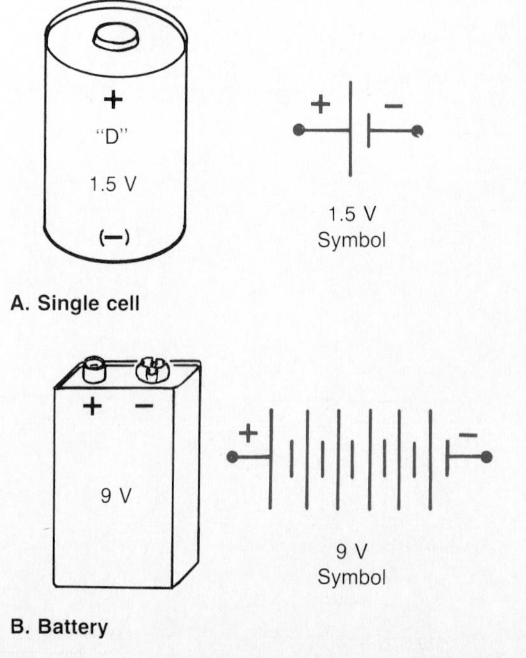

Energy Capacity The battery energy capacity can be rated in *ampere hours* (A·h). This capacity is determined by multiplying the amperes of current the battery will deliver by the number of hours the battery will deliver it. For example, if a battery is capable of delivering 4 A for 8 h, it is said to have a ($4 \times 8 = 32$) 32 A·h (ampere hour) capacity.

The SI unit used to measure the battery energy capacity is the joule (J). It is easy to convert from ampere hours (A·h) to joules if the output voltage of the battery is known. To convert to joules simply multiply the battery A·h rating by its output voltage value and the constant 3 600.

Example
Suppose a 12 V battery is capable of delivering 2 A of current for 8 h (Figure 16-3). The battery energy capacity is then:

$$\text{A·h capacity} = \text{Amperes} \times \text{hours} = (2\text{ A})(8\text{ h}) = 16\text{ A·h}$$

$$\text{Energy capacity} = \text{A·h} \times \text{V} \times 3\,600 = (16\text{ A·h})(12\text{ V})(3\,600) = 691\,200\text{ J} = 691.2\text{ kJ}$$

FIGURE 16-3 BATTERY CAPACITY EXAMPLE

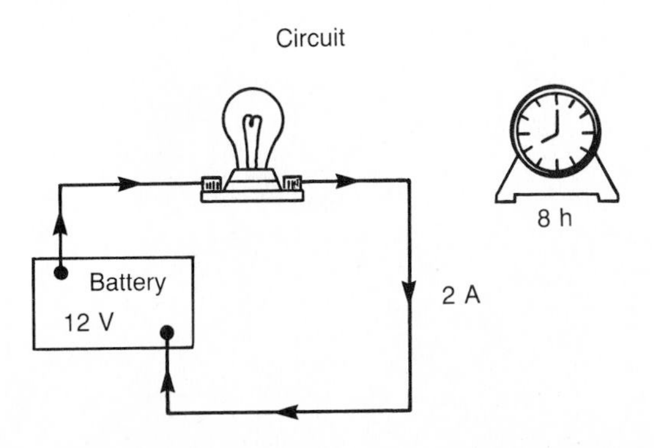

FIGURE 16-4 SIZE DETERMINES BATTERY CAPACITY

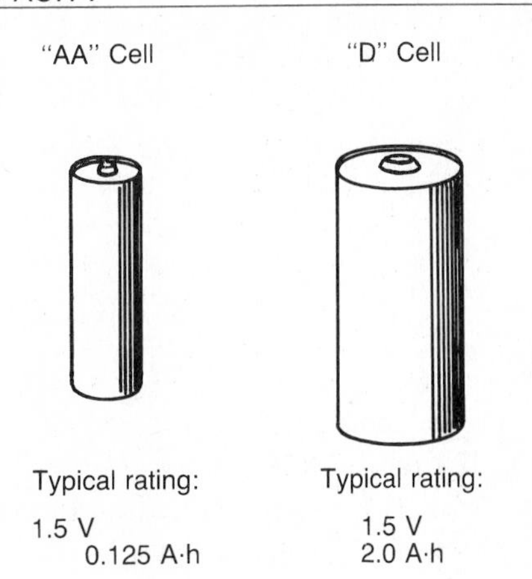

For any given cell or battery type, the energy capacity of the battery is proportional to its physical size (Figure 16-4).

Shelf Life Batteries are also rated for shelf life in years. Even if a cell is not being used, an internal chemical reaction takes place within the cell. Technically called *local action*, this process is at work at all times and will eventually render the cell useless. Shelf life is defined as the time in years a stored battery will produce at least 75% of its initial capacity.

Temperature Batteries are most often rated for a specific output capacity at room temperature or 20°C. Operating them above and below this temperature will reduce their rated output. For example, the automobile battery output drops on cold days making it more difficult to turn the engine over.

16-3 Primary Dry Cells

Carbon-Zinc Cells The voltaic cell is also known as a **wet cell** because it uses a liquid solution for the electrolyte. The danger of spilling the liquid electrolyte from this type

FIGURE 16-5 CARBON-ZINC DRY CELL

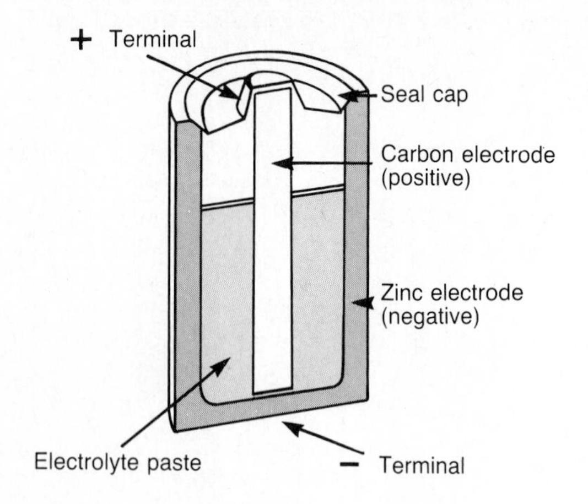

FIGURE 16-6 RELATIVE SIZES OF STANDARD CARBON-ZINC BATTERIES

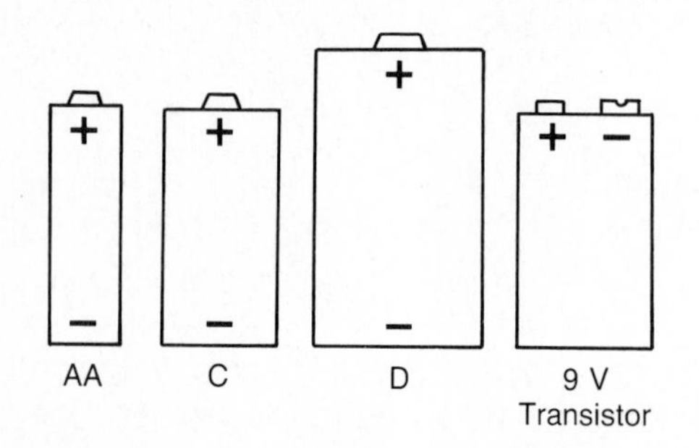

of cell led to the invention of another class of cells called dry cells.

The most common and least expensive type of **dry cell** battery is the *carbon-zinc* type (Figure 16-5). This cell consists of a zinc container which acts as the negative electrode. In the center is a carbon rod which is the positive electrode. The electrolyte takes the form of a moist paste made up of a solution containing ammonium chloride. As with all primary cells, one of the electrodes becomes decomposed as part of the chemical reaction. In this cell the negative zinc container electrode is the one that is used up. As a result, cells left in equipment for long periods of time can rupture, spilling electrolyte and causing damage to the electronic parts.

Carbon-zinc cells are produced in a range of common standard sizes (Figure 16-6). These include 1.5 V AA, C, D cells and 9 V rectangular batteries.

Alkaline Cells Alkaline cells use a zinc container for the negative electrode and a cylinder of potassium hydroxide for the positive electrode. The electrolyte is made up of a solution of potassium hydroxide or alkaline.

Alkaline cells are produced in the same standard sizes as carbon-zinc cells but are more expensive. They have the advantage of being able to supply large currents for a longer period of time. For example, a standard "D" type 1.5 V alkaline cell has a capacity of about 8 A·h compared with about 2 A·h for the carbon-zinc type. A second advantage is that the alkaline cell has a shelf life of about two and one half years as compared to about one year for the carbon-zinc type.

Mercury Cells Mercury cells are most often used in digital watches, calculators, hearing aids and other miniature electronic equipment. They are usually smaller and are shaped differently than the carbon-zinc type (Figure 16-7).

FIGURE 16-7 MERCURY CELL

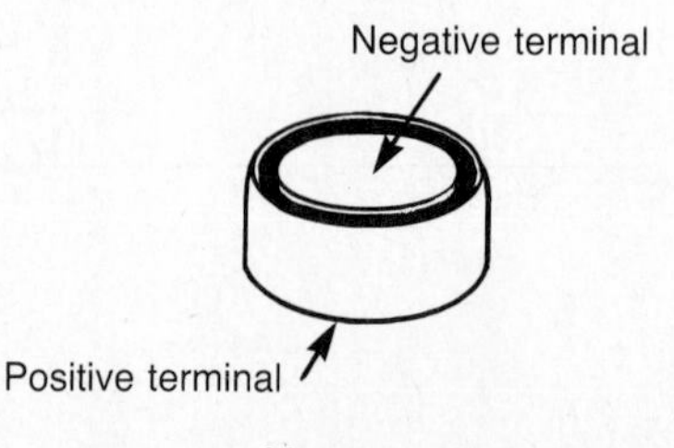

A mercury cell will develop a voltage of 1.34 V by the chemical action between zinc and mercuric oxide. It is expensive to manufacture, but has a decided advantage of producing about 5 times the ampere hour output compared to that of a carbon-zinc cell of the equivalent size. In addition to a longer shelf life (approximately three and one half years) it maintains a constant terminal voltage even as it approaches the end of its usefulness.

16-4 Series and Parallel Cell Connections

Series Connection Often an electronic circuit requires a voltage or current that a single cell is not capable of supplying alone. In this case it is necessary to connect groups of cells in various series and parallel arrangements.

Cells are connected in series by connecting the positive terminal of one cell to the negative terminal of the next cell (Figure 16-8).

Identical cells are connected in series to obtain a higher voltage than is available from a single cell. With this connection of cells the output voltage is equal to the sum of the voltages of all cells. However, the ampere hour (A·h) rating remains equal to that of a single cell.

FIGURE 16-8 SERIES CONNECTION OF CELLS

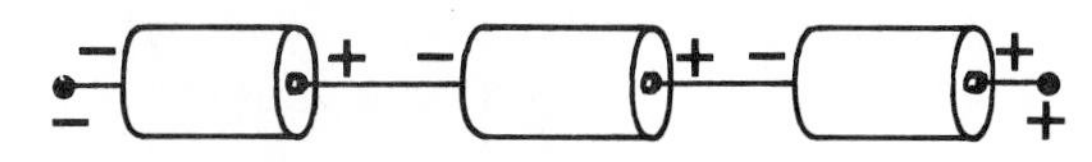

A. Pictorial diagram

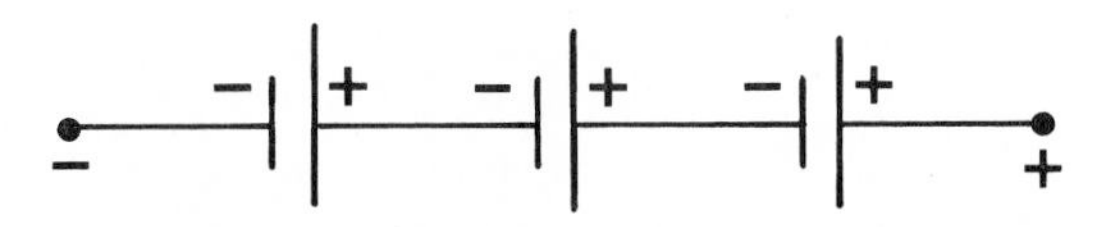

B. Schematic diagram

Example

Suppose three "D" flashlight cells are connected in series (Figure 16-9). Each cell has a rating of 1.5 V and 2 A·h. The voltage and ampere hour rating of this battery would be:

$$\begin{aligned} \text{V Battery} &= \text{V per cell} \times \text{No. of cells} \\ &= (1.5\ \text{V})\ (3) \\ &= 4.5\ \text{V} \end{aligned}$$

$$\begin{aligned} \text{A·h Battery Rating} &= \text{A·h rating of 1 cell} \\ &= 2\ \text{A·h} \end{aligned}$$

FIGURE 16-9 CELLS IN SERIES

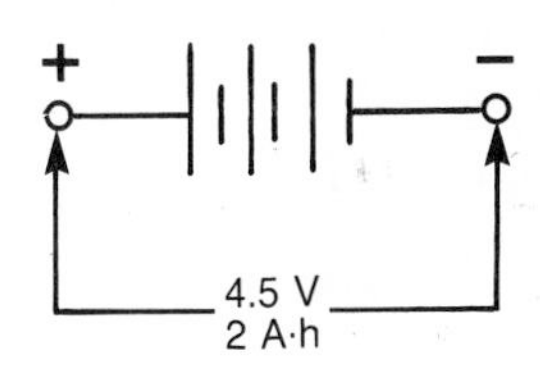

Circuit Schematic

If, by error, one cell connection is reversed in a series group, its voltage will oppose that of the other cells. This will produce a lower than expected battery output voltage.

Example

Suppose that one of the three "D" flashlight cells of the previous example is connected in reverse (Figure 16-10). The output voltage then would be:

$$\begin{aligned} \text{V Battery} &= (1.5\ \text{V}) + (1.5\ \text{V}) - (1.5\ \text{V}) \\ &= (3\ \text{V}) - (1.5\ \text{V}) \\ &= 1.5\ \text{V} \end{aligned}$$

Singer

FIGURE 16-10 SERIES CELL REVERSED

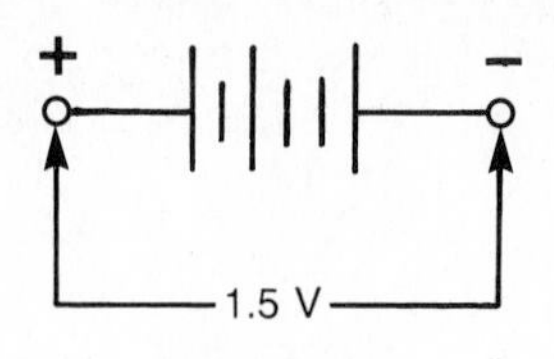

Circuit Schematic

Parallel Connection Cells are connected in parallel by connecting all the positive terminals together and all the negative terminals together (Figure 16-11).

Identical cells are connected in parallel to obtain a higher output current or ampere hour rating. With this connection of cells, the output ampere hour rating is equal to the sum of the ampere hour ratings of all cells. However, the output voltage remains the same as the voltage of a single cell.

FIGURE 16-11 PARALLEL CONNECTION OF CELLS

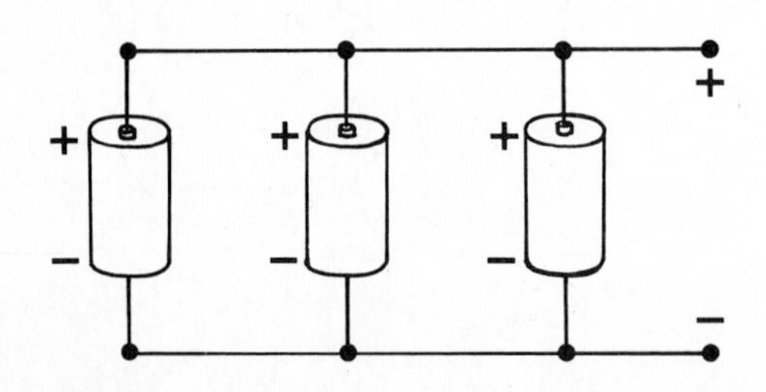

A. Pictorial diagram

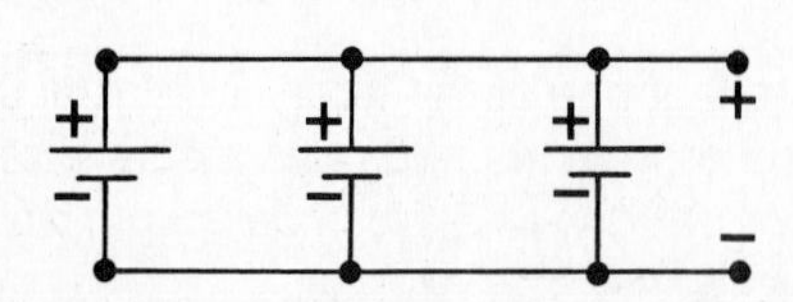

B. Schematic diagram

Example

Suppose four alkaline cells are connected in parallel (Figure 16-12). Each cell has a rating of 1.5 V and 8 A·h. The voltage and ampere hour rating of this battery would be:

V Battery = V rating of 1 cell
= 1.5 V

A·h Battery Rating
= A·h rating per cell × no. of cells
= (8 A·h) (4)
= 32 A·h

FIGURE 16-12 CELLS IN PARALLEL

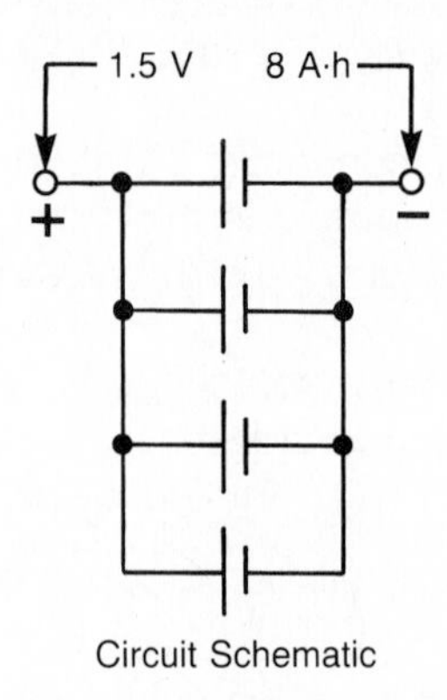

Circuit Schematic

If by error, a cell connection is reversed in a parallel group, it will act as a short circuit. All cells will discharge their energy through this short circuit path. Maximum current will flow and the cells may be permanently damaged.

Series-Parallel Connection Sometimes the requirements of a piece of equipment exceed both voltage and ampere hour rating of a single cell. In this case a series-parallel grouping of cells must be used. The number of cells that must be connected in series is calculated first. Then the number of parallel rows of series connected cells is calculated.

Example

Suppose a battery operated circuit requires 6 V and a capacity of 4 A·h (Figure 16-13). Cells rated at 1.5 V and 2 A·h are available to do the job. The required arrangement of cells would then be:

$$\text{No. of cells in series} = \frac{\text{V required}}{\text{V per cell}} = \frac{(6\ \text{V})}{(1.5\ \text{V})} = 4\ \text{cells}$$

$$\text{No. of parallel rows} = \frac{\text{A·h required}}{\text{A·h per cell}} = \frac{(4\ \text{A·h})}{(2\ \text{A·h})} = 2\ \text{rows}$$

FIGURE 16-13 SERIES-PARALLEL CONNECTION OF CELLS

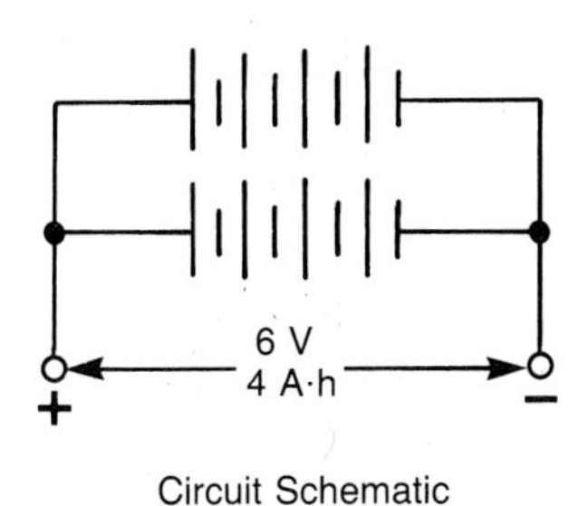

When connecting groups of cells or batteries in parallel, each group must be at the same voltage level. Paralleling two batteries of unequal voltage levels sets up a difference of potential energy between the two. As a result, the higher voltage battery will discharge its current into the other battery until both are at equal voltage value.

16-5 Testing Primary Cells or Batteries

A visual inspection will tell you little about the useful life of a cell or battery unless it has deteriorated to the point where the electrolyte is spilling from the case.

A no-load voltage test of the cell or battery is another poor indication of cell or battery life. This test requires the cell or battery to deliver only a very small amount of current required to operate the voltmeter.

The best method used to check a cell or battery is an in-circuit test of the cell or battery voltage with the normal load connected to it (Figure 16-14). A substantial drop in cell or battery voltage when normal load is applied indicates a bad cell or battery.

FIGURE 16-14 TESTING A BATTERY UNDER LOAD

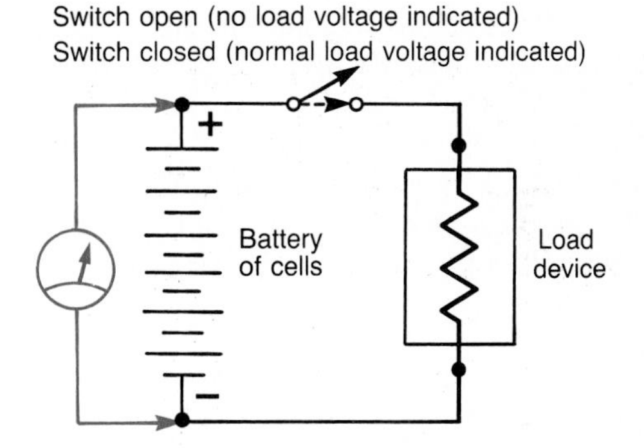

Small cells or batteries can be tested out-of-circuit by calculating the cell's or battery's internal resistance (Figure 16-15). First, the no-load voltage of the cell or battery is measured using a voltmeter. Then, the short circuit current of the cell or battery is measured, using an ammeter of high enough current rating, momentarily connected directly across the cell or battery terminals. The internal resistance is then calculated using Ohm's Law. A high value of internal cell or battery resistance indicates a bad cell or battery.

FIGURE 16-15 CALCULATING CELL INTERNAL RESISTANCE

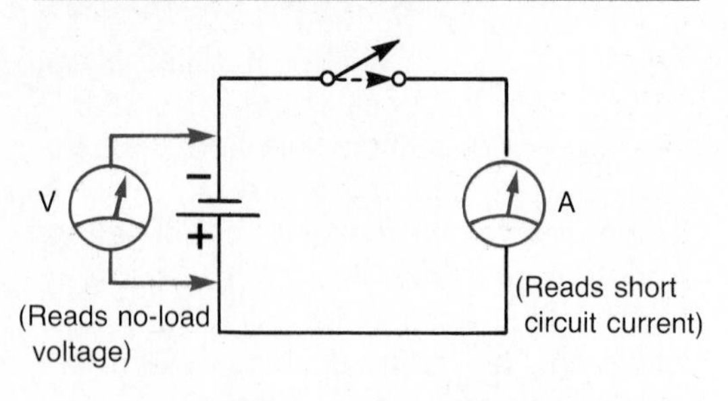

16-6 Lead-Acid Rechargeable Battery

The lead-acid battery is made up of a number of rechargeable or secondary cells (Figure 16-16). Each cell contains two sets of electrode plates and a liquid electrolyte. The large surface area of the plates makes it possible for this battery to deliver heavy surges of current in the 150 A to 500 A range. This feature makes it an ideal battery for starting motors of cars. Each cell produces about 2 V. Six cells connected in series are required to produce the standard 12 V car battery.

Both cell electrodes are basically lead sulphate in an uncharged battery and the electrolyte is a weak solution of sulphuric acid and water. In order for the cell to deliver electrical energy it must have two different electrodes and an active electrolyte. By charging or passing an electric current through the cell, we find that one of the lead sulphate electrodes changes to soft or spongy lead (negative plate) and the other electrode changes to lead peroxide (positive plate) (Figure 16-17). At the same time, we find the electrolyte solution is strengthened and becomes mostly sulphuric acid.

When a load is connected across the charged battery cell, current flows through the load and a reversal of the charging process occurs in the cell. Both electrodes undergo a chemical change and begin to change back to lead sulphate. As part of the same process the amount of sulphuric acid in the electrolyte decreases. When both plates change completely to lead sulphate the current flow to the load stops and the battery is said to be *discharged* (Figure 16-18).

FIGURE 16-16 12 V LEAD-ACID BATTERY

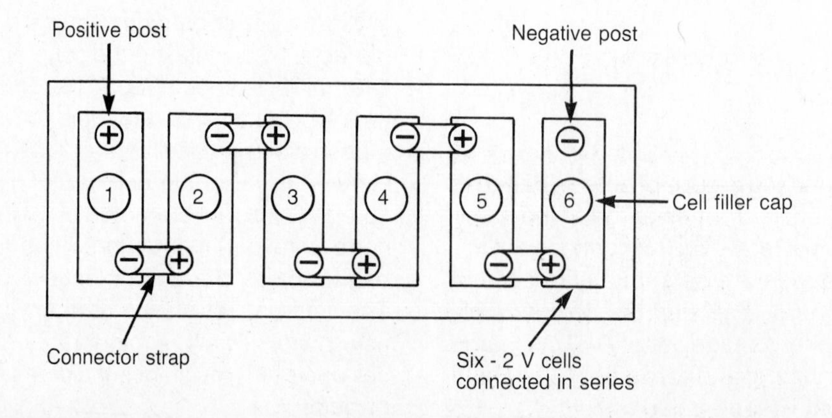

FIGURE 16-17 BATTERY CHARGING CYCLE

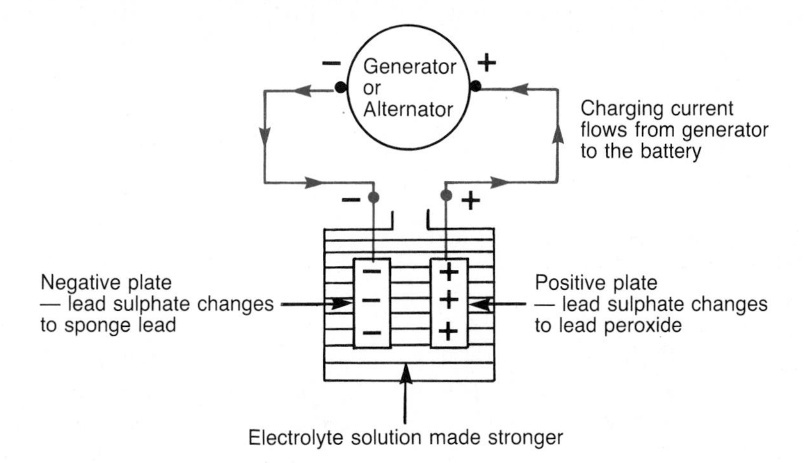

FIGURE 16-18 BATTERY DISCHARGE CYCLE

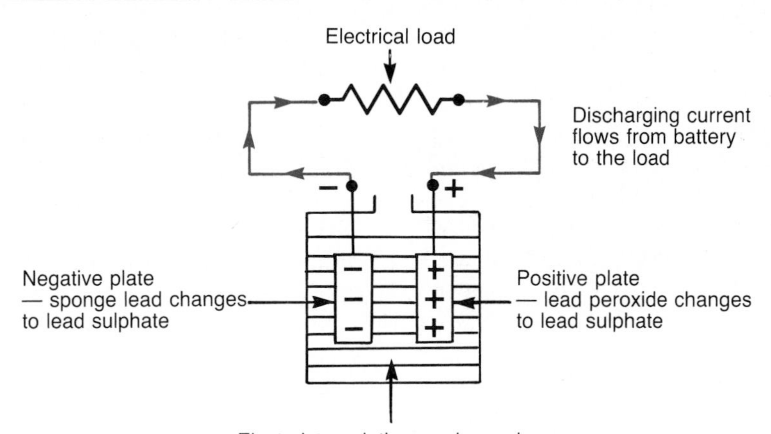

16-7 Testing a Lead-Acid Battery

The state of battery charge can be tested by means of a **battery hydrometer** (Figure 16-19). This instrument measures the relative density of the battery electrolyte. Since the strength of the electrolyte varies directly with the state of charge of each cell, you need only to find what percentage of sulphuric acid remains in each cell electrolyte to determine how much energy is available.

FIGURE 16-19 BATTERY HYDROMETER TESTER

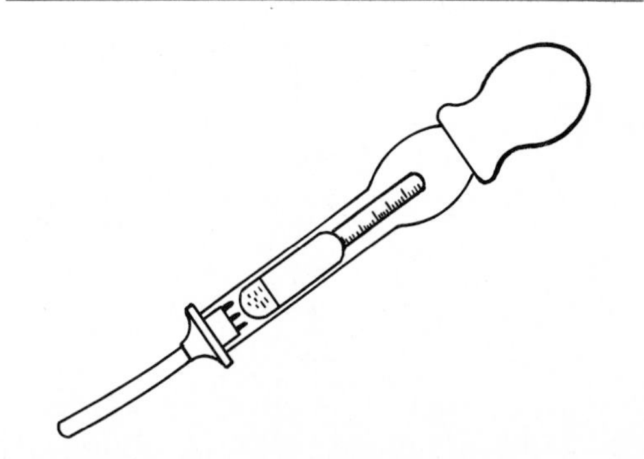

CELL CONDITION	HYDROMETER READING
Full charge	1.26
50% charge	1.20
Discharged	1.15

Voltage tests of lead-acid batteries, like primary cells, should be conducted under load. To make a simple light load voltage test of a car battery check the value of the battery output voltage with and without the headlights on. A maximum load voltage test can be made by metering the battery voltage while operating the starting motor (Figure 16-20). In the case of a 12 V battery, a drop of battery output voltage below 7 V indicates the battery is defective or not fully charged.

FIGURE 16-20 TESTING A BATTERY UNDER LOAD

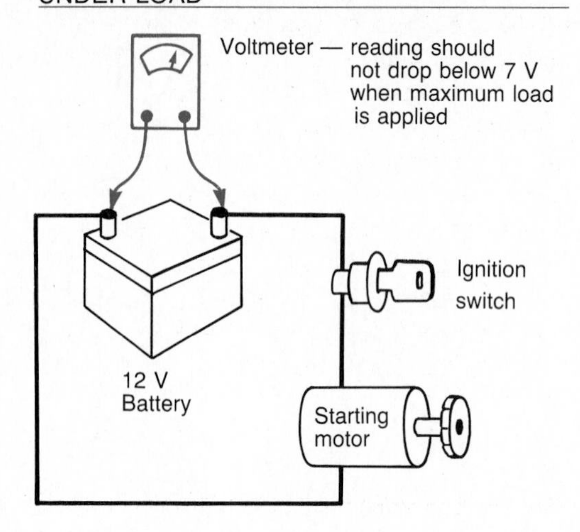

16-8 Nickel-Cadmium Rechargeable Battery

Sealed nickel-cadmium cells are a popular type of smaller rechargeable cell. They use a cadmium container for the negative electrode and a cylinder of nickel for the positive electrode. The electrolyte is made up of a potassium hydroxide paste.

The chemical reaction that occurs within the cell is similar to that of the lead-acid battery. At full charge the output voltage per cell is 1.25 V. When this voltage drops down to 1.1 V, the cell is ready for recharging.

This type of cell is available in the same sizes as the carbon-zinc cell. They are used extensively with portable power tools and electronic equipment (Figure 16-21). Quite often, both the cell and matching charger unit are supplied with the piece of equipment it operates.

FIGURE 16-21 NICKEL-CADMIUM RECHARGEABLE CELL SYSTEM

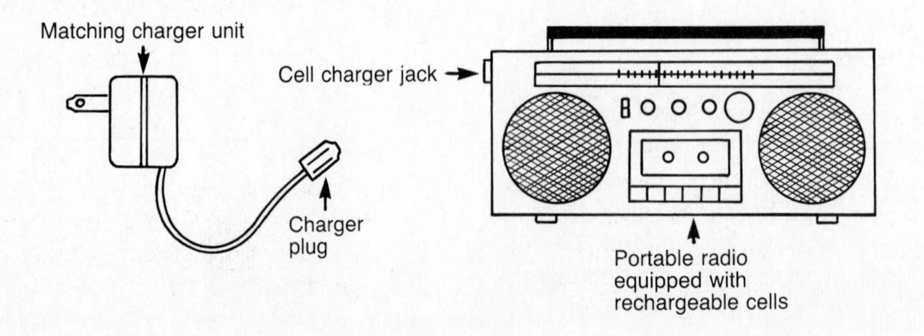

16-9 Battery Chargers

When the chemical reaction in a rechargeable battery has ended, the battery is said to be discharged and can no longer produce the rated flow of electric current. This battery can be recharged, however, by causing direct current from an outside source to flow through it in a direction opposite to that in which it flowed out of the battery. When charging a battery, the *negative* lead of the charger must connect to the *negative* lead of the battery and the *positive* lead of the charger to the *positive* lead of the battery (Figure 16-22). A reversal of these connections will produce a short circuit and may damage both the charger and battery.

FIGURE 16-23 METERING CHARGING CURRENT

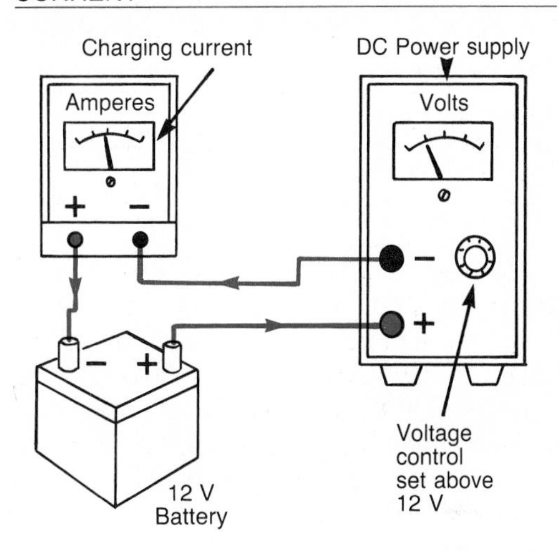

FIGURE 16-22 BATTERY CHARGING AND DISCHARGING CURRENTS

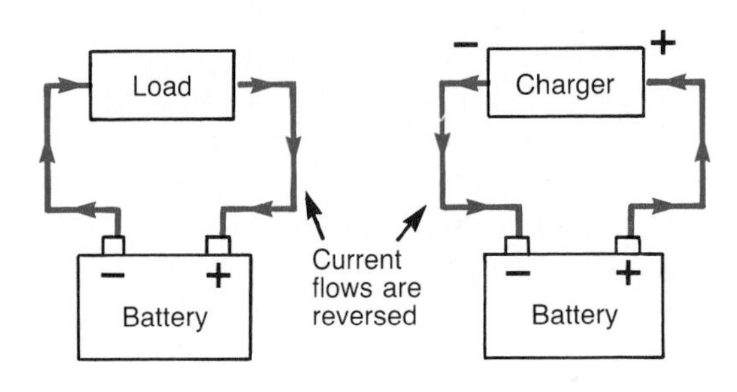

The automobile uses an automatic charging circuit as part of the car's electrical system, which is designed to recharge the battery as required. Out-of-car battery charging is done using large commercial type battery chargers. Smaller type chargers are also available for use on the smaller nickel-cadmium cells. A simple variable-voltage DC power supply works well as a battery charger.

When charging any battery, it is important to set the charging current to a value recommended by the manufacturer. This current is set by adjustment of the output voltage on the charger and read by an ammeter connected in series with the charger and battery (Figure 16-23). When the battery and charger are at the same voltage, no current flows. The charger voltage is set to a value higher than that of the battery to produce a current flow.

16-10 Practical Assignments

1. (a) Obtain five carbon-zinc cells and record the no-load voltage and short circuit current.
   (b) Calculate the internal resistance of each cell tested.
   (c) Rate each cell as being good, fair or poor based on your readings and calculations.

2. (a) Given four size "C" carbon-zinc cells rated for 1.5 V each, draw the schematic for the three battery connections possible using all four cells.
   (b) Calculate the output voltage for each battery connection.
   (c) Wire each battery connection and measure the output voltage.

3. (a) Record the hydrometer reading of each cell of a lead-acid battery.
   (b) Record the no-load voltage of the battery.
   (c) Connect a headlamp across the battery and measure the light load voltage.

4. (a) Draw the wiring diagram of a circuit that could be used to charge a 1.5 V type "AA" nickel-cadmium cell using whatever type of variable DC power supply is available to you.
   (b) Wire the circuit. Set the charging current to 10 mA. Charge the cell for five minutes.

16-11 Self Evaluation Test

1. What is the difference between a primary and secondary cell?

2. By strict definition, what is the difference between a cell and a battery?

3. (a) What is the approximate voltage output range for a single cell?
   (b) What determines the value of this output voltage?

4. (a) State two ways in which the output capacity of a cell or battery is rated.
   (b) For a given cell or battery type, what determines the output capacity?

5. Define the term *shelf-life* as it applies to cell or battery ratings.

6. Explain how cells or batteries are rated with regard to operating temperature.

7. What is the main difference between a wet and dry cell?

8. How does the carbon-zinc cell compare with the equivalent alkaline cell in price, energy capacity and shelf-life?

9. (a) What type of cell is most often used to power digital watches?
   (b) What are the three main features of this type of cell?

10. What battery rating is increased when connecting cells in:
    (a) series
    (b) parallel
    (c) series/parallel

11. Design a battery arrangement that can supply 4.5 V and 8 A·h using cells rated at 1.5 V and 2 A·h each. Draw a schematic diagram of the cell arrangement.

12. An emergency lighting system is supplied by a battery with a rated output of 12 V and 16 A·h. If the lighting system draws 2.5 A at 12 V, calculate the rated length of time that the lights will be operative in case of a power failure.

13. Why is a no-load test of any battery a poor indication of battery life?

14. Outline the procedure that is followed to make an internal resistance check of a small primary cell.

15. Explain how a battery hydrometer is able to test the state of charge of a lead-acid battery.

16. Outline the procedure that is followed to make a maximum load voltage test of a lead-acid automobile battery.

17. (a) Draw a schematic diagram showing the correct connection and metering for a variable DC power supply connected as a battery charger.
    (b) How is the amount of charging current controlled?

17

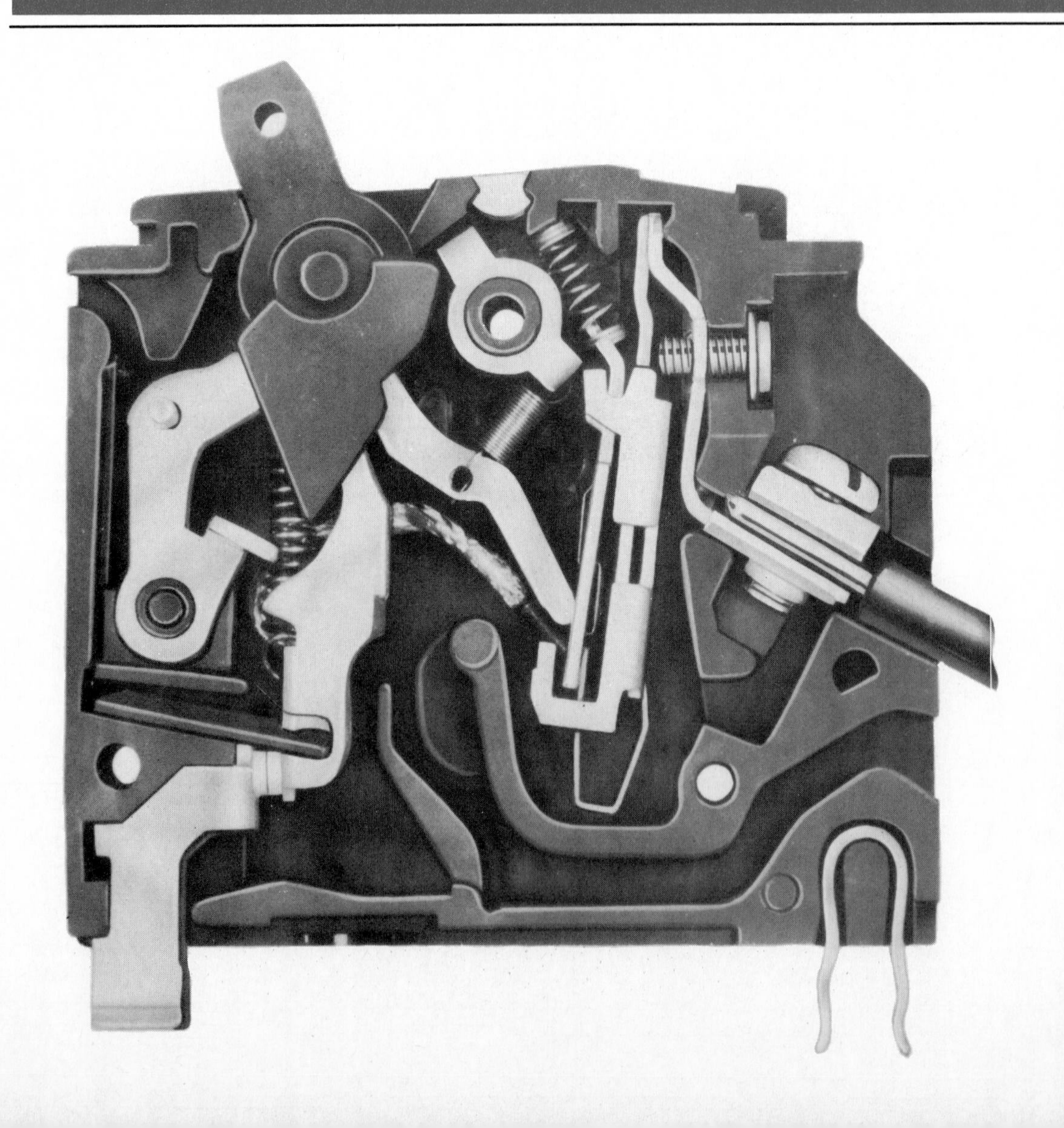

CIRCUIT PROTECTION DEVICES

OBJECTIVES

Upon completion of this unit you will be able to:

- Define the terms *overload* and *short circuit,* and describe the circuit conditions that produce them.
- Compare the basic principle of operation of a fuse and a circuit breaker.
- Explain how fuses and circuit breakers are rated.
- Identify basic fuse types and typical applications.
- Explain the special feature of time-delay fuses.
- Make out-of-circuit tests of fuses.
- Explain how lightning rods and arresters protect electrical equipment.

17-1 Undesirable Circuit Conditions

Overloads An electrical circuit is limited in the amount of current it can handle. The current capacity of the circuit is determined by the size of the wire conductors used. An electrical circuit is said to be *overloaded* when the amount of current flowing through it is more than its rated current capacity.

Overloads can occur in the home when too many lamps and appliances are plugged into the same circuit (Figure 17-1). For example, a general purpose house branch circuit is usually wired for a maximum capacity of 15 A. If the sum of the parallel load currents connected to it exceeds 15 A the circuit is said to be overloaded. The solution to this problem is to remove some of the loads.

In an overloaded circuit, the conductors are required to carry more current than they are safely rated to carry. This results in excessive heat being produced by the conductors, creating a potential fire hazard.

Short Circuits Sometimes a direct path is accidentally created from one side of the voltage source to the other, without passing through a load. This situation is called a *short circuit* (Figure 17-2). When a circuit is shorted, its total resistance is reduced to near zero value. As a result, lots of current will flow! Actual value of current flow will be limited only by the capacity of the source.

FIGURE 17-2 SHORTED CIRCUIT

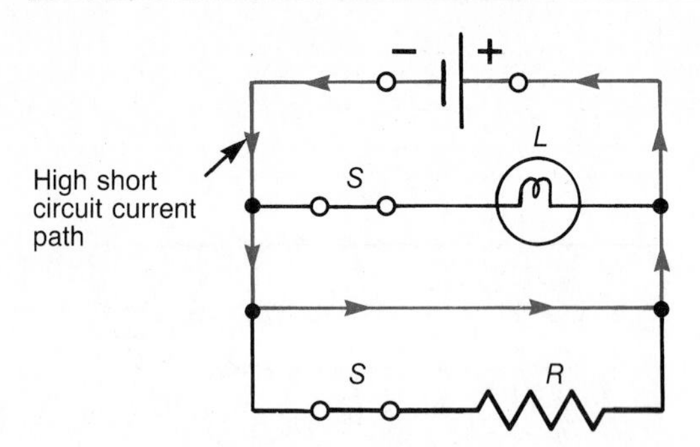

Short circuits can be wired by accident when the circuit is first constructed. When constructing any type of circuit, it is wise to check for shorts before applying voltage.

Any bare or exposed wire spells trouble (Figure 17-3). The fault may be traced to worn insulation or a faulty connection. All that has to happen is for two exposed areas to touch and you have a short circuit. Most short circuits occur in flexible cords, plugs

FIGURE 17-1 OVERLOADED CIRCUIT

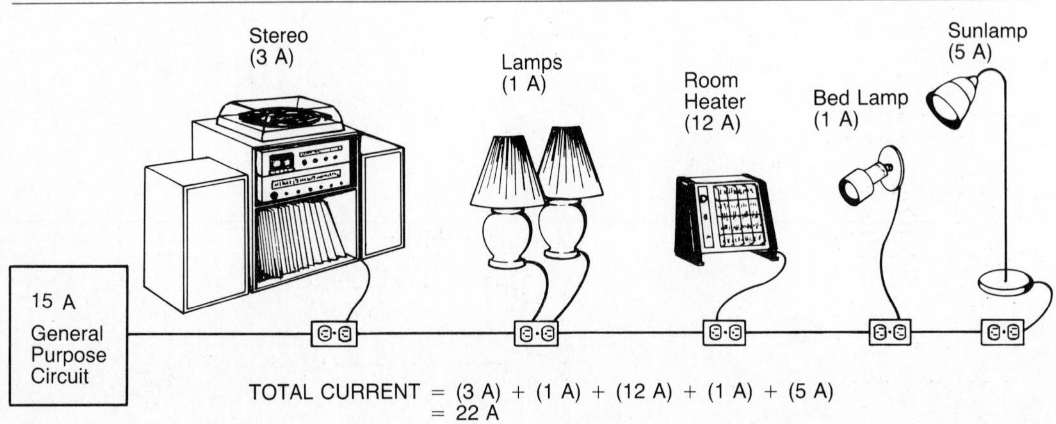

FIGURE 17-3 SHORTED LAMP CORD

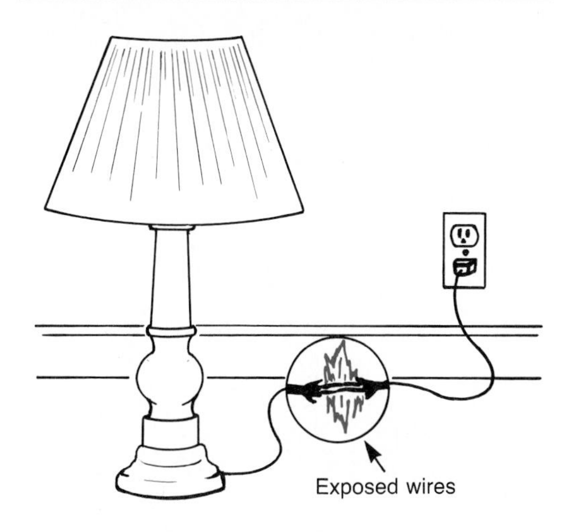

or appliances. Look for black smudge marks on faceplates or frayed or charred cords connected to dead circuits. To correct this problem, simply replace the damaged cord or plug.

A short circuit is one of the most dangerous circuit faults because of the amount of current that flows. Most voltage sources are capable of producing hundreds of amperes of short circuit current that can instantly melt wires and cause fires.

17-2 Fuses and Circuit Breakers

Fuses and circuit breakers are commonly referred to as *overcurrent protection devices* (Figure 17-4). Their purpose is to protect electrical circuits from damage by too much current flow. They are both low resistance devices and are connected in series with the circuit they protect. Whenever wiring is forced to carry more current than it can safely handle, fuses will blow or circuit breakers will trip. These actions open the circuit, disconnecting the supply of electricity.

FIGURE 17-4 OVERCURRENT PROTECTION DEVICES

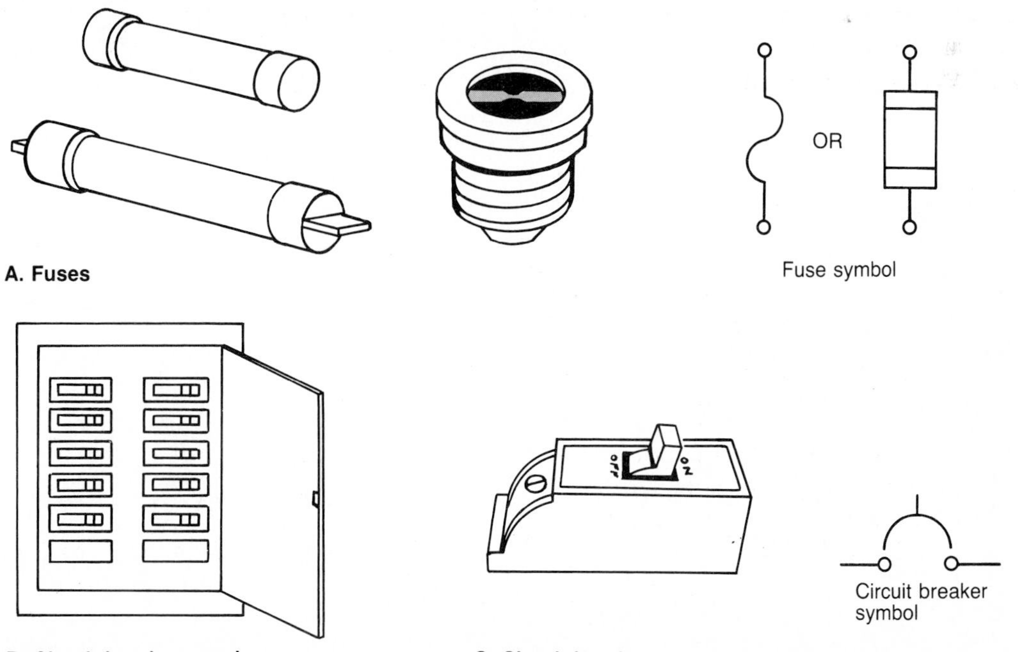

A. Fuses

B. Circuit breaker panel

C. Circuit breaker

Fuses and circuit breakers are rated for both current and voltage. The *continuous current rating* most often referred to represents the maximum amount of current the device will carry without blowing or tripping open the circuit. The continuous current rating is matched to the current rating or ampacity of the circuit conductors they are protecting. For example, a circuit using an AWG #14 copper conductor has a rated capacity of 15 A. The fuse or circuit breaker, therefore, must also be rated for 15 A. Never replace any fuse or circuit breaker with one of a higher current rating!

Fuses and circuit breakers are also rated for *maximum voltage* and *interrupting capacity*. The maximum voltage rating indicates the maximum circuit voltage value under which the device can operate safely. The interrupting capacity is the amount of short circuit current the device can safely interrupt.

17-3 Types of Fuses

Screw-base or Plug Fuse All fuses contain a short metal strip made of an alloy with a low melting point. When installed in a socket or fuseholder, the metal strip becomes a link in the circuit. When the current flow through this link is greater than the rating of the fuse, the metal strip will melt opening the circuit.

The *screw-base* or *plug fuse* is often used to protect branch circuits in a home. It is constructed with a fusible link enclosed in a glass housing (Figure 17-5). This prevents the melted metal link from splattering when the fuse blows. It uses a screw-in base for connection into the fuse holder socket.

Plug fuses have a maximum voltage rating of 125 V. They are available in a number of common current sizes up to a maximum of 30 A. They are used most often in 120 V general house lighting and receptacle circuits. The current rating of the fuse is matched

FIGURE 17-5 THE PLUG FUSE

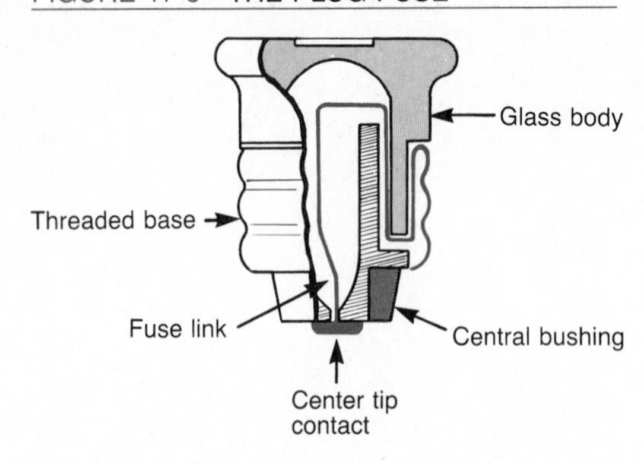

to the maximum current rating of the circuit conductors.

The National Electrical Code requires all plug fuses to be of the *non-interchangeable* type. This type of plug fuse makes it impossible to replace a fuse with one of a higher current rating. Type C plug fuses are designed with color coded central bushings made of different diameters for different current ratings (Table 17-1). The fuse holder is equipped with a plastic rejector ring that will not accept a fuse with a higher current rating.

TABLE 17-1 TYPE "C" PLUG FUSE RATINGS

PLUG FUSE RATING	COLOR CODE OF FUSE AND REJECTOR RING
15 A	Blue
20 A	Pink
25 A	Red
30 A	Green

Fuse links melt for two reasons: either the circuit has developed a short, or it has been overloaded with too many appliances. The see-through glass body of the plug fuse is

FIGURE 17-6 FUSE FAULT INDICATORS

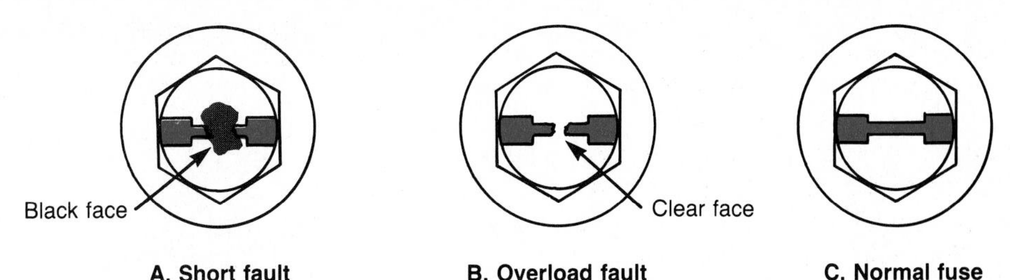

a great help for finding what caused a fuse to blow (Figure 17-6). If the glass front is black, it indicates that there has been a short circuit. A careful check of the circuit should be made before replacing the fuse. If the glass front is clean and clear, it indicates that the circuit is overloaded. In this case some of the appliances should be removed from the circuit before replacing the fuse.

Cartridge Fuses There are two basic types of cartridge fuses. They are the *ferrule-contact* type (Figure 17-7) and the *knife-blade* type (Figure 17-8).

The ferrule-contact type is used to protect 120/240 V heavy duty appliance circuits. These include stoves, clothes dryers and water heaters. They are rated for 250 V and are available in different current sizes up to 60 A. Again, the fuse current rating selected must be matched to the conductor ampacity rating.

The knife-blade cartridge fuse is used for circuit current ratings in excess of 60 A. The contact points of this fuse are larger and more rugged. This allows it to handle higher current flows. Available in current ratings from 65 A to 600 A, they are also suitable for 240 V circuits. Knife-blade fuses are sometimes used in fused main disconnect switches in the home. They provide protection for the consumer's main service wiring.

FIGURE 17-7 FERRULE-CONTACT CARTRIDGE FUSE

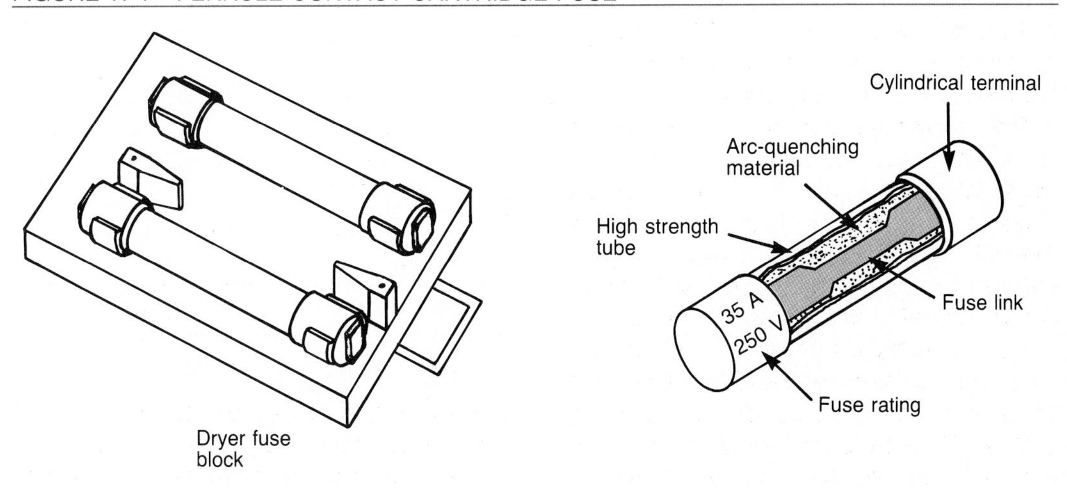

FIGURE 17-8 KNIFE-BLADE CARTRIDGE FUSE

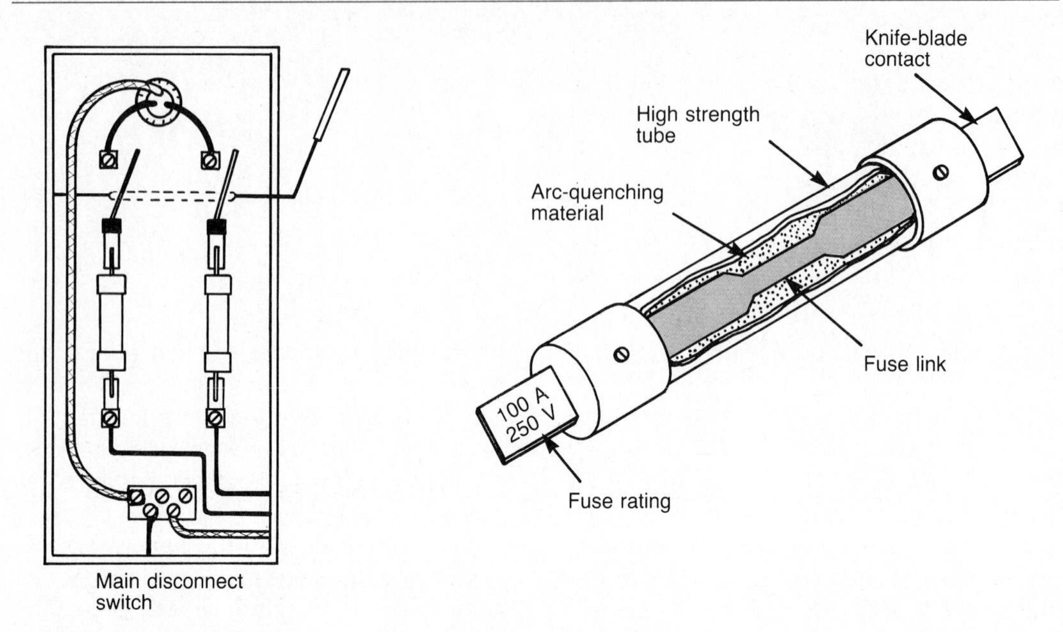

Time-delay Fuses A motor load requires a high surge of current to get it started. Once up to full speed, the motor current drops down to its normal value. Because of this, circuits with motors connected to them will sometimes blow a standard fuse when they are first started. Special *time-delay fuses* are designed to overcome this problem (Figure 17-9 A, B). They are available in both plug and cartridge types. They do not blow like standard fuses on large but temporary overloads. They will, however, blow like standard fuses on small continuous overloads, and instantly on short circuits. This type of fuse is used to protect motor circuits such as the furnace, freezers or water pumps.

The time-delay plug fuse uses a pot of solid solder to hold the fuse link in place. Excessive current flow through the fuse link heats the solder, causing it to melt. The spring then pulls the fuse link away and opens the circuit.

FIGURE 17-9 TIME-DELAY PLUG FUSE

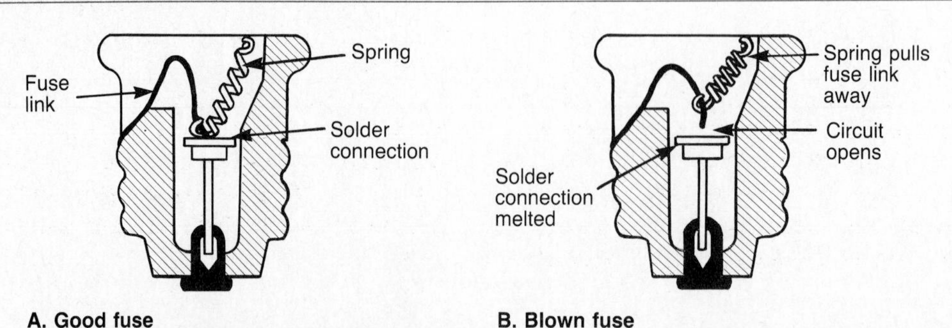

A. Good fuse **B. Blown fuse**

Renewable Fuses Only non-renewable types of fuses are approved by the National Electrical Code for residential use. This type of fuse must be replaced by a new one once it blows. This feature prevents people from tampering with the fuse link.

Renewable cartridge fuses are approved for industrial applications. Unlike the non-renewable type used in the home, these cartridge fuses contain a fuse link that can be replaced once blown (Figure 17-10). Although initially more costly, this fuse will reduce maintenance costs over a long period of time.

FIGURE 17-11 SIMPLE CONTINUITY FUSE TESTER

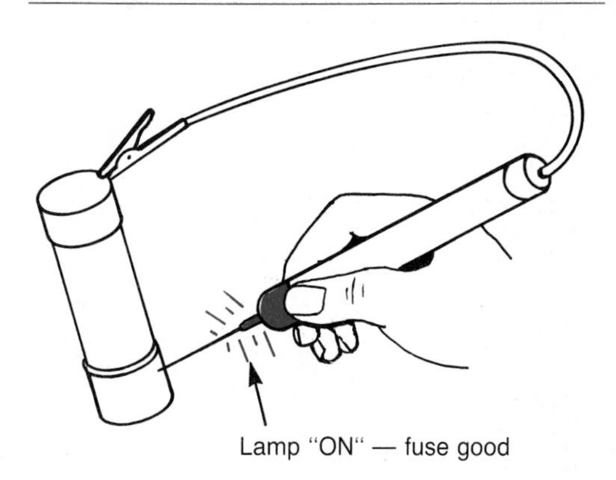

FIGURE 17-10 RENEWABLE CARTRIDGE FUSE

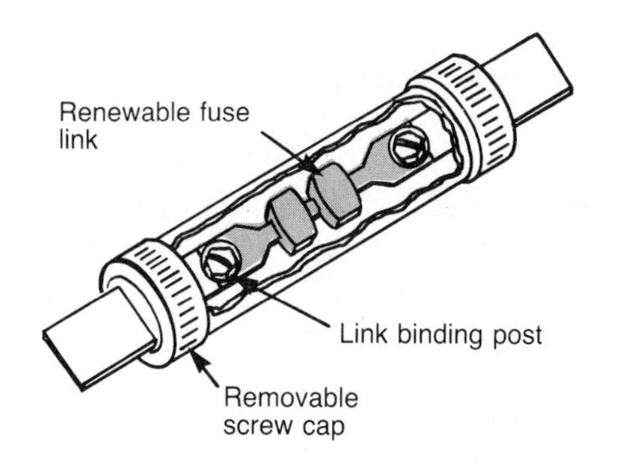

FIGURE 17-12 USING AN OHMMETER TO TEST FUSES

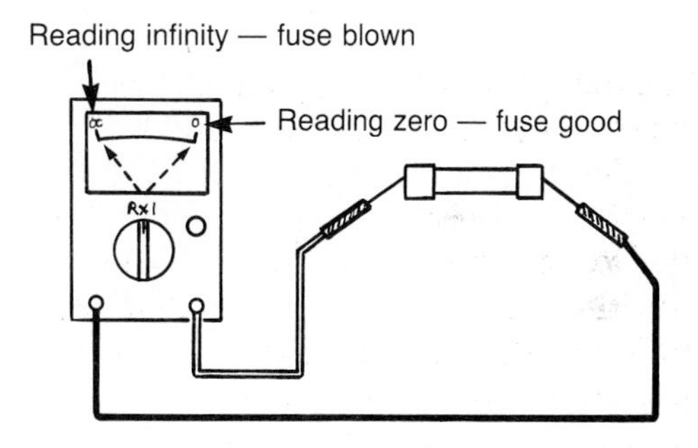

17-4 Testing Fuses

Fuses may be tested out-of-circuit by using a simple *continuity lamp tester* (Figure 17-11). A good fuse connected in series with the test leads will complete the circuit and cause the lamp to turn on.

The ohmmeter can also be used to make an out-of-circuit test of a fuse (Figure 17-12). A good fuse should indicate near zero resistance reading on the meter. A blown fuse will be indicated by an infinite value of resistance.

A voltmeter can be used to make an in-circuit test of a fuse (Figure 17-13). Voltage is checked on the line and load side of the fuse. Full voltage on the line side and zero voltage on the load side indicates a blown fuse.

FIGURE 17-13 TESTING FUSES IN-CIRCUIT WITH A VOLTMETER

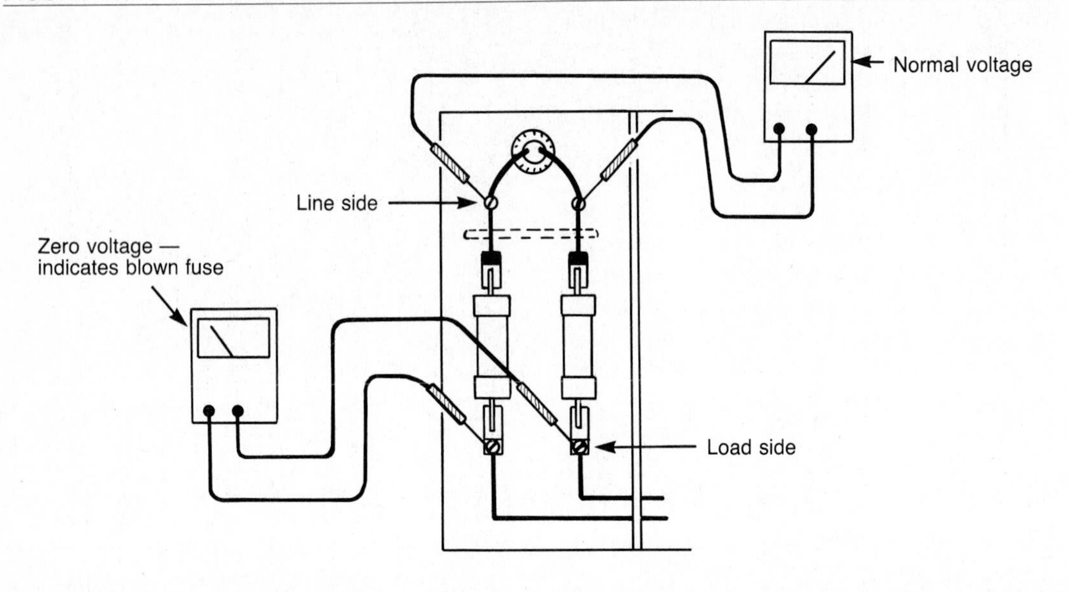

17-5 Circuit Breakers

A *circuit breaker* can be used (in place of a fuse) to protect circuits against overloads and short circuits. Like a fuse, it is connected in series with the circuit it protects. Circuit breakers are rated in a manner similar to that used for fuses. As with fuses, the ampere rating of a breaker must match the ampacity of the circuit it protects.

Resembling an ordinary switch, a circuit breaker serves as both a switch and as a fuse (Figure 17-14). As a switch, a circuit breaker lets you open a circuit (turn switch to off) whenever you want to do work on it.

FIGURE 17-14 CIRCUIT BREAKER CONNECTION

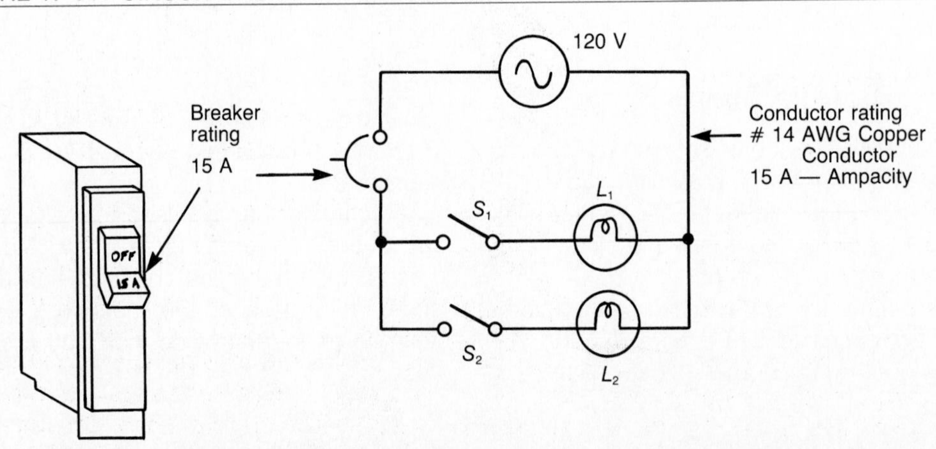

FIGURE 17-15 ACTION OF A CIRCUIT BREAKER

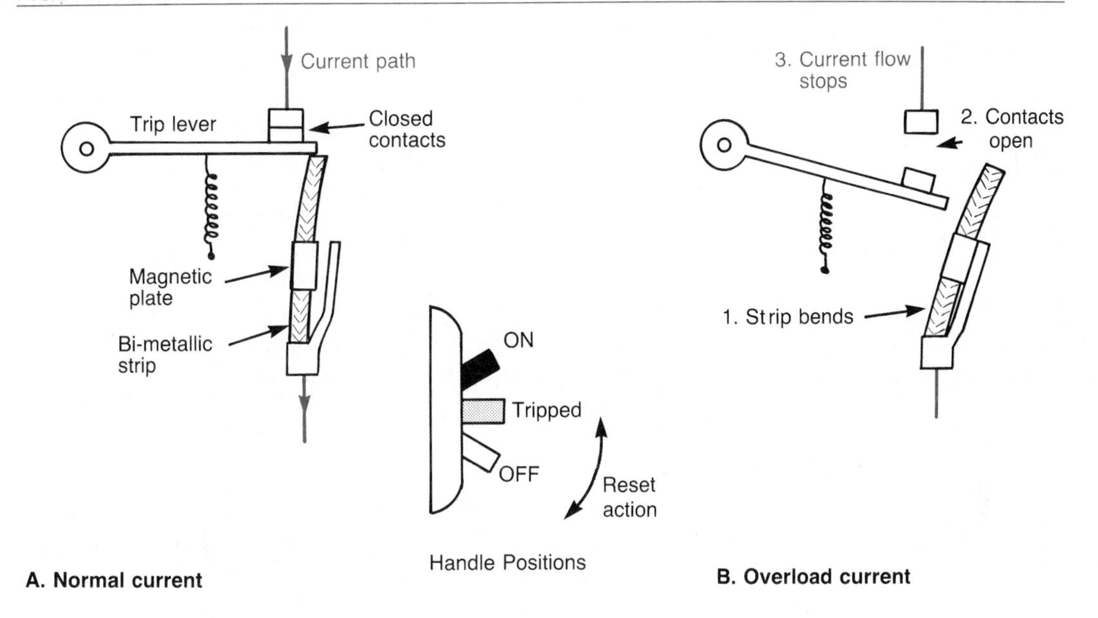

As a fuse, it provides automatic overcurrent protection.

When a breaker is installed in a circuit, a bi-metallic strip inside the breaker becomes the link in the circuit (Figure 17-15 A, B). An overload causes the bi-metallic strip to heat and bend. This action releases the trip lever and opens the breaker's contacts. For short circuits a quicker method is used to release the trip lever of the breaker. The heavy short circuit current creates a magnetic force in a magnetic plate attached to the bi-metallic strip. This releases this trip lever to open the contacts almost instantly.

When the circuit breaker trips, the toggle moves to the "tripped" or intermediate position. Unlike fuses, circuit breakers can be reset (turned back "ON") once they have tripped. To reset the breaker it is first moved all the way to the full "OFF" position and then to the "ON" position.

Although circuit breakers are more expensive to install initially than fuses, they have the following advantages:

- They can be used both as a protective device and an on/off control switch.
- No cost is involved in replacing them when they operate due to over-current.
- It is more convenient and safer to reset them than to replace fuses.
- They can be accurately produced for different time-delay tripping action.

17-6 Lightning Rods and Arresters

Lightning Rods A lightning bolt is the result of an electrical path set up between a charged cloud and the Earth. The cloud and Earth behave as two voltage supply leads with a voltage potential energy in the million volt range. The resistance path between the two can result in short bursts of current in the thousand ampere range. This amount of current flow directed through electrical lines and equipment can do extensive damage to the system.

While lightning-caused damage to electrical installations is quite rare in larger cities, it is rather frequent in rural areas. One protection against a lightning discharge is to provide an alternate direct low resistance path to ground. A *lightning rod* works on this principle. A sharply pointed metal rod is mounted on the highest part of a house or barn and a heavy copper wire is run from it to the ground (Figure 17-16). Any lightning discharge will be directed through this low resistance path thereby protecting the structure and its contents.

FIGURE 17-16 LIGHTNING ROD INSTALLATION

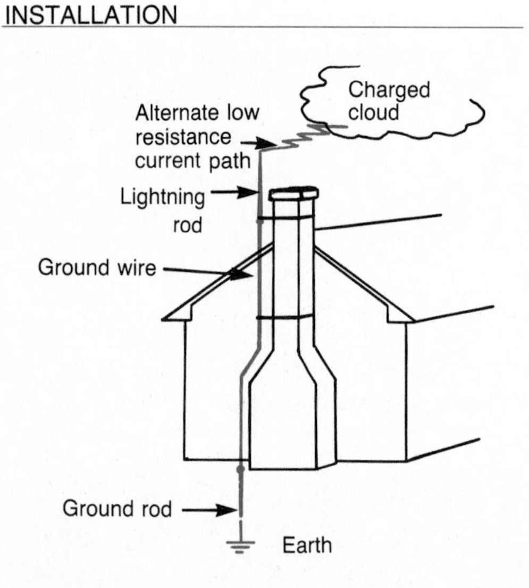

A television antenna mast or other metal support acts the same as a lightning rod and must therefore be well grounded. The steel frame structures of tall commercial and industrial buildings also act as unintended lightning rods and are grounded to provide the same protection.

Lightning Arresters *Lightning arresters* are used to drain the energy from power lines struck by lightning. The lightning arrester works on the spark gap principle like the spark plug in a car. One side of the arrester is connected to ground and the other to the wire to be protected (Figure 17-17). These two points are insulated by the air gap between them under normal circuit voltage conditions. When lightning strikes the line the resulting high voltage ionizes the air and produces a low resistance discharge path to ground. Specially designed arresters are available for use on overhead power lines as well as on signal circuits such as telephone circuits and antenna lead-in wires.

FIGURE 17-17 LIGHTNING ARRESTER

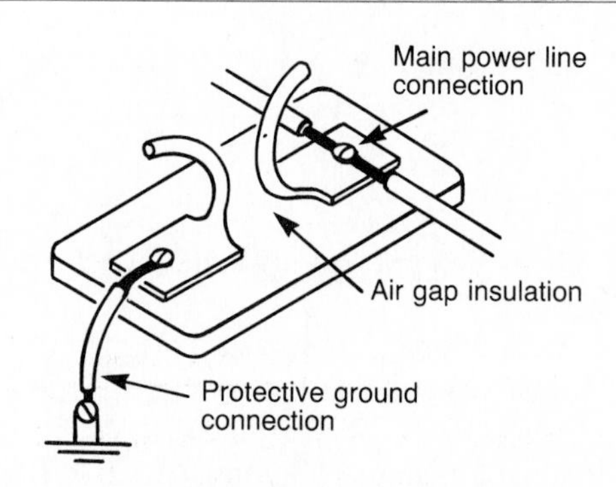

17-7 Practical Assignments

1. (a) Using a 120 V AC supply and a 5 A plug fuse for protection, have the teacher demonstrate:
   (i) normal circuit operating a single lamp load
   (ii) circuit overload
   (iii) short circuit
   (b) Report on the condition of the glass front of the fuse for a circuit overload and a short circuit.
2. (a) Using a 120 V AC supply and a 5 A circuit breaker for protection, have the teacher demonstrate:
   (i) normal circuit operating a single lamp load

(ii) circuit overload
(iii) short circuit
(b) Report on how the breaker is reset after a short circuit or overload.

3. Construct a simple out-of-circuit fuse continuity tester that uses a battery and a lamp. Check a number of identified fuses using this circuit.

4. Given a number of identified fuses, check them using an ohmmeter.

FIGURE 17-18 PRACTICAL ASSIGNMENT 5

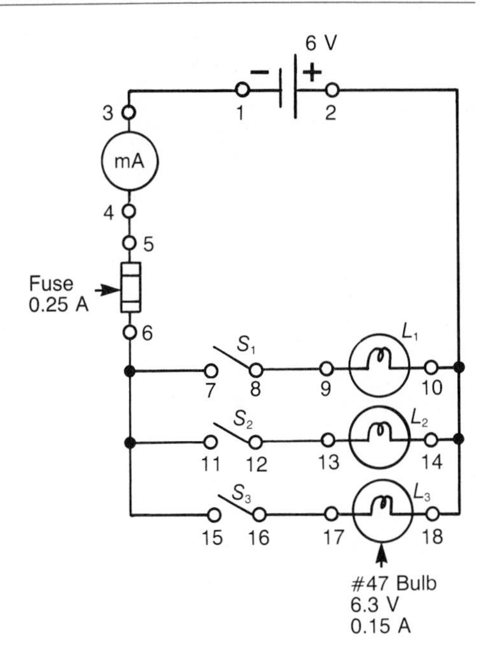

5. (a) Complete a wiring number sequence chart and wiring diagram for the schematic circuit of Figure 17-18.
(b) Close switch S_1. Record the total current flow.
(c) Measure and record the voltage from the positive side of the voltage source to the top and bottom of the fuse.
(d) Close switch S_2. Record the total current flow.
(e) Close switch S_3. Record the total current flow.
(f) Record the length of time it takes for the fuse to blow.
(g) Once the fuse has blown, measure and record the voltage from the positive side of the voltage source to the top and bottom of the fuse.

6. Have the teacher demonstrate the effect of a short circuit on the conductors of an unprotected circuit.

7. Have the teacher demonstrate the action of a bi-metallic strip.

8. Have the teacher demonstrate the spark gap principle used on lightning arresters. Use a high voltage ignition coil or Tesla Coil as the lightning voltage source.

17-8 Self Evaluation Test

1. When is an electrical circuit considered to be overloaded?
2. What is the most common cause of electrical circuit overloads in a home?
3. When is an electrical circuit considered to be short circuited?
4. Why is a short circuit the most dangerous of all circuit faults?
5. What happens when conductors are required to carry more current than they are safely rated to handle?
6. How are fuses and circuit breakers connected with respect to the circuits they protect?
7. How does the resistance of a fuse compare with that of the load device?

8. To what does the continuous current rating of a fuse or circuit breaker refer?
9. How is the continuous current rating of a fuse or circuit breaker determined for a particular circuit?
10. Explain how a fuse operates to protect a circuit.
11. Name the type of fuse used in each of the following circuits:
    (a) 35 A dryer circuit
    (b) 15 A lighting circuit
    (c) 100 A main disconnect switch
12. What is the maximum voltage rating for a plug fuse?
13. (a) Describe the special feature of time-delay fuses.
    (b) In what type of circuit should they be used?
14. State two methods that can be used to check fuses out-of-circuit.
15. (a) Draw the symbols commonly used to represent a fuse.
    (b) Draw the symbol commonly used to represent a circuit breaker.
16. Explain how a circuit breaker trips when the circuit becomes overloaded.
17. List three advantages that circuit breakers have over fuses.
18. Explain how a lightning rod circuit protects an electrical installation.
19. Explain the principle of operation of a lightning arrester.

18

ELECTRICAL POWER

OBJECTIVES

Upon completion of this unit you will be able to:

- Describe the various methods used to generate power in a generating station.
- Outline the method used in transmitting power over long distances.
- Define electrical power.
- Use the power formula to calculate the output power or current of an electrical load device.
- Properly connect a wattmeter to measure the power of an electrical load device.
- Calculate the minimum power rating for resistors connected in series and parallel groups.
- Define power loss.
- Calculate the power loss and transmission efficiency of a simple circuit.

18-1 Electrical Generating Stations

Hydro-electric The electrical energy supplied to our home is produced at a central location called a *generating station*. This station is usually equipped with several large AC generators each of which is driven by a separate prime mover. Generating stations are classified according to the method used to drive the generator. *Hydro-electric generating stations* use water power in the form of waterfalls and water under pressure from giant dams or reservoirs to drive the generators (Figure 18-1). It is the cheapest and most environmentally safe method of producing electricity.

Thermal-electric Unfortunately, there are not enough suitable places for hydro-electric installations to meet the increasing demand for electric energy. As a result, steam power is also used to drive generators. Stations which use generators driven by steam turbines are called *thermal-electric generating stations*. Thermal-electric stations use heat to convert water to steam which forces a turbine wheel to turn (Figure 18-2). The turbine output shaft rotates an AC generator to produce electricity.

Heat used to power thermal-electric generating stations is obtained in one of two ways. The first method is by burning fossil fuels such as coal, oil or natural gas in a furnace. The second method involves splitting uranium atoms or nuclear fusion. In a nuclear station the nuclear reactor performs the same function as the furnace does in fossil-fuelled stations (Figure 18-3). Both use non-renewable resources that can affect the environment.

18-2 Transmitting Electrical Energy

It is not always possible or practical to locate the generating station close to where the electrical energy will be put to use. In most instances the electricity must travel hundreds of miles from the source to the load. Transmitting large amounts of electrical energy over fairly long distances is accomplished most efficiently by using high voltages.

FIGURE 18-1 HYDRO-ELECTRIC GENERATING STATION

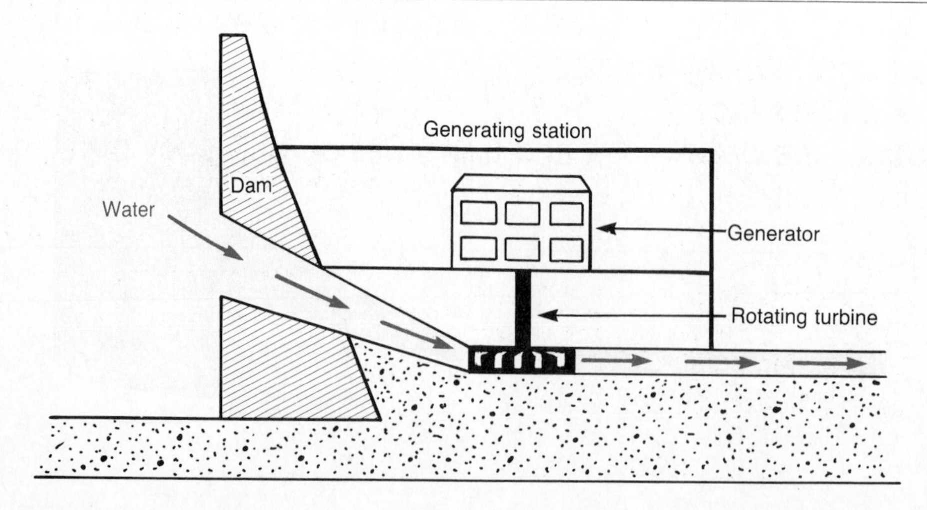

FIGURE 18-2 THERMAL-ELECTRIC CONVENTIONAL (FOSSIL FUEL) STATION

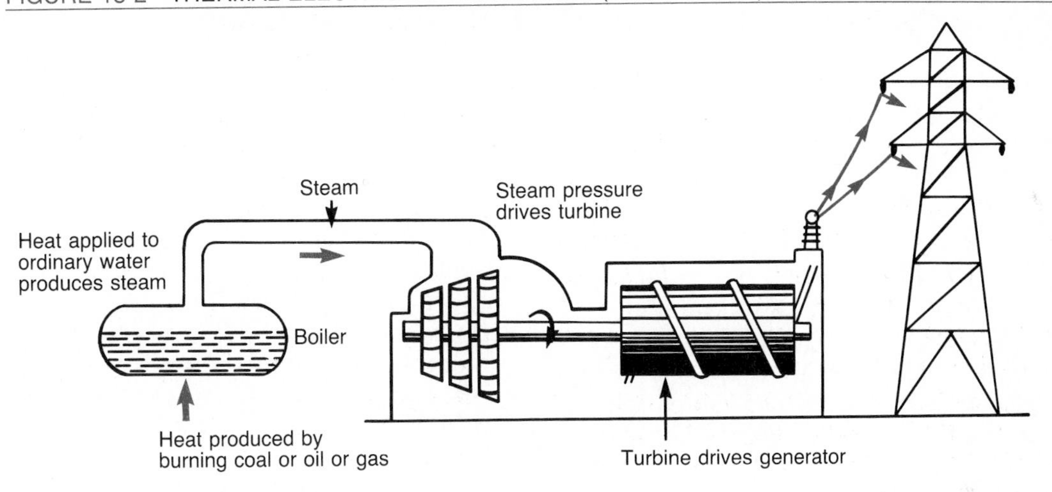

FIGURE 18-3 THERMAL-ELECTRIC NUCLEAR STATION

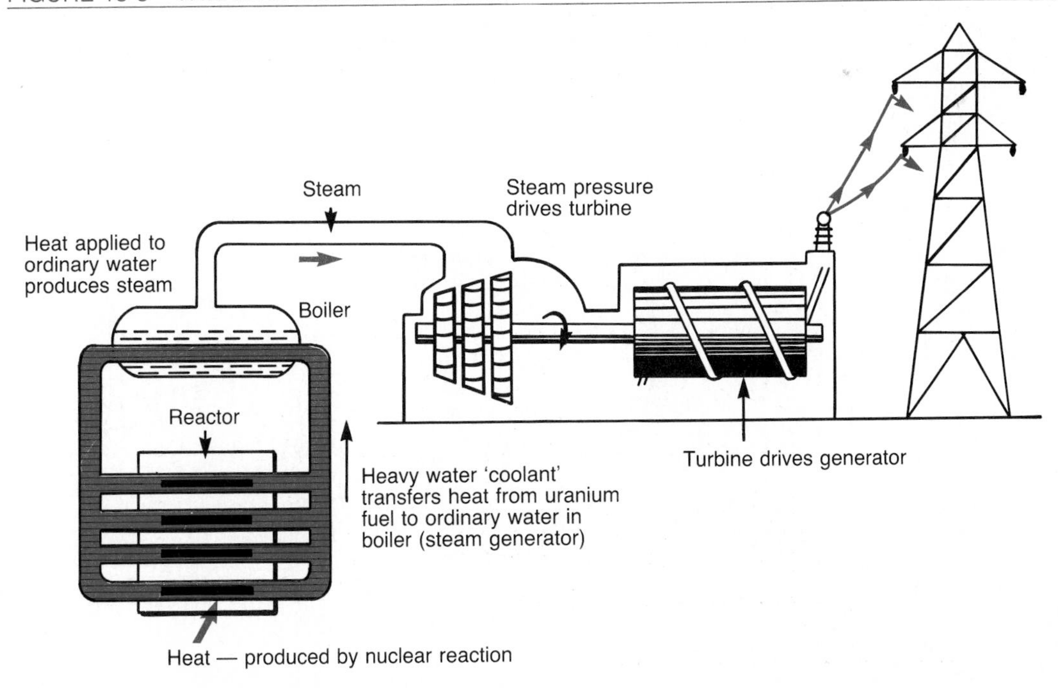

Transmission line losses represent the difference between the primary voltage generated and the voltage delivered to the load. The total electrical power transmitted is equal to the product of the voltage multiplied by the current. Keeping the voltages high has the advantage of lowering the current value for any given amount of power. This, in turn, will reduce the line power losses.

There are certain limitations to the use of high voltage transmission systems. The higher the transmitting voltage, the more difficult and expensive it becomes to safely insulate between line wires as well as from line wires to ground. For this reason the voltages in a typical high voltage grid system are reduced in stages as they approach the area of final use (Figure 18-4).

A network of transmission lines and generating stations are connected together. In this way excess power from one region can be fed to another region in response to demand. During the periods when the demand for electrical power drops, power stations shut down some generators. At times of peak demand, auxiliary equipment is set in operation.

18-3 Electrical Power

Electrical energy is simply the work performed by an electric current. Whenever current exists in a circuit, there is a conversion of electrical energy into other forms of energy. For example, current flow through a lamp filament converts electrical energy into light and heat energy. Electrical power is defined as the rate at which electrical energy is converted. The SI unit of electrical power is the watt (W).

Most electrical load devices carry an electrical power rating. This power rating indicates the rate at which the device converts electrical energy. The *power rating* of a device can be read directly from its nameplate or calculated from other information given (Figure 18-5).

18-4 Calculating Electrical Power

The SI unit used to measure electrical power is the watt. One watt is equivalent to an energy conversion rate of one joule per second

FIGURE 18-4 TYPICAL HIGH VOLTAGE TRANSMISSION SYSTEM

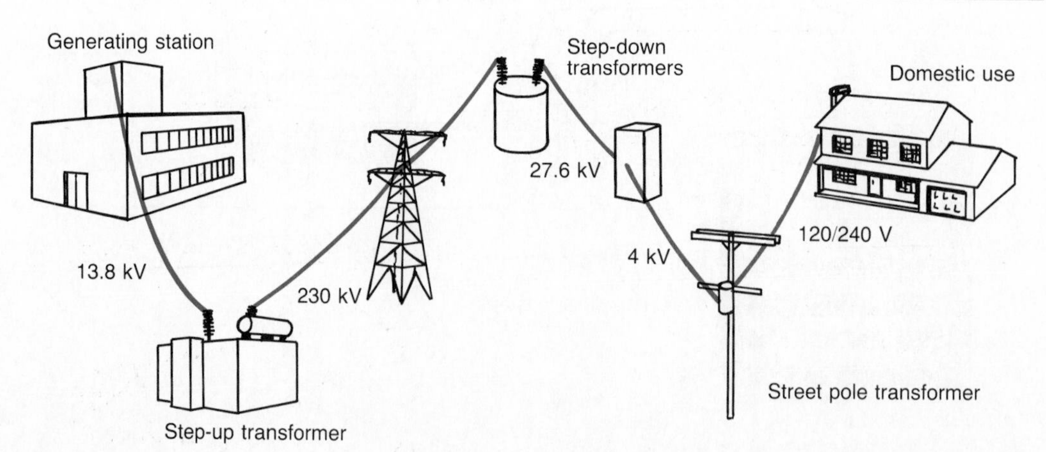

FIGURE 18-5 APPLIANCE POWER RATING

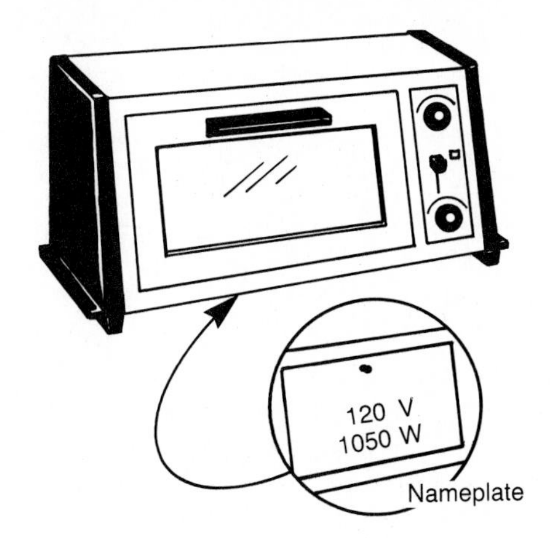

(1 W = 1 J/s). Electrical power can be calculated directly using the measurements of voltage and current. They are related to work and time. To calculate the power in watts you simply multiply the voltage in volts by the current in amperes:

$$P = V \times I$$

where:

P =	power in watts (W)
V =	voltage in volts (V)
I =	current in amperes (A)

Example

Suppose the portable electric heater in Figure 18-6 draws a current of 8 A when connected to its rated voltage of 120 V. The power rating of the heater is then:

Power = Voltage × Current

$P = V \times I$

$P = (120\text{ V})(8\text{ A})$

$P = 960\text{ W}$

FIGURE 18-6 HEATER EXAMPLE

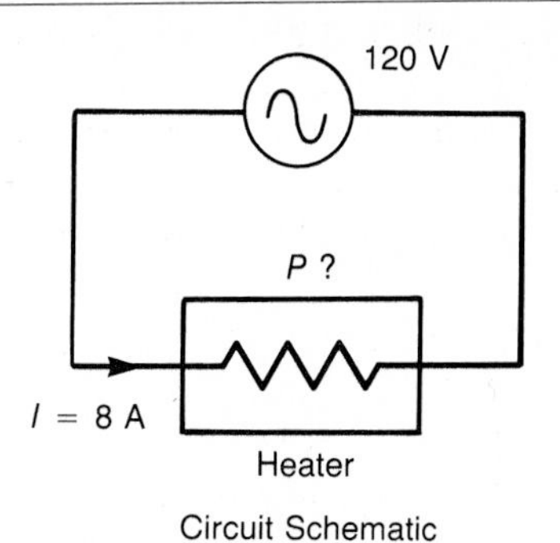

Electrical load devices such as lights and appliances are designed to operate at a specific voltage. They may be marked with both their operating voltage and power output at that voltage. If the current rating is not marked it can be calculated using the rated power and voltage values. To calculate the current, the power is divided by the rated voltage:

$$I = \frac{P}{V}$$

Example

Suppose a 150 W light bulb (Figure 18-7) has a voltage rating of 120 V. When connected to a 120 V source the current flow through the bulb would be:

$$I = \frac{P}{V}$$

$$I = \frac{(150\text{ W})}{(120\text{ V})}$$

$$I = 1.25\text{ A}$$

FIGURE 18-7 LIGHT BULB EXAMPLE

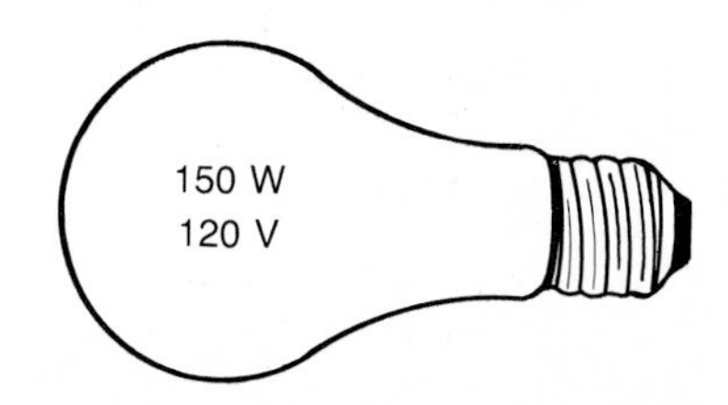

In some cases it is more convenient to calculate power by using the resistance of the material rather than the voltage applied to it. This occurs quite often when calculating the power dissipated in a resistor or lost as heat in a conductor. By substituting the product IR for V in the basic power formula, we get:

$$P = V \times I$$
$$P = (IR)(I)$$
$$P = I^2R$$

Example

Suppose the current flow through a 100 Ω resistor is measured and found to be 0.5 A (Figure 18-8).

(a) How much power is being dissipated as heat?
(b) How does this change if the current is doubled?

(a) $P = I^2 \times R$
$P = (0.5\text{ A})^2 (100\ \Omega)$
$P = 25\text{ W}$

(b) $P = I^2 \times R$
$P = (1\text{ A})^2 (100\ \Omega)$
$P = 100\text{ W}$

(4 times the power)

FIGURE 18-8 RESISTOR EXAMPLE

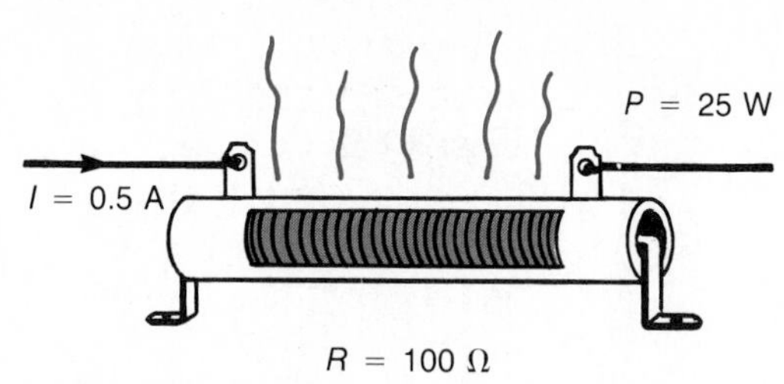

A. Circuit for Part a

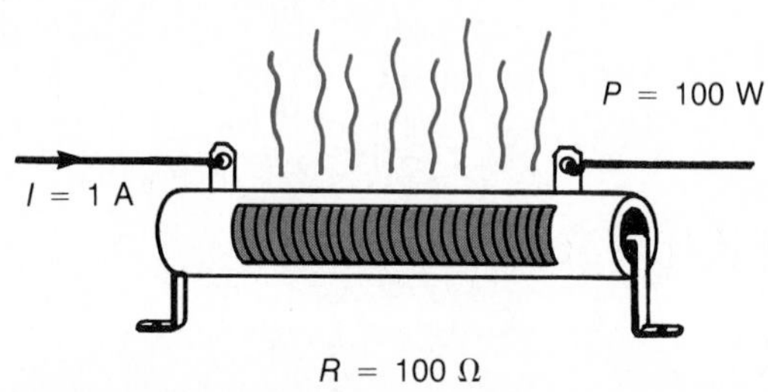

B. Circuit for Part b

18-5 Measuring Electrical Power

A *wattmeter* is an electrical instrument which is used to measure electrical power directly. This meter is a combination of both a voltmeter and an ammeter. It measures both voltage and current at the same time and indicates the resultant power value.

The wattmeter has four terminal connections; two for the voltmeter section and two for the ammeter section (Figure 18-9). The voltmeter section is connected the same as a regular voltmeter, in parallel or across the load. The ammeter section is connected like a regular ammeter, in series with the load. A reverse pointer movement is corrected by reversing the two voltmeter leads or the two ammeter leads but not both.

18-6 Calculating Power in Series and Parallel Circuits

The total power of a group of loads is the sum of the individual power ratings. This rule applies to both the series and parallel circuits.

The total power formula is then:

$$P_T = P_1 + P_2 + P_3 \ldots\ldots$$

FIGURE 18-9 WATTMETER CONNECTION

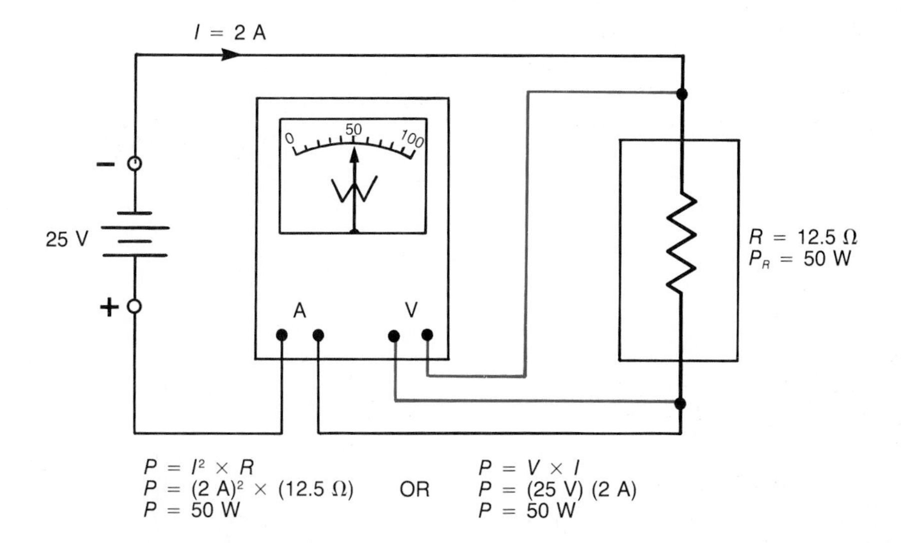

$P = I^2 \times R$
$P = (2\ A)^2 \times (12.5\ \Omega)$
$P = 50\ W$

OR

$P = V \times I$
$P = (25\ V)\ (2\ A)$
$P = 50\ W$

When building circuits using resistor networks, it is important to calculate the maximum power size of resistor required. Using a resistor of a lower power rating than required will cause it to overheat and burn out. The rule of thumb is to calculate the power size of resistor required and then select one of that size or the nearest standard power size above it.

FIGURE 18-10 SERIES RESISTOR EXAMPLE

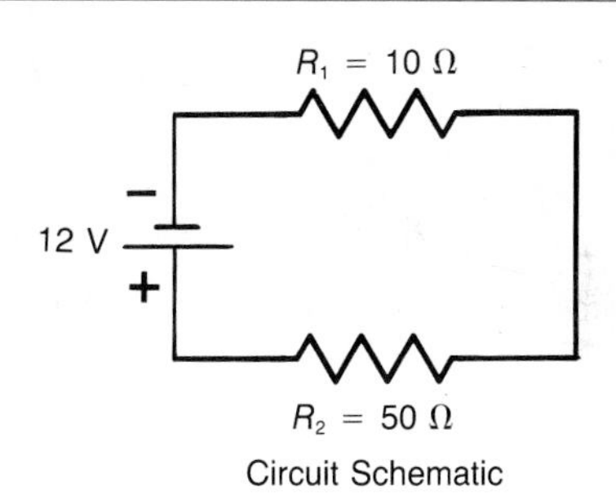

Circuit Schematic

Example

Suppose a 100 Ω resistor (R_1) and a 500 Ω resistor (R_2) are to be connected in series to a 12 V DC source as in Figure 18-10. The minimum power rating of the resistors required would be:

$R_T = R_1 + R_2$
$R_T = (10\ \Omega) + (50\ \Omega)$
$R_T = 60\ \Omega$

$$I_T = \frac{V_T}{R_T}$$

$$I_T = \frac{(12\ V)}{(60\ \Omega)}$$

$I_T = 0.2\ A$

$I_T = I_1 = I_2 = 0.2\ A$

$P_1 = (I_1)^2 \times R_1$
$P_1 = (0.2\ A)^2(100\ \Omega)$
$P_1 = 4\ W$

The minimum power rating for R_1 would have to be 4 W

$P_2 = (I_2)^2 \times R_2$
$P_2 = (0.2\ A)(50\ \Omega)$
$P_2 = 2\ W$

The minimum power rating for R_2 would have to be 2 W

$P_T = P_1 + P_2$
$P_T = (4\ W) + (2\ W)$
$P_T = 6\ W$

The total power dissipated by the circuit would be 6 W

Example
Suppose the same two resistors were to be connected in parallel to a 12 V DC source as in Figure 18-11. Would the minimum power rating required be the same as when they were connected in series? Let's see:

FIGURE 18-11 PARALLEL RESISTOR EXAMPLE

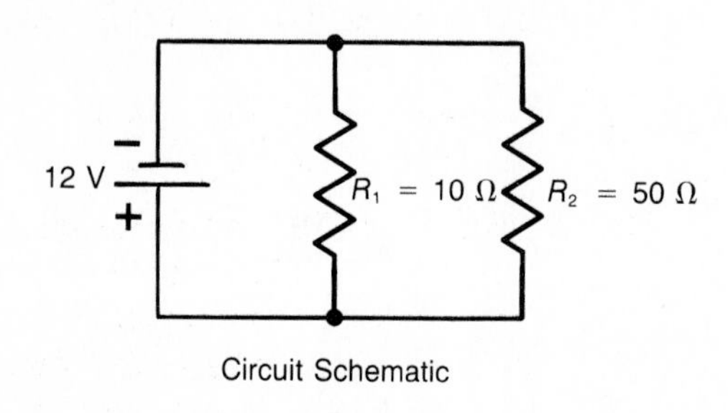

$I_1 = \frac{V_1}{R_1}$

$I_1 = \frac{(12\ V)}{(10\ \Omega)}$

$I_1 = 1.2\ A$

$P_1 = (I_1)^2 \times R_1$
$P_1 = (1.2\ A)^2(10\ \Omega)$
$P_1 = 14.4\ W$

The minimum power rating for R_1 would have to be 14.4 W (compared to 4 W)

$I_2 = \frac{V_2}{R_2}$

$I_2 = \frac{(12\ V)}{(50\ \Omega)}$

$I_2 = 0.24\ A$

$P_2 = (I_2)^2 \times R_2$
$P_2 = (0.24\ A)^2(50\ \Omega)$
$P_2 = 2.88\ W$

The minimum power rating for R_2 would have to be 2.88 W (compared to 2 W)

$P_T = P_1 + P_2$
$P_T = (14.4\ W) + (2.88\ W)$
$P_T = 17.28\ W$

The total power dissipated by the circuit would be 17.28 W (compared to 6 W)

As you can see, the power requirements of a resistor vary with the circuit in which it is connected. To avoid burning out resistors (Figure 18-12), always **CHECK BEFORE YOU CONNECT!**

FIGURE 18-12 OVERHEATED RESISTOR

18-7 Power Loss

In all electrical circuits, some power is always lost in the line wires or conductors. This is due to the fact that the conductors themselves possess a certain amount of resistance. Power is consumed in forcing the current through this resistance. This power is converted into heat by the conductor and is dissipated or given off into the surrounding air.

The power loss in conductors is easily calculated using the power formula: $P = I^2R$. In fact, power loss is often referred to as the I^2R loss. The resistance value used is the total resistance of the conductors. The current value used is the amount of current flow through the conductors.

Example

An AC generator delivers 60 A of current to a load (Figure 18-13). The voltage across the load is 230 V. The total resistance of the line wires is 0.25 Ω. Calculate the power loss, power received by the load, power delivered by the generator and efficiency of transmission.

FIGURE 18-13 POWER LOSS EXAMPLE

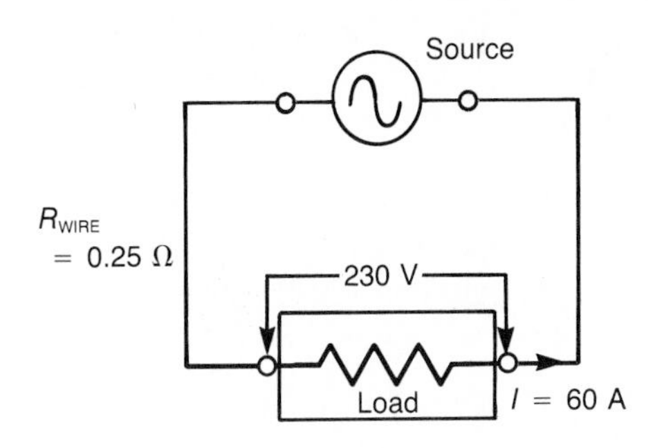

$P_{Loss} = I^2R$
$P_{Loss} = (60\ A)^2(0.25\ \Omega)$
$P_{Loss} = 900\ W$

Power loss is 900 W

$P_{Load} = V \times I$
$P_{Load} = (230\ V)\ (60\ A)$
$P_{Load} = 13\,800\ W$

Power received by the load is 13 800 W

$P_{Total} = P_{Loss} + P_{Load}$
$P_{Total} = (900\ W) + (13\,800\ W)$
$P_{Total} = 14\,700\ W$

Power delivered by the generator is 14 700 W

$$\%\ \text{efficiency} = \frac{P_{Load} \times 100}{P_{Total}}$$

$$\%\ \text{efficiency} = \frac{(13\,800\ W)(100)}{(14\,700\ W)}$$

% efficiency = 93.9 %

Efficiency of transmission is 93.9 %

Power loss is wasted power but still has to be provided by the generating source. Careful control of power loss is essential for an efficient distribution system. High efficiency is obtained by keeping conductor resistance and circuit currents as low as possible.

18-8 Practical Assignments

1. (a) Record the voltage, current and power ratings from the nameplate of five different electrical load devices.
   (b) Calculate any missing quantity values of current or power.
2. (a) Calculate the current flow for each of the following 120 V light bulbs: 40 W, 60 W, 150 W and 200 W.
   (b) Have the teacher wire each bulb, in turn, to a 120 V source. Measure and record the actual current flow for each.
3. (a) Have the teacher wire a voltmeter, ammeter and wattmeter to a single load device. Record the readings on all three meters.
   (b) Calculate the power from the voltmeter and ammeter readings. Compare this calculated power value with that indicated by the wattmeter.

FIGURE 18-14 PRACTICAL ASSIGNMENT 4

4.

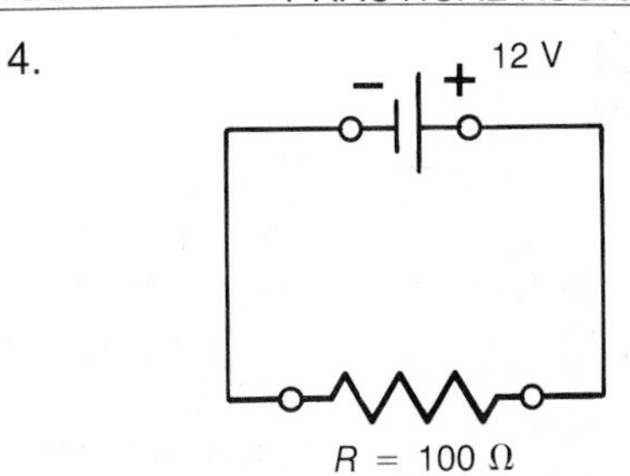

(a) Calculate the minimum power rating of resistor required for the schematic circuit of Figure 18-14.

(b) Have the teacher wire the circuit using each of the following carbon resistors, in turn:
100 Ω – 2 W
100 Ω – 1 W
100 Ω – 0.5 W
100 Ω – 0.25 W
Observe what happens to the resistor after a few minutes.

18-9 Self Evaluation Test

1. Explain how electricity is generated using the hydro-electric process.
2. (a) Explain how electricity is generated using the thermal-electric process.
   (b) State the two primary sources of heat used in the thermal-electric process.
3. (a) Why are high voltages used when transmitting electrical power over long distances?
   (b) What limitation is there to the use of high voltage transmission systems?
4. Define electrical power.
5. What is the basic SI unit used to measure electrical power?
6. Calculate the power rating of a range element that draws 7.5 A when connected to its rated voltage of 230 V.
7. How many amperes will a 1 250 W toaster draw when connected to its rated voltage of 120 V?
8. How much power is dissipated in the form of heat when 0.5 A of current flows through a 50 Ω resistor?
9. Draw the schematic diagram of a wattmeter properly connected to measure the power of a light bulb fed from a 120 V AC supply.
10. A 50 Ω resistor (R_1) and a 100 Ω resistor (R_2) are connected in series to a 24 V DC voltage source
    (a) Calculate the minimum power rating for resistor R_1
    (b) Calculate the minimum power rating for resistor R_2
    (c) Calculate the total power dissipated by the circuit
11. An 80 Ω resistive load (R_1) and a 120 Ω resistive load (R_2) are connected in parallel to a 120 V AC voltage source.
    (a) Calculate the power dissipated by R_1
    (b) Calculate the power dissipated by R_2
    (c) Calculate the total power dissipated by the circuit
12. What is meant by the "power loss" of an electrical circuit?

FIGURE 18-15 SELF EVALUATION TEST 13

13.

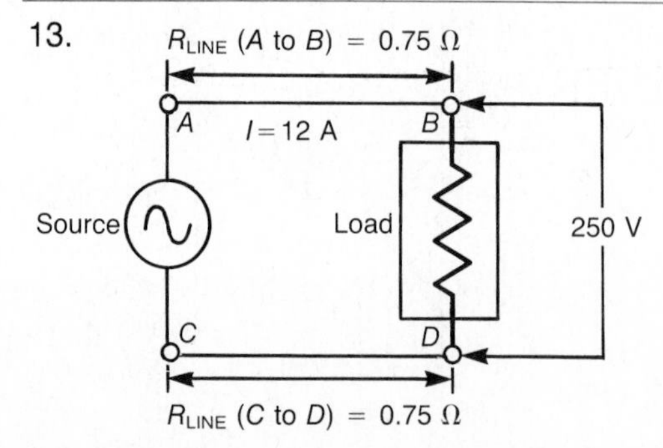

For the schematic circuit of Figure 18-15 calculate:
(a) Total resistance of the two line wires
(b) Power loss
(c) Power received by the load
(d) Power delivered by the source
(e) Efficiency of transmission

19

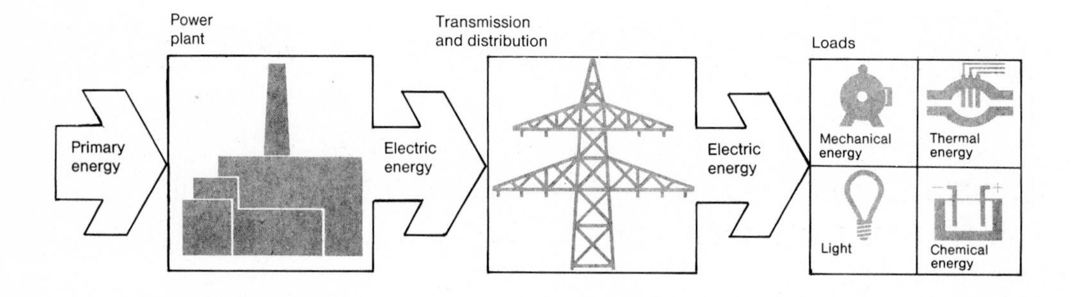
Power
plant
Transmission
and distribution
Loads
Primary
energy
Electric
energy
Electric
energy
Mechanical
energy
Thermal
energy
Light
Chemical
energy

ELECTRICAL ENERGY

OBJECTIVES

Upon completion of this unit you will be able to:

- Define electrical energy.
- Calculate energy consumption both in joules and kilowatt-hours.
- Correctly read a dial type kilowatt-hour meter.
- Calculate the energy cost of an electrical device.
- Calculate the electrical energy cost for a residence over a billing period.
- Outline the problems associated with meeting electrical energy demands and possible solutions.
- Rank common electrical devices according to average yearly energy conversion.

19-1 Energy

Energy is defined as the ability to do work. It exists in many different forms. These include electrical, heat, light, mechanical, chemical and sound energy. Electrical energy is the energy carried by moving electric charges.

Every kind of energy can be measured in joules (J). One joule of electrical energy is equivalent to the energy carried by one coulomb of electric charge being propelled by a force of one volt.

One of the most important laws of science states that energy can neither be created nor be destroyed. It can only be changed from one form to another. For example, an incandescent lamp transforms electrical energy into useful light energy. However, not all the electrical energy is converted into light energy. About 95% of it is converted into wasted heat (Figure 19-1). In this process the important thing to note is that while the total energy may take on different forms, in each case the total amount of energy converted remains constant.

FIGURE 19-1 ENERGY CONVERSION IN A LIGHT BULB

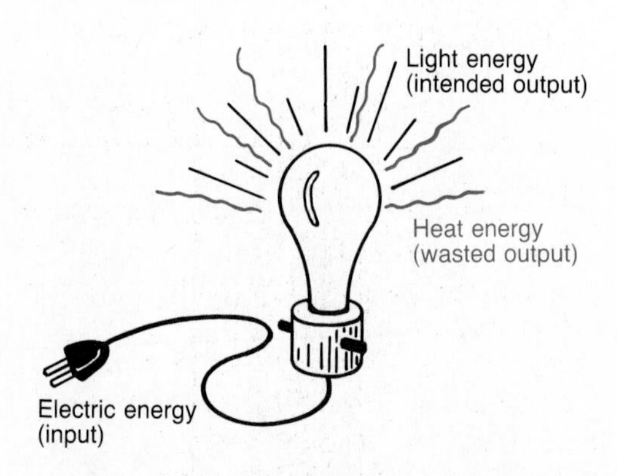

19-2 Calculating Electrical Energy

The SI unit used for measuring electrical energy is the joule (J). One joule represents a very small amount of energy. The energy used by a device is most easily calculated from its power rating. To calculate energy in joules, we multiply the power in watts (W) times the time the device is used in seconds (s):

$$E = Pt$$

where:

E =	energy in joules (J)
P =	power in watts (W)
t =	time in seconds (s)

Example

Suppose the 100 W light bulb in Figure 19-2 is operated for 5 min. The amount of energy converted in joules is then:

$$\text{Energy} = \text{Power} \times \text{time}$$
$$E = Pt$$
$$E = (100\ \text{W})(5 \times 60\ \text{s})$$
$$E = 30\ 000\ \text{J}$$
$$E = 30\ \text{kJ}$$

FIGURE 19-2 LIGHT BULB EXAMPLE

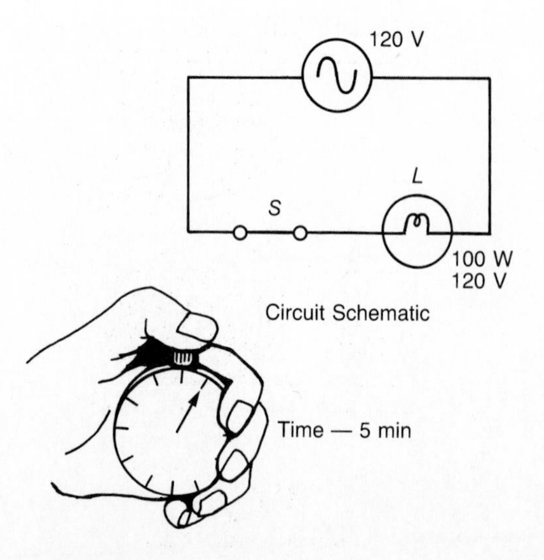

The practical non-SI unit for measuring electrical energy is the kilowatt-hour (kW·h). To find the energy used by a device in kilowatt-hours the same basic energy formula is used. However, in this case, the power rating is expressed in kilowatts (kW) and the time in hours (h):

$$E = Pt$$

where:

E =	energy in kilowatt-hours (kW·h)
P =	power in kilowatts (kW)
t =	time in hours (h)

Example

Suppose an electric coffee maker is rated for 900 W of power (Figure 19-3). It is used an average of six hours per month. The average monthly energy conversion in kilowatt-hours is then:

$$\text{Energy} = \text{Power} \times \text{time}$$
$$E = Pt$$
$$E = \left(\frac{900}{1\,000}\,\text{kW}\right)\left(6\ \text{h}\right)$$
$$E = 5.4\ \text{kW·h}$$

FIGURE 19-3 COFFEE MAKER EXAMPLE

Electrical energy can be expressed in terms of *joules* or *kilowatt-hours*. Both are acceptable and used for different applications. Therefore, it is important to be able to think of energy in both terms. To convert from kilowatt-hours to megajoules, we can use the equation:

$$1\ \text{kW·h} = 3.6\ \text{MJ}$$

19-3 The Energy Meter

The kilowatt-hour (kW·h) meter is used to measure the amount of electrical energy used in your house. It is usually found on the wall outside the house (Figure 19-4A). This meter is read regularly and is the basis for determining your electric bills.

The kilowatt-hour meter records the energy used in the home in the same way as the odometer of a car records the distance travelled. Some meters will give a direct reading on a digital type display while others use a series of dials that must be read in order to determine the value recorded (Figure 19-5).

Steps to follow in reading a dial type meter

1. Read the order of the dials from left to right. Note that some dials increase their readings in a counter-clockwise direction while others increase in a clockwise direction.
2. When the dial pointer is between two numbers, read the smaller of the two numbers.
3. When the pointer rests almost squarely on a number refer to the dial to the immediate right to determine which number you record.

Your meter may have four or five dials. The kilowatt-hour reading can be obtained by reading the dials and by using the multiplier (if applicable) shown on the meter.

The kilowatt-hour meter gives a total cumulative reading. To determine the amount of energy used in a specific time period you would calculate the difference between two successive readings (Figure 19-6).

FIGURE 19-4 ENERGY METER

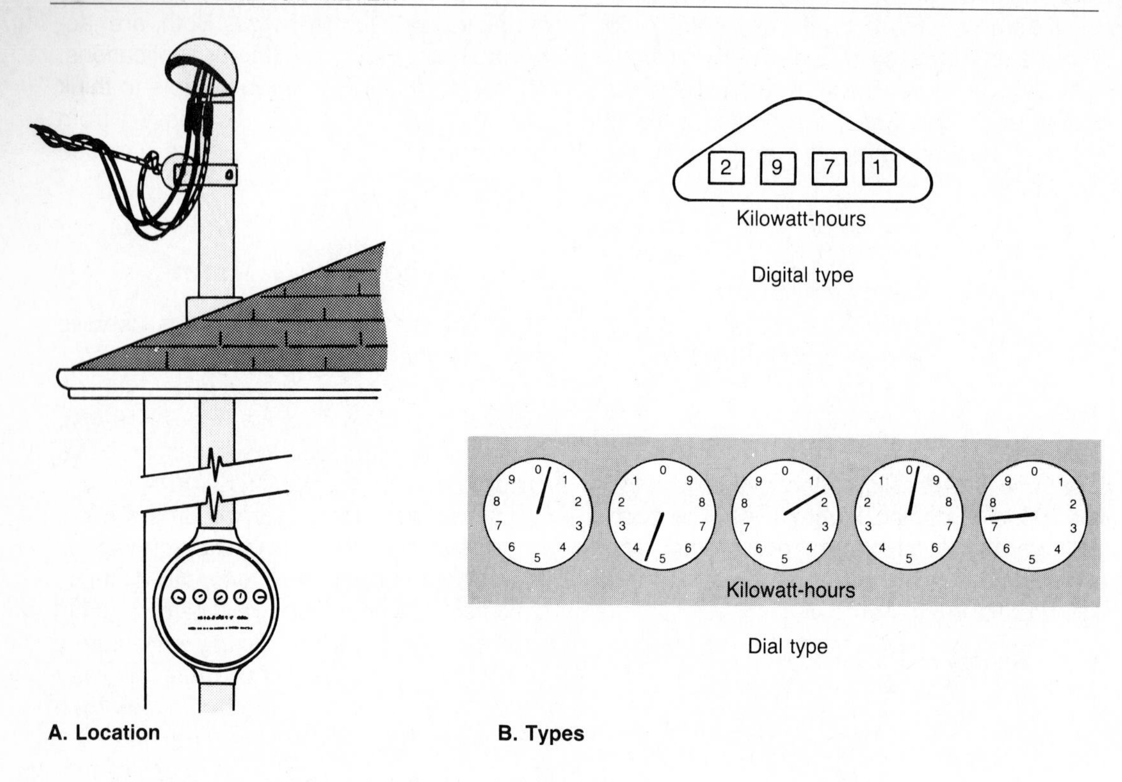

FIGURE 19-5 FOUR DIAL METER

(Multiply by 10)

Read: 89 520 kW·h (89 525 is also correct)

FIGURE 19-6 DETERMINING ENERGY CONSUMPTION

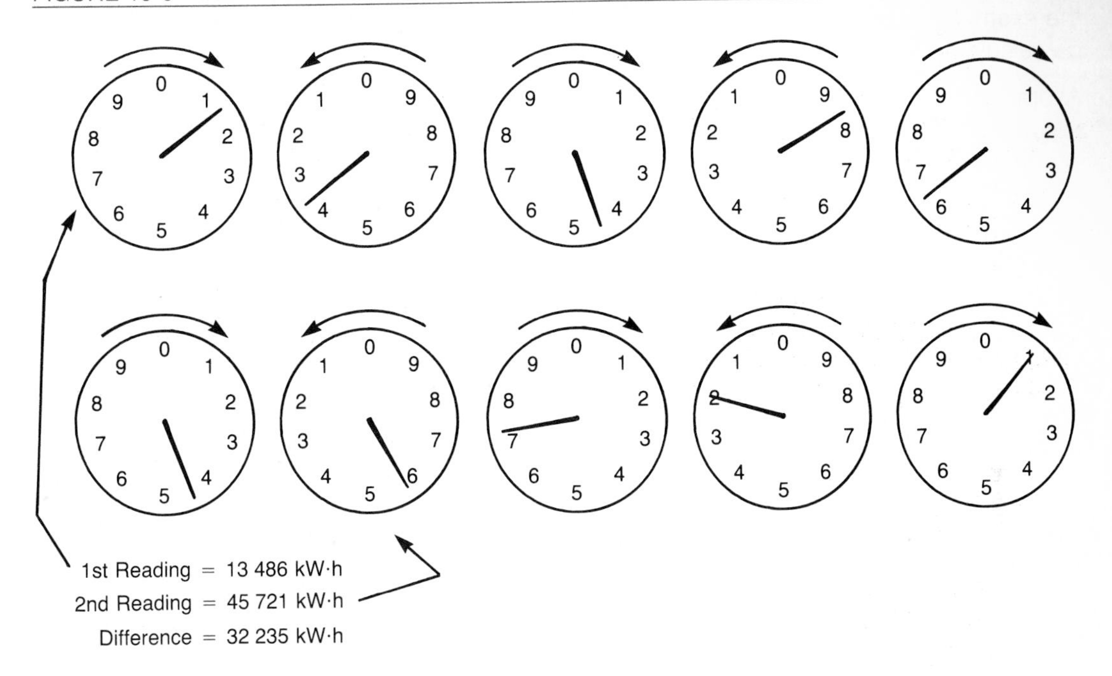

19-4 Energy Costs

Electrical energy is sold by the kilowatt-hour (kW·h) converted. Although the rate structure varies, residential electrical energy bills generally cover a one- or two-month period and are calculated as shown in Tables 19-1A and B:

TABLE 19-1A SAMPLE RATE STRUCTURE

BI-MONTHLY ENERGY RATE STRUCTURE					
1st	100 kW·h	at	9¢	per	kW·H
next	300 kW·h	at	5.4¢	per	kW·h
all	additional	at	3.8¢	per	kW·h

TABLE 19-1B SAMPLE BILLING CALCULATION

BILLING CALCULATION	
100 kW·h at 9.00 ¢ / kW·h =	$ 9.00
300 kW·h at 5.40 ¢ / kW·h =	$16.20
1 160 kW·h at 3.8 ¢ / kW·h =	$44.08
1 560 kW·h	$69.28
for two months	

It is often useful to know the average cost of 1 kW·h of energy when estimating the cost of operating various electrical devices. To calculate the average cost of 1 kW·h of energy you simply divide the total cost by the number of kW·h being charged for. In the example given the average cost would be calculated as follows:

$$\text{Average cost per kW·h} = \frac{\text{Total Cost}}{\text{Total Energy Used}}$$
$$= \frac{(6\ 928\ ¢)}{(1\ 560\ \text{kW·h})}$$
$$= 4.44\ ¢\ \text{per kW·h}$$

Learning to calculate energy costs can help you keep close track of your energy conversion. It can also serve as a basis for a planned personal energy conservation program.

Example

The electric clothes dryer in Figure 19-7 is rated for 4.2 kW. It is being used on the average of twenty hours per month. The average cost of the energy converted is 4.5¢ per kW·h. What is the cost of operating it for a one-month period.

FIGURE 19-7 CALCULATING ENERGY CONVERSION

Energy converted — 168 kW·h
Two-month cost — $7.56

$$E = Pt$$
$$E = (4.2\ \text{kW})\ (20)$$
$$E = 84\ \text{kW·h}$$
$$\text{Cost} = E \times \text{rate per kW·h}$$
$$\text{Cost} = (84\ \text{kW·h})\ (4.5¢)$$
$$\text{Cost} = \$3.78$$

19-5 Practical Assignments

1. (a) Record the amount of energy consumed in your home during one week by taking a set of readings on the kilowatt-hour meter.
   (b) Compare the energy used by your classmates by finding a class average.
   (c) Suggest reasons for very high or low consumptions.
2. (a) Obtain a copy of your electrical energy bill and record the rate schedule.
   (b) Calculate the cost of 5 480 kW·h of energy based on your rate schedule.

19-6 Self Evaluation Test

1. Define electrical energy.
2. (a) What is the basic SI unit used to measure electrical energy?
   (b) What is the non-SI unit often used to measure electrical energy?
3. A 150 W light bulb is on for a total of 16 h.
   (a) How many kilowatt-hours of energy is converted?
   (b) How many megajoules of energy is converted?
4. (a) A 2 500 W water heater is operated for a total of 42 h over a one-month period. Calculate the kilowatt-hours of energy converted.
   (b) Calculate the energy converted in megajoules.

FIGURE 19-8 SELF EVALUATION TEST 5

(c) Calculate the cost of operating the heater for this time period at an average rate of 8¢ per kW·h.

5. (a) Two successive readings of a kilowatt-hour meter are shown in Figure 19-8. What are the readings?
   (b) How much energy was used?
   (c) Calculate the cost of the energy converted using an average rate of 6¢ per kW·h.

6. Calculate the electrical energy bill for a household that converted 2 580 kW·h over a two-month period. The rate schedule is as follows:
   1st 50 kW·h at 8.2¢ per kW·h
   next 300 kW·h at 6.5¢ per kW·h
   all additional at 5.5¢ per kW·h

20

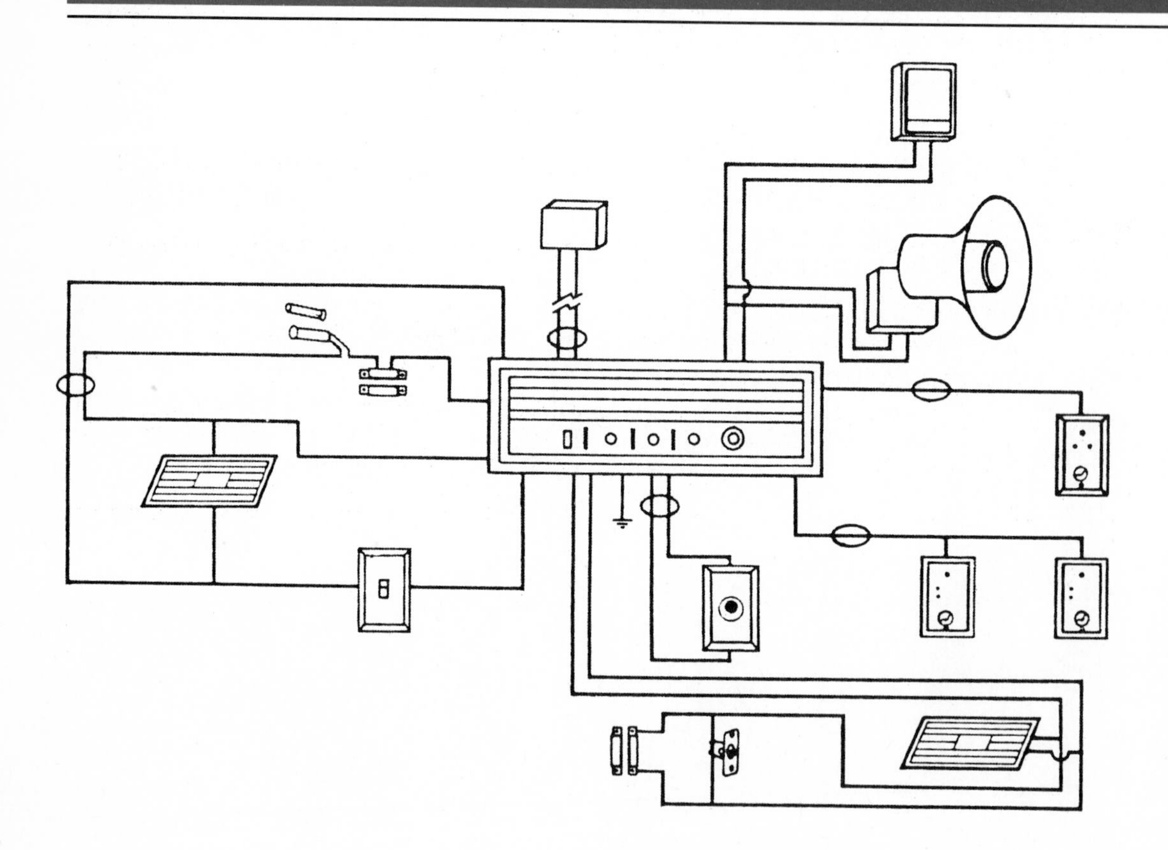

SIGNAL SYSTEMS

OBJECTIVES

Upon completion of this unit you will be able to:

- Outline the general National Electrical Code requirements for residential signaling systems.
- Draw the circuit schematics for typical signal systems from a set of written instructions.
- Complete a wiring layout for a signal system using the schematic diagram.
- Wire typical signal systems in accordance with the planned wiring layout.

20-1 Signal Circuit Requirements

Signal circuits operate at much lower voltages and currents than standard power and lighting circuits. Article 725 of the 1987 National Electrical Code outlines the rules that apply to these circuits. Listed below are some of the general requirements that apply to residential signaling systems:

Power Supply Unit Power is supplied by a transformer approved for this purpose. Approved transformers have built-in overcurrent protection. Their rated output power can not exceed 100 VA.

Voltage Levels The voltage level of the circuit must not exceed 30 V. A bell transformer permanently connected to the 120 V AC line steps the voltage down to a lower level (Figure 20-1). Standard bell transformer secondary voltages are 6-10 V AC and 12-18 V AC.

FIGURE 20-1 TRANSFORMER CONNECTION

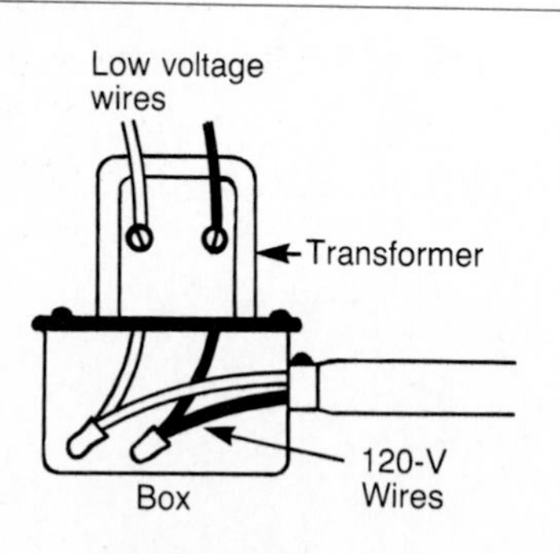

Transformer connection

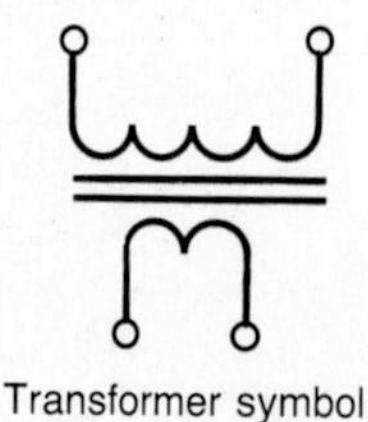

Transformer symbol

Conductor Material and Size A No. 18 AWG copper conductor with thermoplastic insulation is approved for use on these circuits (Figure 20-2). The conductors are usually enclosed in a thermoplastic sheath forming a cable. Both 2- and 3-conductor cables are used. Each conductor is color coded for identification.

FIGURE 20-2 TWO-CONDUCTOR CABLE

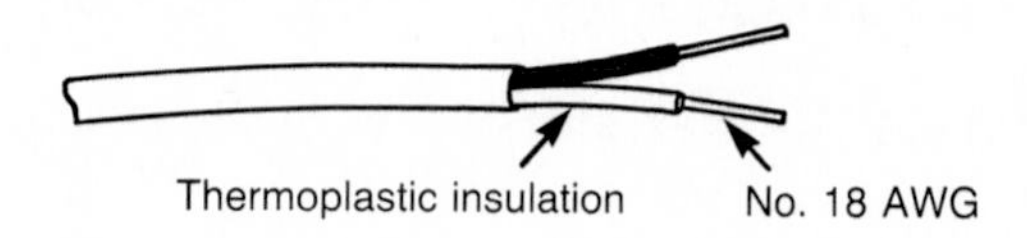

Cable Installation Cables are usually supported with special insulated staples. They must be run completely apart from electric lighting or power circuits.

20-2 Signal Devices

The Door Chime The door chime is a very popular signal device used in homes today. It produces a sound more pleasing than that of the old bell and buzzer systems.

A typical two-tone door chime allows you to identify signals from two locations. It is made up of two 16 V electrical solenoids and two tone bars. A **solenoid**, as you may recall, is an electromagnet that has a movable core or plunger. When voltage is momentarily applied to the *front* solenoid its plunger strikes both tone bars. When voltage is momentarily applied to the *back* solenoid its plunger strikes only the one-tone bar. Thus a double tone is produced by a signal from the front door and a single tone from the back door (Figure 20-3).

The terminal board of the chime unit usually has three terminal screws (Figure 20-

FIGURE 20-3 TWO-TONE DOOR CHIME

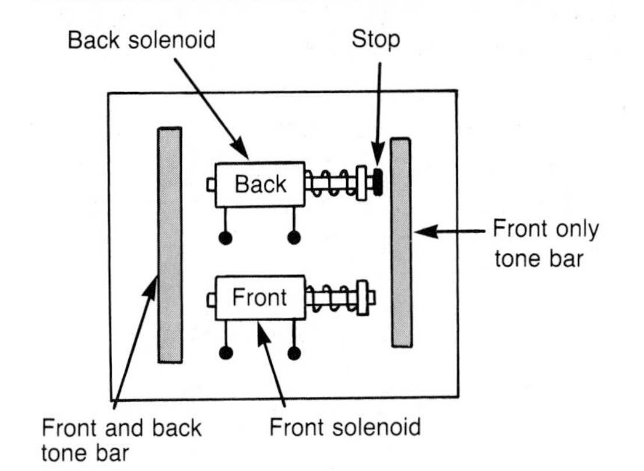

FIGURE 20-5 ANNUNCIATOR SIGNALING SYSTEM

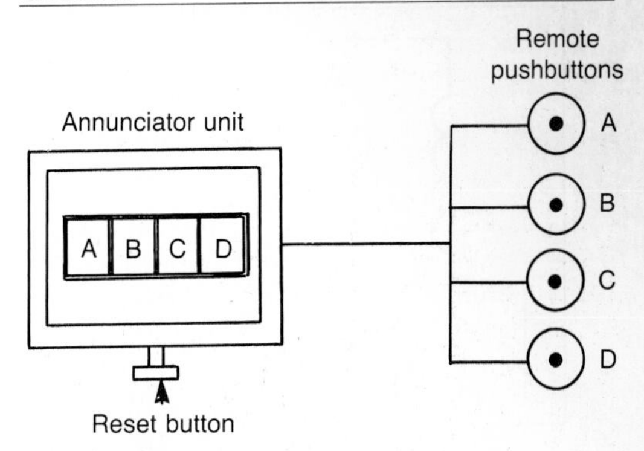

4). The one marked "F" connects to one side of the front door solenoid. The one marked "B" connects to one side of the back door solenoid. The one marked "T" connects to both remaining leads of the solenoids. The "T" terminal is then common to both solenoids.

The Annunciator An annunciator is used to identify signals from several different locations. For example, maintenance personnel working out of a central shop may have to be summoned from different locations within the building. An annunciator installed in the shop is used to indicate where service is required. Pushbuttons installed at different locations within the building are used to signal for service (Figure 20-5).

The annunciator unit itself consists of a number of independently operated electromagnets (Figure 20-6). Momentarily pressing one of the remote pushbuttons sends a current through one of the electromagnets. This releases a letter tab indicating the area from which the signal was sent.

FIGURE 20-4 INTERNAL CHIME WIRING

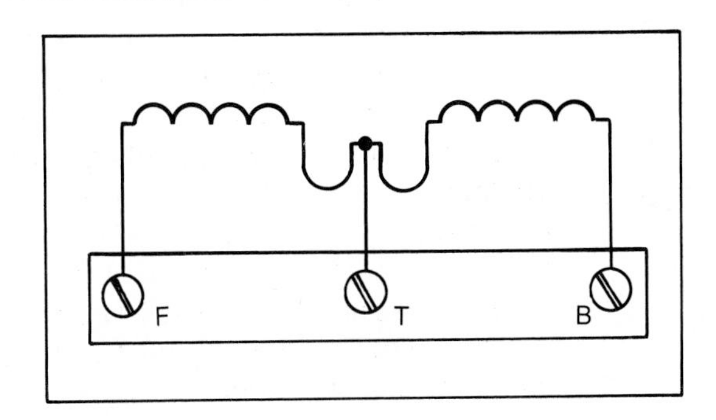

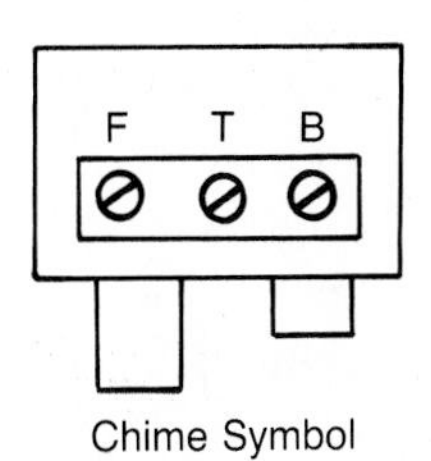

FIGURE 20-6 FOUR-POINT ELECTRO-MECHANICAL ANNUNCIATOR

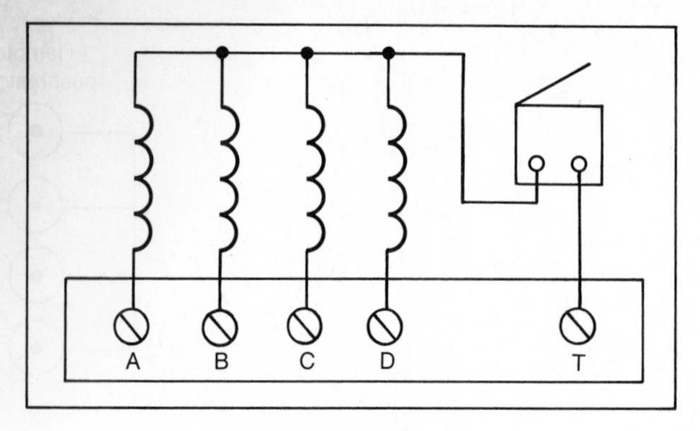

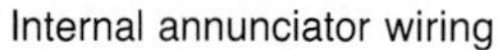
Internal annunciator wiring

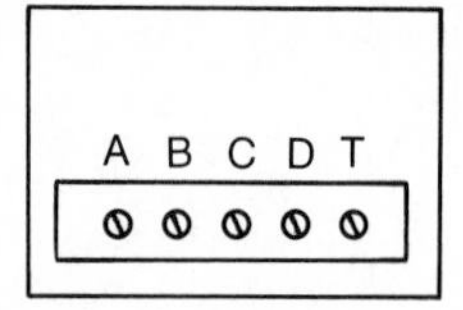

Annunciator Symbol

A buzzer, wired in series with the electromagnet also sounds calling attention to the annunciator. One handy feature of this unit is that the letter remains visible after the pushbutton is released. It is reset by means of a hand control located on the bottom of the annunciator unit. This feature allows calls to be registered in the absence of any personnel in the area. This type of annunciator is known as a gravity-drop type. Also available are the more modern electronic types of annunciators that use *solid state devices* and *lights* in place of the electro-mechanical devices.

20-3 Wiring the Door Chime Circuit

The complete schematic and wiring number sequence chart for a door chime circuit is drawn in Figure 20-7. A 120 V/16 V bell transformer is used as the power supply unit. The schematic circuit can be easily read to show how the circuit operates. Pushing the appropriate pushbutton will complete the circuit to the front or back solenoid. Pushbuttons are used instead of switches so that the circuit will remain operative only as long as the button is being pressed. A two-tone chime indicates the signal is from the front door location. A single-tone chime indicates the signal is from the back door location.

When this circuit is wired in a house, the circuit layout is not as shown in the schematic. The various components are not located in the positions shown in the schematic. Wires are not run singly but in groups of 2 or 3 within a cable sheath.

There can be any number of different types of wiring layouts and cable runs for the same schematic. The one drawn in Figure 20-8 is designed to simulate one typical of a house layout. The transformer is usually located in the basement of the house with its 120 V primary permanently wired into the house's electrical system. Three cable runs are used. One, 2-conductor cable is installed from the transformer to each of the doors and a three conductor cable from the transformer to the chime. The chime is located in a central location on the first floor. Note that the components have been numbered according to the numbering sequence used on the schematic. The buttons and transformer are represented pictorially.

The wiring diagram is completed by connecting the terminals according to the wiring

FIGURE 20-7 CHIME SCHEMATIC DIAGRAM

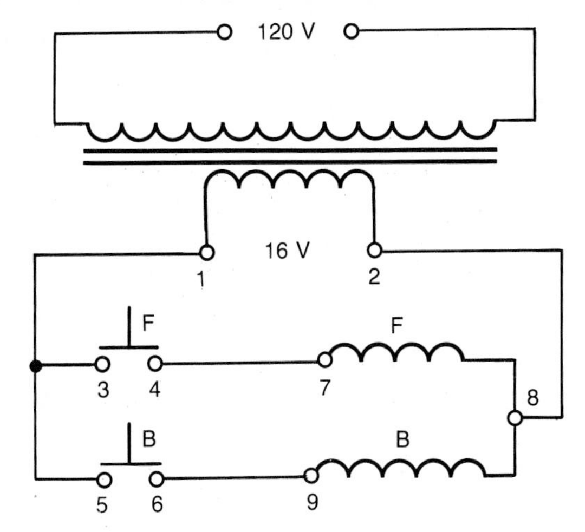

A. Chime schematic diagram

1, 3, 5
2, 8
4, 7
6, 9

B. Wiring number sequence chart

FIGURE 20-8 CHIME WIRING DIAGRAM

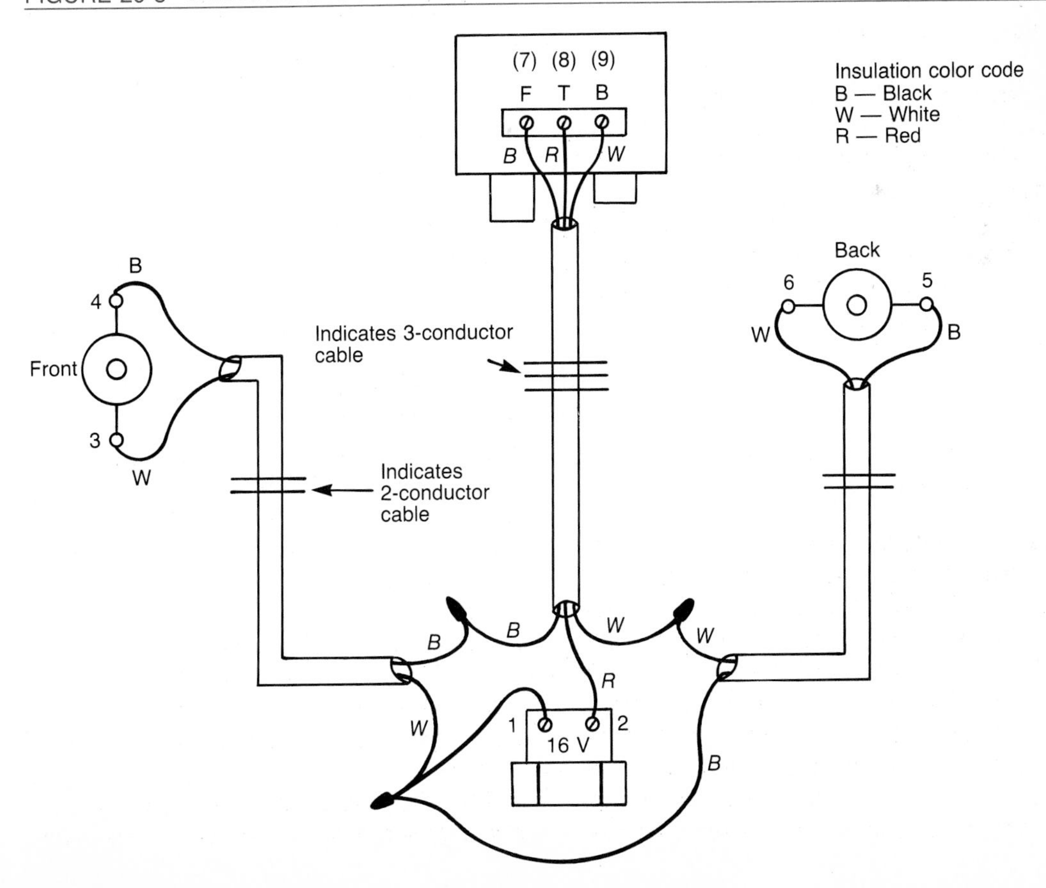

sequence chart. Use the wire insulation color code to properly identify cable wire ends. A two conductor cable usually contains a white and black wire. The three conductor cable wires are usually color coded white, black and red.

20-4 Wiring an Annunciator System

The complete schematic and wiring number sequence chart for an annunciator system is shown in Figure 20-9. This particular system can identify signals from four different locations. Pushbutton *A* controls the current to electromagnet *A*. The same applies to the other sets of buttons and electromagnets. A buzzer is connected in series with all three electromagnets. As a result, it operates when any of the four buttons are pressed.

A typical wiring layout for this annunciator system is drawn in Figure 20-10. For this installation the annunciator and transformer are located adjacent to each other in the basement of the building. One button is located on each of the four floors. A two-conductor cable is run from each pushbutton to the annunciator unit.

FIGURE 20-9 ANNUNCIATOR SCHEMATIC DIAGRAM

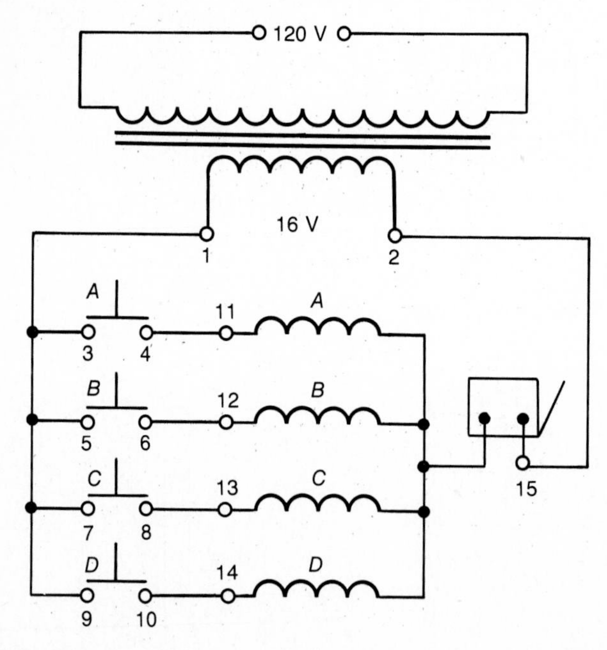

A. Annunciator schematic diagram

1, 3, 5, 7, 9
2, 15
4, 11
6, 12
8, 13
10, 14

B. Wiring number sequence chart

FIGURE 20-10 ANNUNCIATOR WIRING DIAGRAM

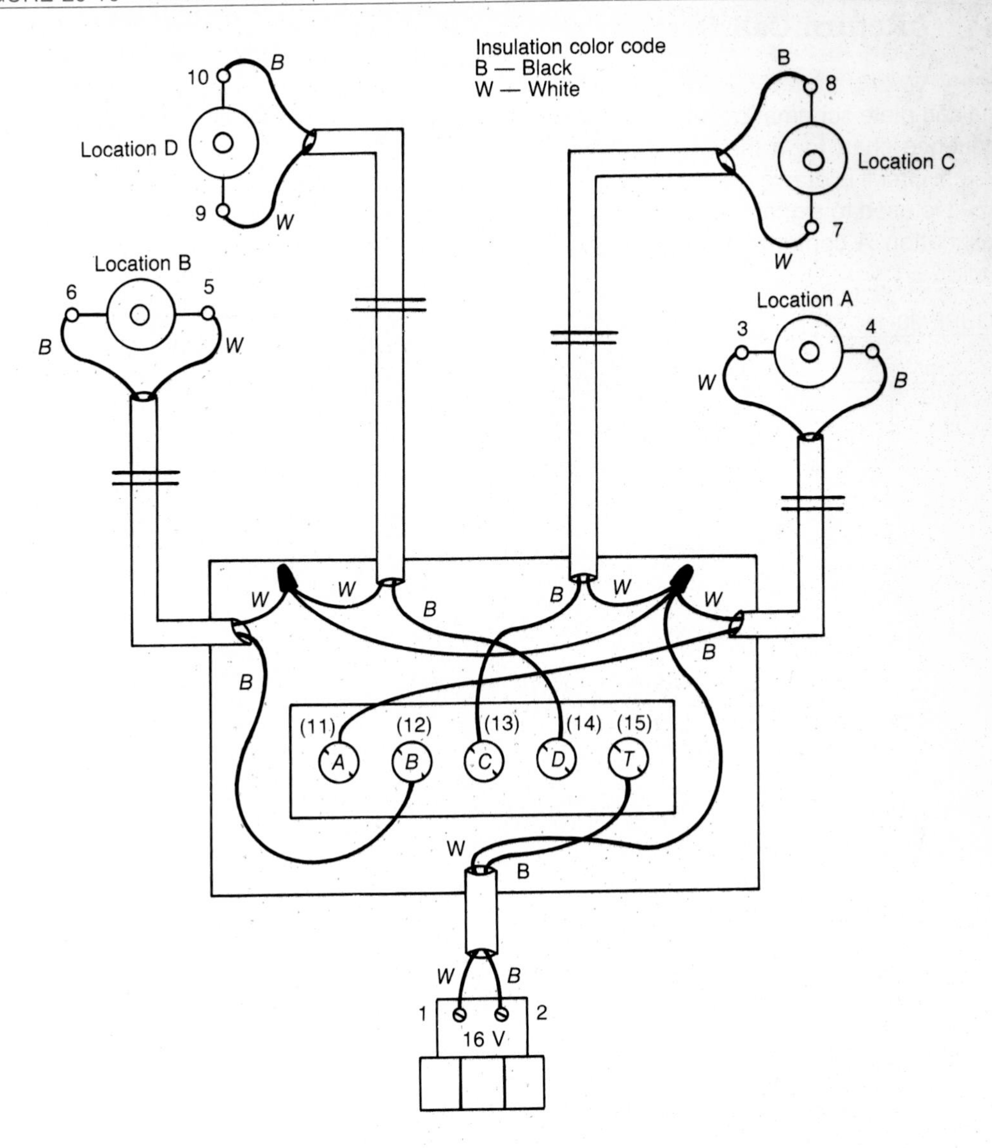

20-5 Wiring a Three-Wire Return Call System

The complete schematic and wiring number sequence chart for a three-wire return call system circuit is drawn in Figure 20-11. This circuit is used to signal between two points. Pushbutton *A* controls the current through the buzzer. Pushbutton *B* controls the current through the bell. The arrangement of components is such that the circuit can be wired with a minimum of three conductors connecting the two signal points.

A typical wiring layout for this three-wire return call system is drawn in Figure 20-12. Notice that only one three-conductor cable is required between the two signalling points.

FIGURE 20-11 SCHEMATIC DIAGRAM FOR A THREE-WIRE RETURN CALL SYSTEM

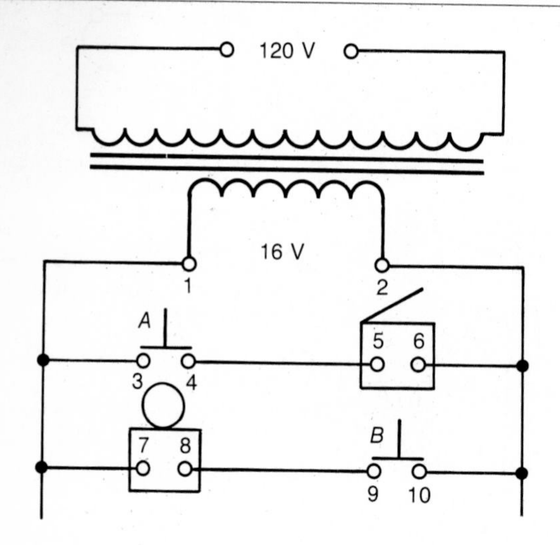

1, 3, 7
2, 6, 10
4, 5
8, 9

Wiring number sequence chart

FIGURE 20-12 THREE-WIRE RETURN CALL SYSTEM WIRING DIAGRAM

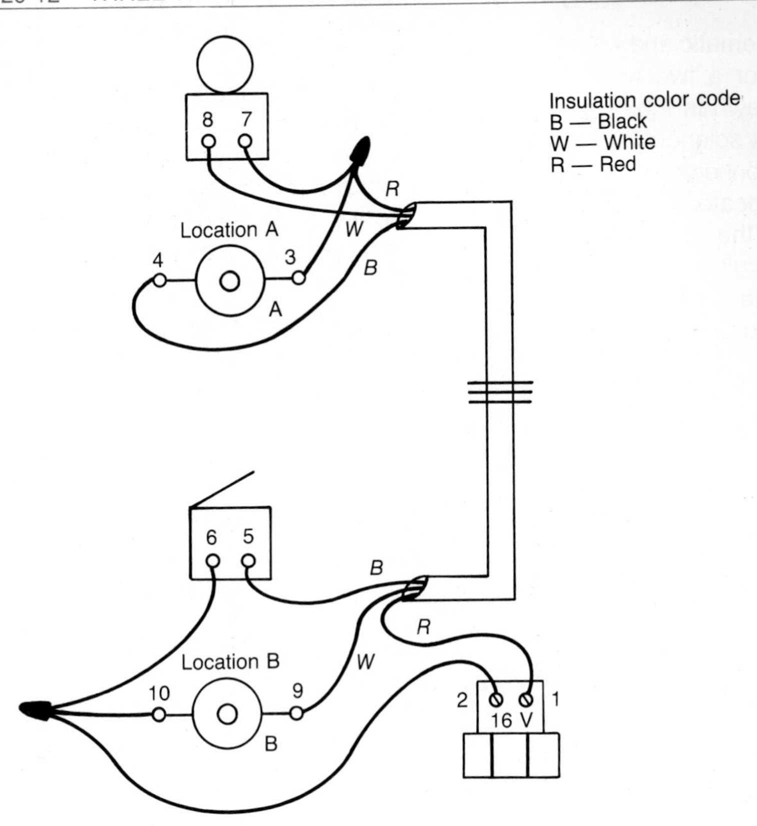

20-6 Apartment Lock System

The complete schematic and wiring number sequence chart for a two apartment door lock system is drawn in Figure 20-13. The current to the lock solenoid is controlled by two pushbuttons connected in parallel. These pushbuttons are located in each apartment. Pressing one or the other of the buttons completes the circuit to the door lock solenoid. Current to the buzzer located in apartment *A* is controlled by pushbutton A_2 located in the lobby. Similarly, current to the buzzer located in apartment *B* is controlled by pushbutton B_2 located in the lobby.

A typical wiring layout for the two apartment door lock circuit is drawn in Figure 20-14. In this installation, terminal boards are used for making the necessary connections. These terminal boards are located in the basement next to the transformer. The routes of the conductor cable runs are as shown on the diagram.

FIGURE 20-13 APARTMENT LOCK SCHEMATIC DIAGRAM

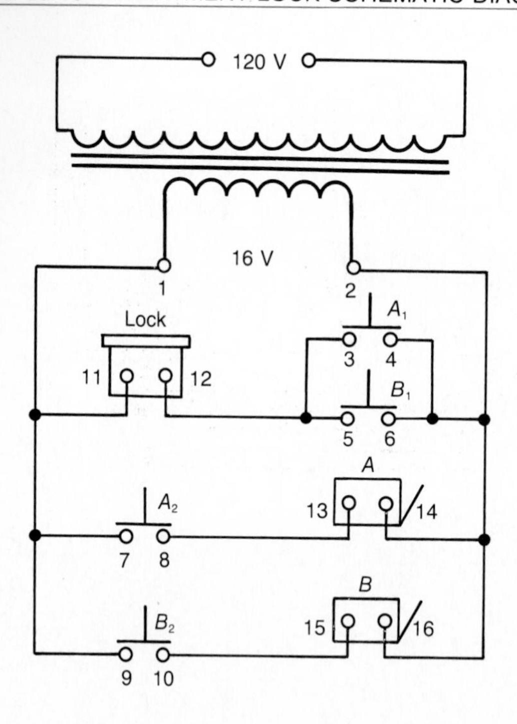

1, 11, 7, 9
2, 4, 6, 14, 16
12, 3, 5
8, 13
10, 15

Wiring number sequence chart

FIGURE 20-14 APARTMENT LOCK WIRING DIAGRAM

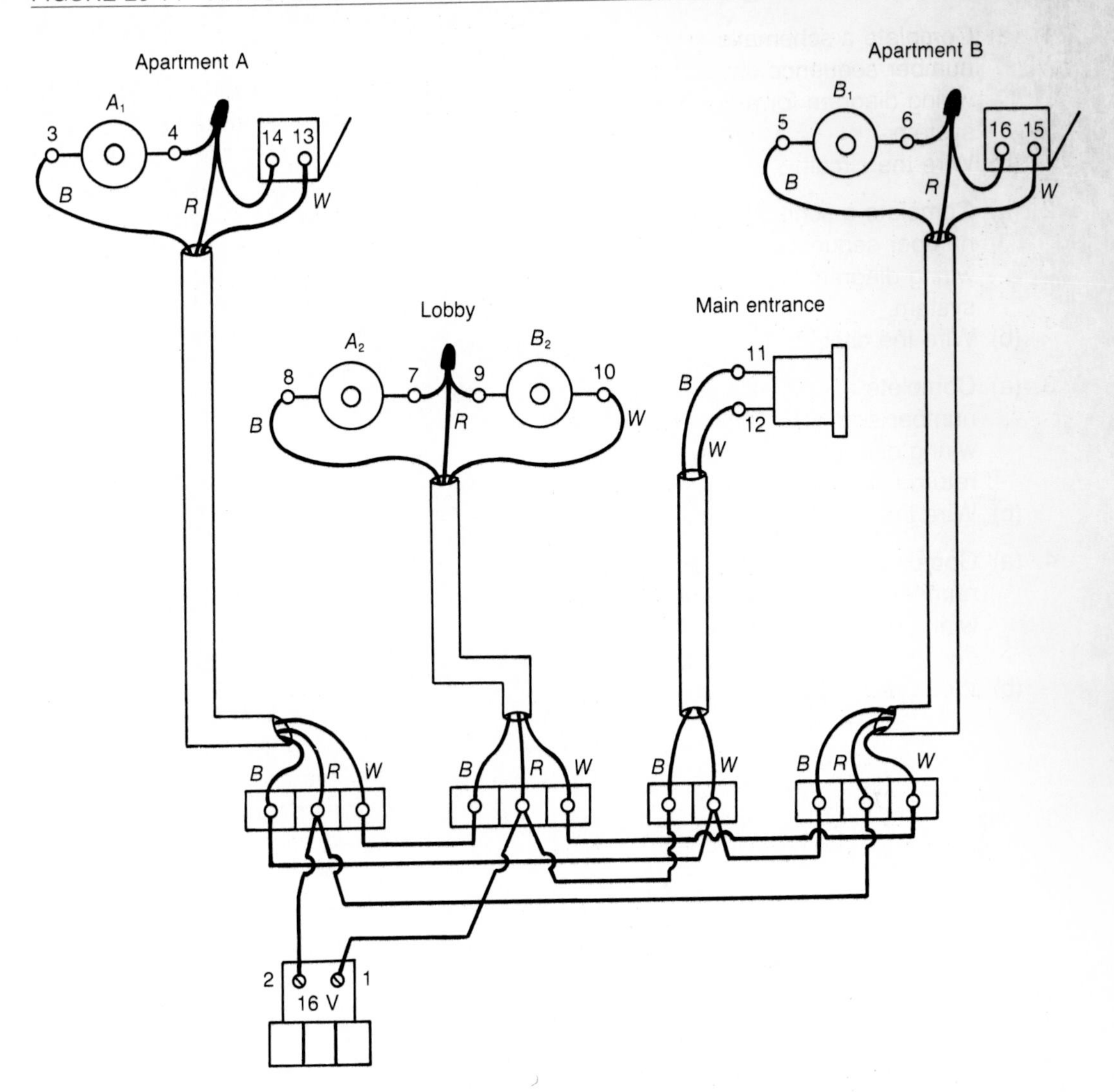

20-7 Practical Assignments

1. (a) Complete a schematic, wiring number sequence chart and wiring diagram for a door chime system.
   (b) Wire the circuit.
2. (a) Complete a schematic, wiring number sequence chart and wiring diagram for an annunciator system.
   (b) Wire the circuit.
3. (a) Complete a schematic, wiring number sequence chart and wiring diagram for a three-wire return call system.
   (b) Wire the circuit.
4. (a) Complete a schematic, wiring number sequence chart and wiring diagram for an apartment door lock system.
   (b) Wire the circuit.

20-8 Self Evaluation Test

1. What type of voltage supply unit is usually used for signal circuits?
2. (a) What is the maximum voltage that can be used on signal circuits?
   (b) At what voltage does a house door chime circuit usually operate?
3. (a) What approved AWG size and type of wire is most often used when wiring a door chime circuit?
   (b) Why is this size of wire not approved for home lighting or power circuits?
4. When installing signal circuit cables they must be kept apart from regular lighting or power circuit cables. What is the reason for this rule?
5. On work sheets provided, complete a schematic, wiring number sequence chart and wiring diagram for each of the circuits listed below:
   (i) door chime circuit
   (ii) annunciator system
   (iii) three-wire return call system
   (iv) apartment lock system

APPENDIX

Conversion of Units

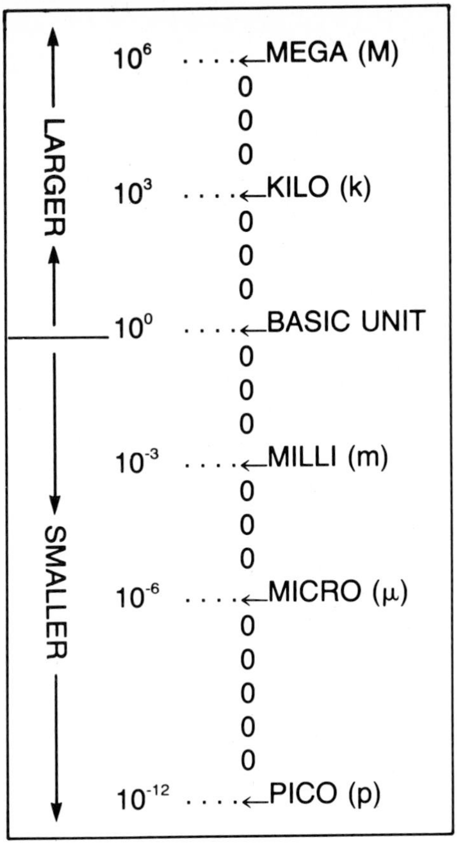

↑ LARGER ↑	10^6	.... ←	MEGA (M)
		0	
		0	
		0	
	10^3	.... ←	KILO (k)
		0	
		0	
		0	
	10^0	.... ←	BASIC UNIT
↓ SMALLER ↓		0	
		0	
		0	
	10^{-3}	.... ←	MILLI (m)
		0	
		0	
		0	
	10^{-6}	.... ←	MICRO (μ)
		0	
		0	
		0	
		0	
		0	
	10^{-12}	.... ←	PICO (p)

Common SI prefixes

Tera	T	one trillion (10^{12})
Giga	G	one billion (10^9)
Mega	M	one million (10^6)
kilo	k	one thousand (10^3)
hecto	ha	one hundred (10^2)
deca	da	ten (10^1)
deci	d	one-tenth (10^{-1})
centi	c	one-hundredth (10^{-2})
milli	m	one-thousandth (10^{-3})
micro	μ	one-millionth (10^{-6})
nano	n	one-billionth (10^{-9})
pico	p	one-trillionth (10^{-12})

Common Electrical SI Units and Symbols

Quantity	Symbol	Basic SI Unit	SI Unit Abbreviation
Electric potential energy difference	V	volt	V
Electric charge	Q	coulomb	C
Electric current	I	ampere	A
Resistance	R	ohm	Ω
Electric power	P	watt	W
Electric energy	E	joule	J
Frequency	f	hertz	Hz
Capacitance	C	farad	F
Inductance	L	henry	H
Capacitive reactance	X_C	ohm	Ω
Inductive reactance	X_L	ohm	Ω
Impedance	Z	ohm	Ω

DC Circuits
Formula circle

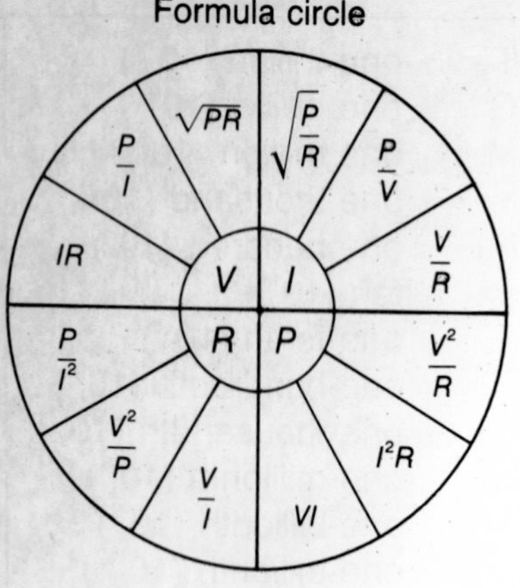

Formulas

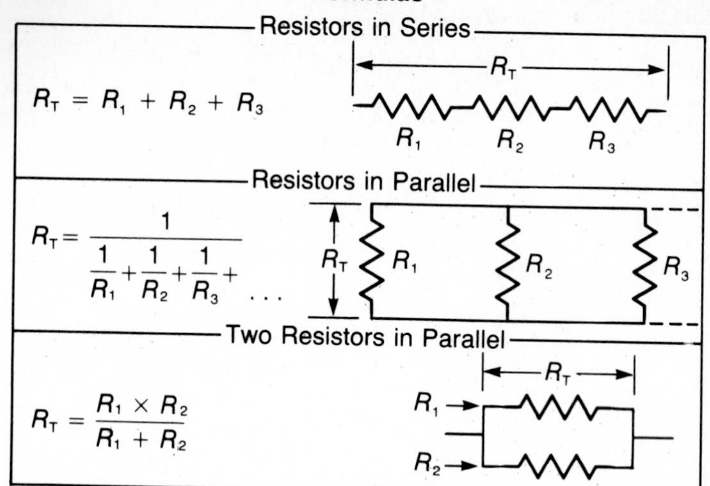

Resistor Color Code

COLOR	1st NUMBER	2nd NUMBER	MULTIPLIER	TOLERANCE (PERCENT)
Black	0	0	1	
Brown	1	1	10	
Red	2	2	100	
Orange	3	3	1 000	
Yellow	4	4	10 000	
Green	5	5	100 000	
Blue	6	6	1 000 000	
Violet	7	7	10 000 000	
Gray	8	8	100 000 000	
White	9	9	1 000 000 000	
Gold			0.1	5
Silver			0.01	10
None				20
	BAND 1	BAND 2	BAND 3	BAND 4

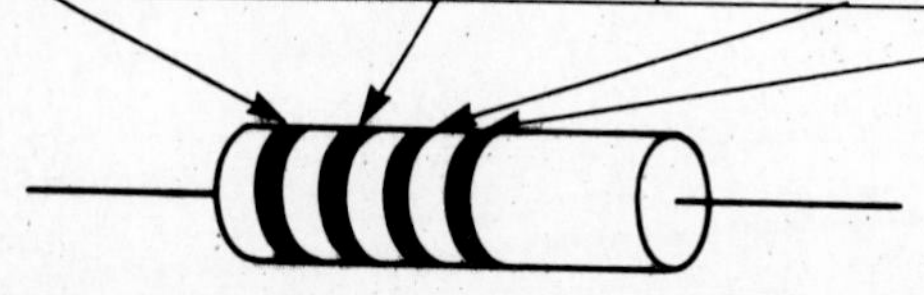

Common Schematic Symbols

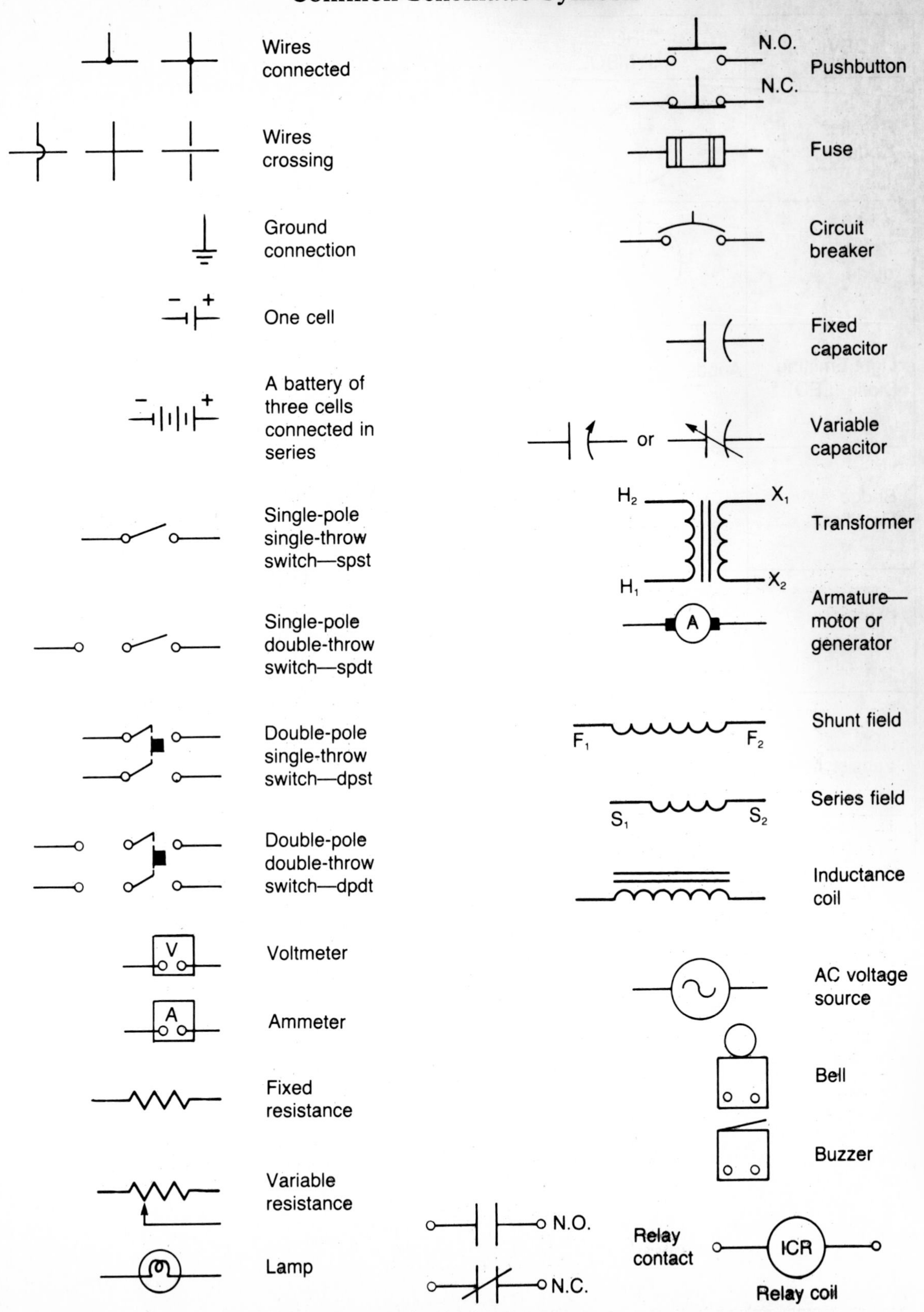

Semiconductor devices and symbols.

DEVICE	CIRCUIT SYMBOL
Rectifier diode	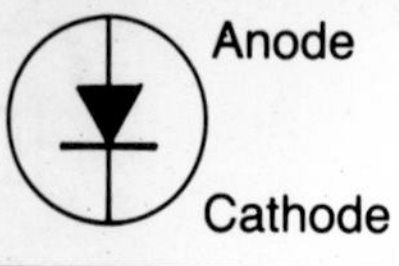 Anode, Cathode
Zener diode	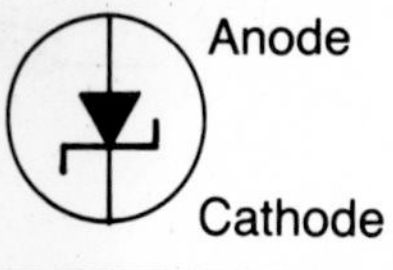 Anode, Cathode
Light Emitting Diode (LED)	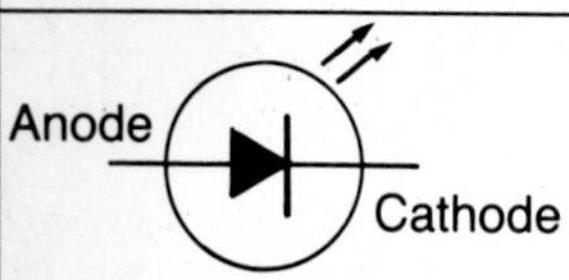 Anode, Cathode
Bridge Rectifier	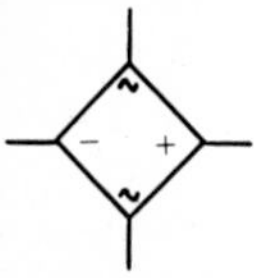 ~ − + ~
PNP transistor	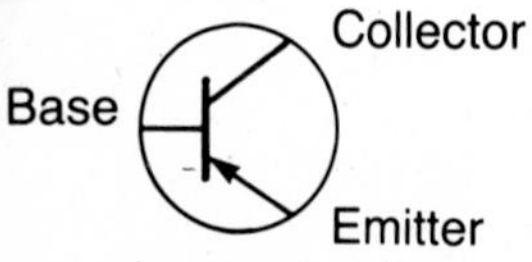 Base, Collector, Emitter
NPN transistor	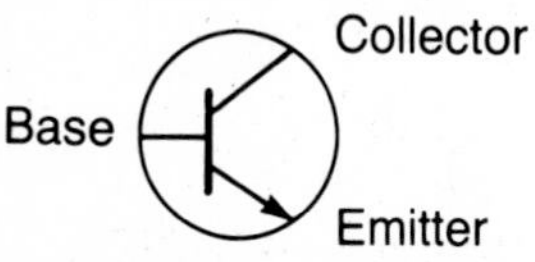 Base, Collector, Emitter
N – channel JFET	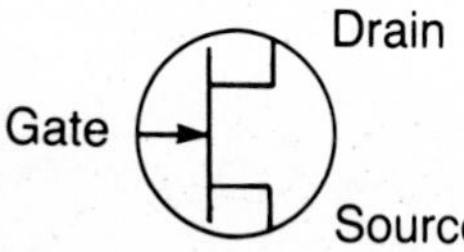 Gate, Drain, Source
P – channel JFET	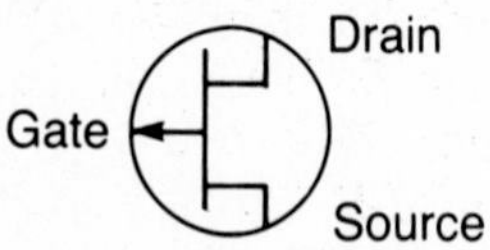 Gate, Drain, Source
P – channel MOSFET	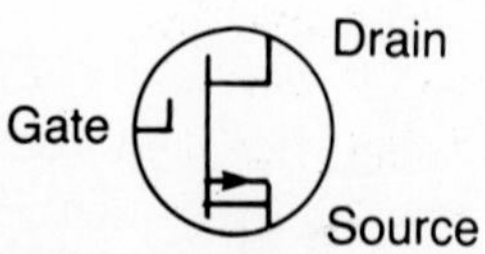 Gate, Drain, Source
DIAC	

DEVICE	CIRCUIT SYMBOL
N – channel MOSFET	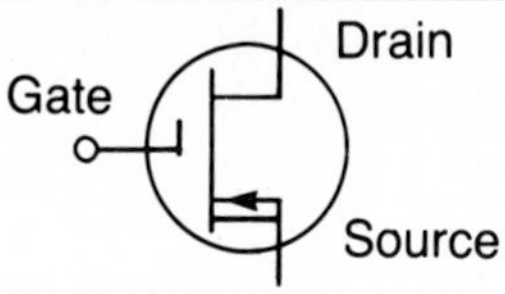 Gate, Drain, Source
Unijunction transistor (UJT)	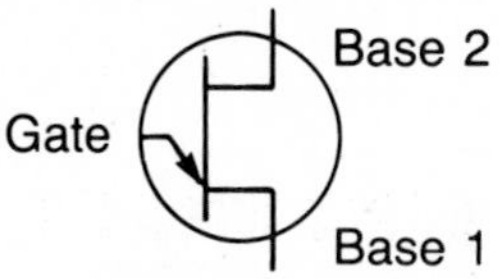 Gate, Base 2, Base 1
Silicon controlled rectifier (SCR)	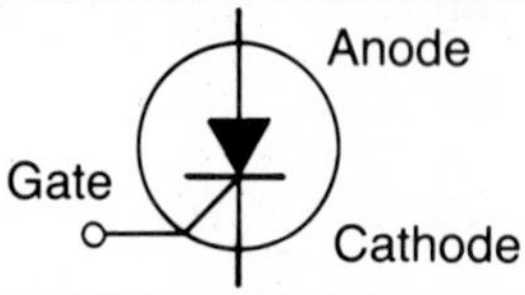 Anode, Gate, Cathode
Triac	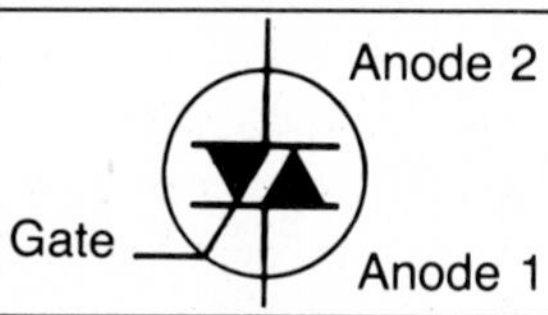 Anode 2, Gate, Anode 1
Integrated circuit (IC)	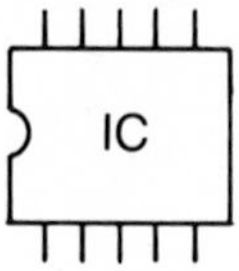 IC
Operational amplifier (OP-AMP)	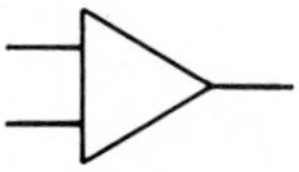

Logic Gates

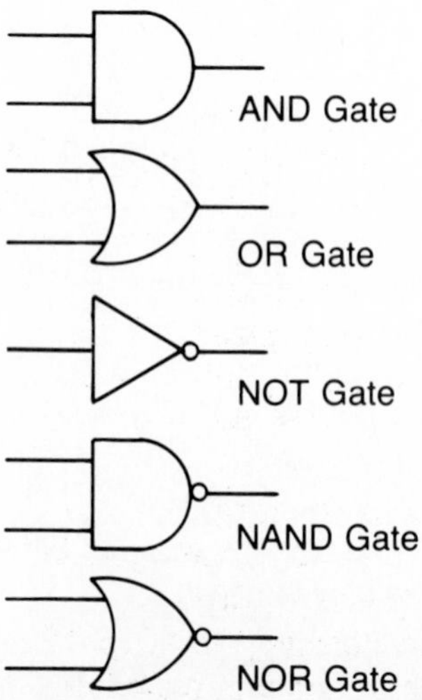

AND Gate

OR Gate

NOT Gate

NAND Gate

NOR Gate

INDEX

BOOK I